Fundamental Interactions in Low-Energy Systems

ETTORE MAJORANA INTERNATIONAL SCIENCE SERIES

Series Editor:
Antonino Zichichi
European Physical Society
Geneva, Switzerland

(PHYSICAL SCIENCES)

Recent volumes in the series:

A Continuation Order Plan is available for this series. A continuation order will bring delivery of each new volume immediately upon publication. Volumes are billed only upon actual shipment.

Fundamental Interactions in Low-Energy Systems

Edited by

P. Dalpiaz

INFN and Department of Physics
University of Ferrara
Ferrara, Italy

G. Fiorentini
G. Torelli

INFN and Department of Physics
University of Pisa
Pisa, Italy

Plenum Press • New York and London

Library of Congress Cataloging in Publication Data

International School of Physics of Exotic Atoms (4th: 1984: Erice, Italy)
  Fundamental interactions in low-energy systems.

  (Ettore Majorana international science series; 23)
  "Proceedings of the fourth course of the International School of Exotic Atoms, held
March 31—April 6, 1984, at the Ettore Majorana Center for Scientific Culture, Erice,
Sicily, Italy"—T.p. verso.
  Bibliography: p.
  Includes index.
  1. Nuclear reactions—Congresses. 2. Exotic atoms—Congresses. I. Dalpiaz, P. II.
Fiorentini, G. III. Torelli, G. IV. Series: Ettore Majorana international science series.
Physical sciences; v. 23.
QC793.9.I595   1984   539.7'5   85-6335
ISBN 978-1-4684-4969-3

ISBN 978-1-4684-4969-3   ISBN 978-1-4684-4967-9 (eBook)
DOI 10.1007/978-1-4684-4967-9

Proceedings of the Fourth Course of the International School of Exotic Atoms, held
March 31—April 6, 1984, at the Ettore Majorana Center for Scientific Culture,
Erice, Sicily, Italy

©1985 Plenum Press, New York

Softcover reprint of the hardcover 1st edition 1985

A Division of Plenum Publishing Corporation
233 Spring Street, New York, N.Y. 10013

PREFACE

The fourth course of the International School of Physics of Exotic Atoms took place at the "Ettore Majorana" Center for Scientific Culture, Erice, from March 31 to April 6, 1984.

As tradition, exotic atoms have been a tool for studying electromagnetic, weak and strong interactions at low energies. We felt it appropriate to have a full course devoted to a discussion of the information to be gained on the fundamental interactions from the study of low energy systems.

In this kind of physics, which is characterized experimentally by very intense particle sources and very sensitive apparatuses, one can search for rare events and can perform precise measurements. Sensitive tests of the predictions of current theories of electro-weak and strong interactions can thus be achieved.

The course was attended by 54 participants from 23 institutes in 9 countries. The morning lectures reviewed the achievements of the field in the last few years and the afternoon seminars dealt with new results and projects.

The lecturers were kind enough to edit their notes so that we are now able to present the procedings of the course. In section I the electro-weak force at low energy is discussed. Following an introduction to the status of the theory, the results of recent experiments in the field are presented. Section II is devoted to the strong interactions: two theoretical papers discuss the phenomenology of Quantum Chromodynamics at low energies and the first experimental results from the LEAR machine are reported. In section III the search for new particles and interactions, which are predicted in several schemes of unified theories, are discussed. In section IV instrumentation developments are considered in connection with superconducting junctions. Some different facets of the physics of muonic atoms and molecules are discussed in the final section.

The course was sponsored by the European Physical Society, the Italian Ministry of Education, the Italian Ministry of Scientific and

Technological Research, the Italian National Institute of Nuclear
Physics (INFN) and the Sicilian Regional Government.

The Center for Scientific Culture was a very efficient and
pleasant host for this course, and our sincere thanks go to its
Director, Prof. A. Zichichi.

P. Dalpiaz
G. Fiorentini
G. Torelli

CONTENTS

II
STRONG INTERACTIONS

III
SEARCH FOR HYPOTHETICAL PARTICLES AND INTERACTIONS

WEAK INTERACTIONS AT LOW ENERGY

J. Bernabéu

Department of Theoretical Physics
University of Valencia
Burjasot, Valencia, Spain

1. INTRODUCTION

In this series of lectures, I would like to discuss the basic aspects of the standard electroweak theory, as well as some alternatives. The phenomenology associated with charged and neutral currents weak interaction will show the success of the standard theory in giving a quantitative account of the existing data. The framework of most of our considerations will be that of an effective theory at low momentum transfer, low with respect to the masses of the intermediate vector bosons, 80-90 GeV. As an extension of the standard theory for the leptonic sector, I will contemplate the possibility of massive neutrinos being described by Majorana fields, leading to exotic phenomena induced by lepton number violation.

2. THE STANDARD THEORY

In this section I wish to illustrate the gauge theory dogma, considering Q.E.D. as an abelian gauge theory being constructed from the principle of local gauge invariance. The next step corresponds to the way how a locally gauge invariant theory of charged vector bosons can appear through Yang-Mills fields. The theory based on the SU(2) x U(1) group is able to describe charged current interactions and two neutral current interactions, one of which can be chosen to be the electromagnetic interaction. But Yang-Mills fields are massless, a situation which does not correspond to the behavior of the short range weak interactions. The way out of this problem goes under the name of spontaneous symmetry breaking, based on the possibility that a fully symmetric Lagrangian can give rise to non-symmetric solutions for the ground state.

2.1 QED as an Abelian Gauge Theory

We construct the simplest example of a gauge theory employing the principle of local gauge invariance. Consider a massless fermion described by the field ψ (x). The free lagrangian density is

$$L_o(x) = i\bar{\psi}(x)\gamma^\mu \partial_\mu \psi(x) \tag{1}$$

which is invariant under the global gauge transformation

$$\psi(x) \rightarrow \psi'(x) = e^{iQ\Lambda}\psi(x) \tag{2}$$

This invariance leads to a conserved current j^μ (x), the electromagnetic current if Q is the fermion electric charge.

Suppose that the Λ-parameter of the U(1) transformation (2) is made local, $\Lambda \rightarrow \Lambda$ (x). Then L_o (x) is no more invariant. To restore it, one needs a new field A_μ (x) appearing in the covariant derivative

$$\partial_\mu \rightarrow \partial_\mu - ieQA_\mu(x) \tag{3}$$

As long as the transformation properties of A_μ (x) are given by

$$A_\mu(x) \rightarrow A'_\mu(x) = A_\mu(x) + e^{-1}\partial_\mu\Lambda(x) \tag{4}$$

the new Lagrangian obtained from the replacement (3) into equation (1) is left invariant. This new Lagrangian describes two pieces: a) the kinematic term for the fermion, b) the interaction of the charged fermion with the A_μ field, say the photon.

In order to have a complete theory we must add a kinematic term for the photon field which must be gauge invariant by itself. This term is

$$L_o^{(\gamma)} = -\frac{1}{4} F_{\mu\nu}(x) F^{\mu\nu}(x)$$

with

$$F_{\mu\nu} = \partial_\mu A_\nu - \partial_\nu A_\mu \tag{5}$$

Notice that the photon mass must vanish, because a term like $A_\mu A^\mu$ violates the gauge principle and therefore it is forbidden.

2.2 Yang-Mills Fields

The basic idea is to generalize the abelian single parameter group U(1) of local phases to more general unitary groups. This allows[1] the construction of a local gauge invariant theory of charged vector bosons.

Let us take the group SU(2), like the isospin group, and $\psi(x)$ a spin-$\frac{1}{2}$ field with isospin T and free Lagrangian density $L_o(x)$. Consider the local isospin gauge transformation

$$\psi(x) \rightarrow \psi'(x) = S(x) \, \psi \, (x) \tag{6}$$

where $S(x)$ is a $(2T + 1)$ unitary unimodular matrix. If $S(x)$ is dependent of x, $L_o(x)$ is invariant; otherwise, it is not. Again one needs additional fields $B_\mu(x)$ appearing in the covariant derivative

$$\partial_\mu \rightarrow \partial_\mu - igB_\mu(x) \tag{7}$$

where $B_\mu(x)$ denote $(2T + 1) \, (2T + 1)$ real matrices.

For infinitesimal gauge transformations depending on the parameters $\vec{\alpha}(x)$, one has

$$S(x) = I + i\vec{T}\cdot\vec{\alpha}(x) \; ; \; B_\mu(x) = 2\vec{b}_\mu(x)\cdot\vec{T} \tag{8}$$

where $\vec{T}$ are the matrices of $(2T + 1)$ dimensions representing the generators of the group. In order to keep the new Lagrangian

$$L_1(x) = i\bar{\psi}(x)\gamma^\mu[\partial_\mu - i \, g \, 2\vec{b}_\mu(x)\cdot\vec{T}]\psi(x) \tag{9}$$

invariant under (6), the transformations of the $\vec{b}_\mu$-fields have to be given by

$$\vec{b}'_\mu(x) = \vec{b}_\mu(x) + \vec{b}_\mu(x)\times\vec{\alpha}(x) + \frac{1}{2g}\partial_\mu\vec{\alpha}(x) \tag{10}$$

The last inhomogeneous term is analogous to the one appearing in equation (4) and it shows that the source of the field is isospin. But now there is also an homogeneous term in equation (10) that shows that $\vec{b}_\mu(x)$ is an isovector. As $\vec{b}_\mu$ has a non vanishing isospin, it must couple to itself. This will be apparent below. In fact, to construct the kinematic term for the gauge fields, one needs the antisymmetric tensor with a unit of isospin

$$\vec{f}_{\mu\nu}(x) = \partial_\mu\vec{b}_\nu(x) - \partial_\nu\vec{b}_\mu(x) + 2g \, \vec{b}_\mu(x)\times \vec{b}_\nu(x)$$

able to provide a term

$$-\frac{1}{4} \vec{f}_{\mu\nu}(x)\cdot\vec{f}^{\mu\nu}(x) \tag{11}$$

with the required properties. It is worth noting that $\vec{f}_{\mu\nu}(x)$ is not gauge invariant, contrary to $F_{\mu\nu}(x)$ in equation (5), but covariant. All that is needed is the invariance of the Lagrangian and equation

(11) satisfies this requirement. This term contains self-couplings
with three and four gauge fields

with coupling strengths that depend only of g. The unique coupling
constant g for self-couplings of gauge fields and for the coupling
of these fields to fermions is a consequence of the local gauge
invariance under this non abelian group of transformations.

2.3 Standard Theory

I would like to consider now the SU(2) x U(1) theory of
Glashow[2], Weinberg[3] and Salam[4] to describe weak and electromagnetic interactions. Take two fermions f and f' such that their
charges satisfy $Q_f = Q_f' + 1$. With the use of the notation

$$\psi_1(x) \equiv \begin{Bmatrix} f_L \\ f'_L \end{Bmatrix} \quad , \quad \psi_2(x) \equiv f_R \quad , \quad \psi_3(x) \equiv f'_R \tag{12}$$

where $f_{L,R}(x) \equiv \tfrac{1}{2}(1 \pm \gamma_5) f(x)$, the kinematic term can be written
as

$$L_o(x) = i \sum_{j=1}^{3} \bar{\psi}_j(x) \gamma^\mu \partial_\mu \psi_j(x) \tag{13}$$

We require invariance of the Lagrangian under the gauge group SU(2) x
U(1) of transformations

$$\psi_j(x) \rightarrow \psi_j'(x) = \exp\{i\vec{\alpha}(x) \cdot \tfrac{\vec{\tau}}{2}\} \exp\{iy_j \beta(x)\} \psi_j(x) \tag{14}$$

where $\vec{\alpha}(x)$ are the three real arbitrary parameters for SU(2), τ are
the matrix representations of the SU(2) generators and $y_j \beta(x)$ is the
phase for U(1). Assume that ψ_1 is a doublet under SU(2), whereas ψ_2
and ψ_3 are singlets. The free term (13) is not invariant under (14),
so we replace the derivative by the covariant derivative.

$$\partial_\mu \rightarrow D_\mu^j = \partial_\mu - ig \frac{\vec{\tau}}{2} \cdot \vec{W}_\mu(x) - ig' B_\mu(x) y_j \tag{15}$$

where four gauge fields have been introduced: three $\vec{W}_\mu(x)$ non
abelian fields and one $B_\mu(x)$ abelian field. Notice that $\vec{\tau}\psi_2 = \vec{\tau}\psi_3 = 0$, so the $\vec{W}_\mu(x)$ fields are only coupled to the left handed fermion
doublet.

The transformation properties of the gauge fields can be read off the equations of section 2.1 and 2.2, as well as the kinematic terms for $\vec{W}_\mu$ and B_μ and the self-couplings of the non abelian fields $\vec{W}_\mu$. Up to this level all particles are massless, in particular the vector bosons in order to keep gauge invariance.

2.4 Charged Current Interaction

There is a piece in the theory just constructed which couples charged vector bosons to left handed fermions. This component mediated by charged currents is

$$L_{cc} = g\bar{\psi}_1(x)\gamma^\mu \tfrac{1}{2}[\tau_1 W^1_\mu(x) + \tau_2 W^2_\mu(x)]\ \psi_1(x) \tag{16}$$

With the definition of the field

$$W_\mu(x) \equiv (W^1_\mu(x) + iW^2_\mu(x))/\sqrt{2}$$

which creates W^- and annihilates W^+, equation (16) can be expanded as

$$L_{cc} = \frac{g}{2\sqrt{2}}\ \{\bar{f}(x)\gamma^\mu_5(1+\gamma_5)f'(x)W_\mu(x) + h.c.\} \tag{17}$$

which is precisely the interaction of the intermediate vector boson theory for charged currents, corresponding to the vertex

2.5 Neutral Current Interactions

The interaction of fermions with vector bosons mediated by neutral currents is

$$L_{nc} = \sum_{j=1}^{3} \bar{\psi}_j(x)\gamma^\mu[g\frac{\tau_3}{2}W^3_\mu + g'y_j B_\mu]\psi_j(x) \tag{18}$$

Since W^3_μ and B_μ are massless, neutral and as yet unphysical, we may form two orthogonal linear combinations

$$W^3_\mu = \cos\theta_W Z_\mu + \sin\theta_W A_\mu$$

$$B_\mu = -\sin\theta_W Z_\mu + \cos\theta_W A_\mu \tag{19}$$

and separate the interaction (18) into the ones with the A_μ and Z_μ fields.

In order to include electromagnetism in the theory, we wish to identify $A_\mu(x)$ with the photon field. This is possible, with the following consequences

$$L_{nc}^A \equiv L_{em} \rightarrow \begin{cases} e = g \sin\theta_W = g' \cos\theta_W \\ \\ y_1 = Q_f - \tfrac{1}{2} , \quad y_2 = Q_f , \quad y_3 = Q_f , \end{cases} \tag{20}$$

so this attitude leaves only two free parameters, say e and θ_W. The weak hypercharges y_j have to be chosen connected with electric charges.

The interaction of fermions with the neutral vector boson Z_μ is given by

$$L_{nc}^Z = \frac{e}{\sin\theta_W \cos\theta_W} \{\bar{f}(x)\gamma^\mu [(1/4 - Q_f \sin^2\theta_W) + 1/4\gamma_5]f(x)$$
$$+ \bar{f}'(x)\gamma^\mu [(-1/4 - Q_{f'} \sin^2\theta_W) - 1/4\gamma_5]f'(x)\}Z_\mu(x) \tag{21}$$

This is a parity violating interaction but not, in general, of the V-A type, unless $Q_f = 0$.

It is interesting to discuss the relation among the currents appearing in this theory, independent of their explicit forms in terms of the fermion fields. The new neutral current, the one between brackets in equation (21) satisfies:

$$j_Z^\mu(x) = j_3^\mu(x) - \sin^2\theta_W \, j_{em}^\mu(x) \tag{22}$$

where $j_3^\mu(x)$ is the rotated current, under SU(2), of $j_{cc}^\mu(x)$, the one for charged currents. Equation (22) allows the connection between different processes, in particular for electroweak interactions of hadrons.

2.6 Spontaneous Symmetry Breaking

Let us assume that a Lagrangian is invariant under the group G of transformations. If there is a unique ground state, it must be a singlet under G. But there could be a degenerate set of states with minimal energy which transform under G as the members of a given multiplet. The choice of one of these states as the ground state of the theory corresponds to the so-called spontaneous symmetry breaking[5].

6

For a Lagrangian describing the behavior of a complex scalar field, which is invariant under a global gauge transformation G, one can find situations in which

$$< 0 \, |\phi(x) \, | \, 0 > = \lambda/\sqrt{2} \neq 0 \tag{23}$$

so the redefinition of the field from the one of minimal energy, $\phi \equiv 1/\sqrt{2} \, [\lambda + \phi_1(x) + i\phi_2(x)]$, leads to a massless Goldstone boson ϕ_2 and a massive field ϕ_1. Massless bosons exist[6] as long as there is a solution of the theory which does not share the continuous global symmetry of the Lagrangian.

The consequences change drastically when the spontaneous symmetry breaking mechanism is applied to a local gauge symmetry G. In that case[7], the Goldstone boson disappears in the physical theory, and its degree of freedom appears as the third helicity state of the gauge vector boson, which in this process acquires a mass. Apart from this, a physical massive neutral scalar field remains, the so-called Higgs particle.

This strategy can be applied to the standard theory in order to give masses to the $W^{\pm}$ and Z° vector bosons responsible of weak interactions, which we know phenomenologically that are short-ranged. One needs[3] a complex scalar field

$$\phi(x) = \begin{Bmatrix} \phi^{(+)}(x) \\ \phi^{(o)}(x) \end{Bmatrix} \tag{24}$$

which is an isodoublet with respect to the SU(2) group of the theory. When the spontaneous symmetry breaking (23) is applied to the neutral field $\phi^{(o)}(x)$, the local gauge invariance of the theory allows the replacement

$$\phi^{(+)}(x) \rightarrow 0 \; ; \; \phi^{(o)}(x) \rightarrow 1/\sqrt{2}[\lambda + \chi(x)] \tag{25}$$

where $\chi(x)$ is the real physical Higgs field. The missing degrees of freedom have gone to the vector bosons, which have now a mass

$$M_W = \tfrac{1}{2} \, \lambda \, g \; ; \; M_Z = \tfrac{1}{2} \, \frac{\lambda g}{\cos\theta_W} \tag{26}$$

whereas the photon remains massless.

If one considers charged current weak interactions at low momentum transfer, the amplitude

tum transfer, the amplitude ⟩⟨ W is proportional to

$$(\frac{g}{2\sqrt{2}})^2 \, \frac{1}{M_W^2 - q^2} \simeq \frac{g^2}{8 \, M_W^2}$$

The identification with the Fermi coupling constant G gives

$$G/\sqrt{2} = \frac{g^2}{8\,M_W^2} = \frac{1}{2\lambda^2} \tag{27}$$

where the last equality comes from equation (26). Using equation (20) too, one gets a definite prediction for the masses

$$M_W = \left(\frac{\pi\,\alpha}{G\sqrt{2}}\right)^{\frac{1}{2}} \frac{1}{|\sin\theta_W|} \;,\; M_Z = M_W \,/\, \cos\theta_W \tag{28}$$

The vector bosons $W^{\pm}$ and Z° have been discovered[8] at CERN using the Sp$\bar{\text{p}}$S collider at the predicted masses of about 80 GeV and 90 GeV, respectively, for [9] $\sin^2\theta_W \simeq 0.23$.

Apart from this, the Higgs particle χ has to exist, with unpredicted mass, leading to additional interactions represented diagramatically by

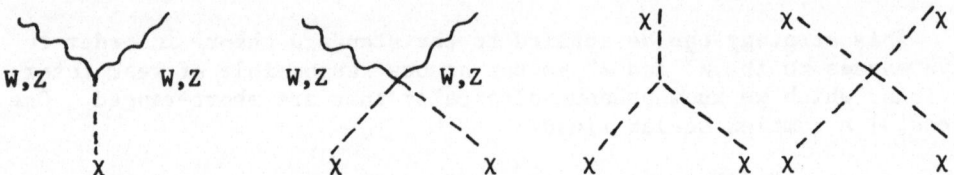

with well defined couplings.

2.7 Fermion Masses

Starting from massless fermions, one can use the scalars of the theory to introduce Yukawa couplings with fermions. The terms which are gauge invariant under $SU(2)_L \times U(1)$ are

$$L(x) = c_f, [\bar{f}_L, \bar{f}'_L] \left\{ \begin{matrix} \phi^{(+)} \\ \phi^{(o)} \end{matrix} \right\} f'_R + c_f [\bar{f}_L, \bar{f}'_L] \left\{ \begin{matrix} \phi^{(o)+} \\ \phi^{(+)+} \end{matrix} \right\} f_R \tag{29}$$
$$+ \text{ h.c.}$$

This can be applied as such for one generation of leptons or quarks, let us say $\{^\nu_e\}$ or $\{^\mu_d\}$ respectively. For neutrinos, there is no right-handed piece in the standard model, so the second term of equation (29) is absent for the leptonic couplings.

After spontaneous symmetry breaking of the gauge invariance, one rewrites

$$L(x) = 1/\sqrt{2} \, [\lambda + \chi(x)] \, \{c_f \bar{f}f + c_f, \bar{f}'f'\} \tag{30}$$

We notice in equation (30) two ingredients: i) fermions have ac-
quired a mass

$$m_f = - c_f \ \lambda/\sqrt{2} \ , \ m_{f'} = - c_{f'} \lambda/\sqrt{2}$$

ii) there is a Yukawa coupling of the fermions to the physical Higgs
boson, with a characteristic coupling proportional to the mass of the
fermion

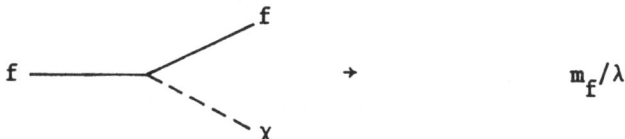

$$\rightarrow \qquad m_f/\lambda$$

In this theory, neutrinos are then massless.

When several generations are present, as

$$\left\{\begin{matrix} u \\ d \end{matrix}\right\} \qquad\qquad \left\{\begin{matrix} c \\ s \end{matrix}\right\} \qquad\qquad \left\{\begin{matrix} t \\ b \end{matrix}\right\}$$

for quarks, the fields of definite mass do not necessarily coincide
with the fields with definite transformation properties under the
electroweak $SU(2)_L$ x $U(1)$ gauge group. This is a consequence of the
fact that equation (30) has to be replaced by a matrix of couplings,
which implies a mass matrix which has to be diagonalized to find the
stationary states.

The mixing between weak states and mass eigenstates is irrelevant for all neutral currents, leading[12] to natural flavor
conservation. However, if the charged current is written in terms
of mass eigenstates of quarks, this mixing leads to the Cabbibo-
Kobayashi-Maskawa unitary transformation[11], which can be parametrized in the form

$$L_{cc}(x) = \frac{g}{2\sqrt{2}} \ [\bar{u},\bar{c},\bar{t}] \ \gamma_\mu \ (1+\gamma_5) \ \begin{pmatrix} 1 & S_1 & 0 \\ -S_1 & 1 & S_3+S_2e^{i\delta} \\ 0 & -S_2-S_3e^{i\delta} & e^{i\delta} \end{pmatrix} \ \left\{\begin{matrix} d \\ s \\ b \end{matrix}\right\} \quad (31)$$

with three angles and one phase. In writing equation (31) we have
assumed small mixings, so that $s_i \equiv \sin\theta_i \ll 1$. The phase δ can be
responsible of the CP-violation observed in some weak processes.

3. ALTERNATIVE MODELS

The success of the standard theory in explaining the experimental data is impressive. Nevertheless, small deviations, either in

the charged current interaction or neutral current interaction, are allowed. There are several lines of alternative ideas in the current literature: other gauge group; other fermion multiplets; quarks, leptons and W's as composite objects; supersymmetric ideas,.... I will comment on the extension of the $SU(2)_L \times U(1)$ gauge group to the left-right symmetric $SU(2)_L \times SU(2)_R \times U(1)$ model[10].

In this alternative model, the right-handed components f_R, f'_R of the fermion fields, which are singlets under $SU(2)$, transform as doublets under $SU(2)_R$. Gauge invariance of the theory imposes the existence of seven gauge vector bosons, which we call $\vec{W}L$, $\vec{W}R$ and $\vec{B}_\mu$. The coupling constant g of the standard theory for the coupling of W_μ the W_μ to fermions is now replaced by two independent couplings g_L and g_R. The theory is manifestly left-right symmetric if we impose $g_L = g_R = g$, i.e. it is parity invariant. As we shall see, in this theory parity violation is a low energy phenomenon related to the spontaneous symmetry breaking of the gauge symmetry.

It is immediate to write the charged current interaction in terms of the fields defined above. One gets

$$L_{cc}(x) = \frac{g}{2\sqrt{2}} \, [\bar{f}\gamma^\mu(1 + \gamma_5)f'W^L_\mu + \bar{f}\gamma^\mu(1 - \gamma_5)f'W^R_\mu + \text{h.c.}] \qquad (32)$$

But the vector boson fields of the same charge are massless, and therefore degenerate, at this level, so one can redefine

$$\begin{Bmatrix} W^L_\mu \\ \\ W^R_\mu \end{Bmatrix} = \begin{pmatrix} \cos \xi & \sin \xi \\ \\ -\sin \xi & \cos \xi \end{pmatrix} \begin{Bmatrix} W^1_\mu \\ \\ W^2_\mu \end{Bmatrix} \qquad (33)$$

new fields W^1_μ and W^2_μ, for which the coupling to fermions is given by the currents:

$$: \frac{g}{2\sqrt{2}} \, \gamma^\mu \, \{\cos\xi(1+\gamma_5) - \sin\xi(1-\gamma_5)\}$$

$$(34)$$

$$: \frac{g}{2\sqrt{2}} \, \gamma^\mu \, \{\sin\xi(1+\gamma_5) + \cos\xi(1-\gamma_5)\}$$

The interest of these new fields W^1 and W^2 is that they are the physical fields with definite mass, obtained from the spontaneous symmetry breaking of the local gauge symmetry. If $M_{W^1} \neq M_{W^2}$ parity invariance is also broken in this process.

One sees from equation (34) that the charged current structure is a general combination of vector and axial vector currents.

The neutral current structure is obtained from the Lagrangian

$$L_{nc}(x) = L_{em}(x) + L_{nc}^{Z^1 + Z^2} \qquad (35)$$

where Z^1, Z^2 and the photon field A are the physical fields, with definite mass, coming from a unitary transformation of W^{L^3}, W^{R^3} and B, the old gauge fields. The identification of A with the electromagnetic field leads to precise relations of the couplings and mixings of this unitary transformation. The resulting couplings for Z^1 and Z^2 are vector and axial vector, but not of (V–A) type.

The way that leptons and quarks are incorporated into the model is straightforward at the level of one generation. One has for leptons

$\underline{SU(2)_L \times U(1)}$ $\qquad\qquad\qquad$ $\underline{SU(2)_L \times SU(2)_R \times U(1)}$

$$\begin{Bmatrix} \nu_L \\ e_L \end{Bmatrix} \; ; \; e_R \qquad\qquad \rightarrow \qquad \begin{Bmatrix} \nu_L \\ e_L \end{Bmatrix} \; ; \; \begin{Bmatrix} \nu_R \\ e_R \end{Bmatrix}$$

$(\tfrac{1}{2},-\tfrac{1}{2}) \; ; \; (0,-1)$ $\qquad\qquad \rightarrow \qquad$ $(\tfrac{1}{2},0,-\tfrac{1}{2}) \; ; \; (0,\tfrac{1}{2},-\tfrac{1}{2})$

where the parentheses under the multiplets denote the representation of the group to which they belong.

For quarks one introduces the multiplets:

$\underline{SU(2)_L \times U(1)}$ $\qquad\qquad\qquad$ $\underline{SU(2)_L \times SU(2)_R \times U(1)}$

$$\begin{Bmatrix} u_L \\ d_L \end{Bmatrix} \; ; \; u_R \; ; \; d_R \qquad \rightarrow \qquad \begin{Bmatrix} u_L \\ d_L \end{Bmatrix} \; ; \; \begin{Bmatrix} u_R \\ d_R \end{Bmatrix}$$

$(\tfrac{1}{2},1/6);(0,2/3);(0,-1/3) \; \rightarrow \;$ $(\tfrac{1}{2},0,1/6); \; (0,\tfrac{1}{2},1/6)$

One notices that in the left-right symmetric model the weak hypercharge is fixed by the lepton number or baryon number, depending on whether we consider leptons or quarks, respectively. In general, $y = (B-L)/2$.

4. EFFECTIVE LAGRANGIAN

We are interested in the low momentum transfer limit of the electroweak interaction. We shall write the charged current and neutral current interactions in a general form of vector and axial vector currents coupled as a contact interaction. The effective couplings will be given in the left-right symmetric model described above, and then particularized in the standard theory.

4.1 Charged Current Interaction

The limit can be understood in terms of the relevant diagrams for the semileptonic process $\ell + u \rightarrow \nu_\ell + d$

under the conditions $|q^2| \ll M^2_1$, M^2_2. One has [13]

$$L_{cc}(x) = -\frac{G}{\sqrt{2}} [C_V(\bar{\nu}\gamma^\mu \ell)(\bar{d}\gamma_\mu u) + C_A(\bar{\nu}\gamma^\mu \gamma_5 \ell)(\bar{d}\gamma_\mu \gamma_5 u)$$

$$+ C'_V(\bar{\nu}\gamma^\mu \gamma_5 \ell)(\bar{d}\gamma_\mu u) + C'_A(\bar{\nu}\gamma^\mu \ell)(\bar{d}\gamma_\mu \gamma_5 u)] \qquad (36)$$

In the left-right symmetric model one has the current-current couplings

$$\begin{aligned}
C_V &= C\,[(1 - \sin2\xi) + \Delta(1 + \sin2\xi)] \\
C_A &= C\,[(1 + \sin2\xi) + \Delta(1 - \sin2\xi)] \\
C'_V &= C\,(1-\Delta)\,\cos2\xi \\
C'_A &= C\,(1-\Delta)\,\cos2\xi
\end{aligned} \qquad (37)$$

where ξ is the mixing angle of the rotation from W^L, W^R to W^1, W^2, and $\Delta = (M_1/M_2)^2$. The parity invariant limit is obtained for the degenerate case $\Delta=1$. This implies $C'_V = C'_A = 0$ and the mixing ξ is irrelevant for C_V, C_A. This is not chosen by nature. A different limit is the current-current lagrangian, obtained for $\Delta=0$. This implies the factorization $C_V C_A = C'_V C'_A$ and a current of the form $(V-A) + \varepsilon (V+A)$. The standard limit, compatible with all present data on charged currents is a particular case of the current-current lagrangian, in which not only $\Delta = 0$, but also $\xi = 0$. The current is $(V-A)$, and $C_V = C_A = C'_V = C'_A = 1$.

12

4.2 Neutral Current Interaction

For the semileptonic sector of electrons and quarks, it is interesting to write an effective lagrangian for the parity violating piece. The parity conserving piece is masked by the electromagnetic interaction, which dominates at low energies.

With the same procedure as above

the general form gives[14] at $|q^2| \ll M_{Z_1}^2$, $M_{Z_2}^2$

$$L_{pv}^{e-q}(x) = - \frac{G}{\sqrt{2}} [\bar{e}\gamma^\mu \gamma_5 e \ (\tilde{\alpha}V_\mu^3 + \tilde{\gamma}V_\mu^0)$$

$$+ \bar{e}\gamma^\mu e \ (\tilde{\beta}A_\mu^3 + \tilde{\delta}A_\mu^0)] \tag{38}$$

where V_μ^3, V_μ^0, A^3 and A^0 are the isovector vector, isoscalar vector, isovector axial and isoscalar axial currents of quarks.

In the left-right symmetric model, the couplings are:

$$\tilde{\alpha} = - f_{pv}(1-\sin^2\theta) \ , \ \tilde{\gamma} = \frac{1}{3}f_{pv}\sin^2\theta$$

$$\tilde{\beta} = - f_{pv}(1-\sin^2\theta) \ , \ \tilde{\delta} = 0 \tag{39}$$

where the global parity violating coupling is

$$f_{pv} = \frac{\sqrt{2}e^2 \ \sin\phi\cos\phi}{\sin\theta\sin2\theta \ G} \ (\frac{1}{M_{Z_1}^2} - \frac{1}{M_{Z_2}^2}) \tag{40}$$

and (θ, ϕ) are the angles of the rotation of the neutral gauge vector bosons to the ones of definite mass.

The parity invariant limit corresponds to $M_{Z_1} = M_{Z_2}$, whereas the current-current limit of the lagrangian is obtained for $(M_{Z_1}/M_{Z_2})^2 \to 0$. In the standard limit, one has to make the replacement $\sin^2\theta \to 2 \sin^2\theta_W$ and $\cos\phi \to (\sqrt{2} \cos\theta_W)^{-1}$, where θ_W is the weak unification angle of the standard theory.

5. LOW ENERGY PHENOMENOLOGY

5.1 Charged Currents

There are many old tests of the (V-A) nature of the charged weak couplings. Recently, there has been new impetus to improve the precision.

- Measurements of the e^+ polarization and ξ parameter in μ-decay, $\mu^+ \to e^+ \nu \bar{\nu}$. They are described[15] by Fetscher and Burkard in their lectures at this School.
- The simultaneous measurement[16] of the average polarization and the longitudinal polarization of recoil in the semileptonic process $\mu^- + {}^{12}C \to \nu_\mu + {}^{12}B$. The ETH-Louvain collaboration analyzed their experiment to conclude that the helicity of the neutrino emitted from muon capture is $h_\nu = -1.08 \pm 0.11$. The analysis assumes, like the Goldhaber experiment in electron capture, that $|h_\nu| = 1$, so the sign is measured. In order to look for deviations of the longitudinal polarization of the neutrino from one, one has to use a more general framework. This is possible[17] measuring both the longitudinal polarization P_L of recoil and the asymmetry α in the angular distribution $\Lambda(\theta) = 1 + \alpha(\vec{P}_\mu \cdot \hat{p})$, where $\vec{P}_\mu$ is the incoming muon polarization and $\hat{p}$ the recoil direction. From the use of rotational invariance only, one has

$$h_\nu = 2P_L - \alpha \tag{41}$$

so h_ν follows.
- The helicity conservation predicted by vector and axial currents has been tested, measuring[18] the μ^+ polarization in the deep inelastic process $\bar{\nu}_\mu + Fe \to \mu^+ + X$. One gets $P_\mu = 0.80 \pm 0.07 \pm 0.14$, where the first error is statistical and the second systematic, to be compared with the prediction $P_\mu = +1$.

5.2 Neutral Currents

- The coupling of neutrinos to u- and d-quarks has been obtained from inclusive neutrino nucleon scattering. For isoscalar targets, $\nu_\mu + N \to \nu_\mu + X$, one can extract[19] from the use of neutrino and antineutrino processes the following combination of couplings

$$u_L^2 + d_L^2 = 0.305 \pm 0.013$$

$$u_R^2 + d_R^2 = 0.036 \pm 0.013 \tag{42}$$

where the notation is self-explanatory. The results (42) are compatible with the standard theory prediction, for a value

$$\sin^2\theta_W = 0.220 \pm 0.014 \tag{43}$$

- The coupling of electrons to u- and d-quarks has been determined incoherently from the asymmetry of longitudinally polarized electrons on deuterons, and coherently from parity violation in atoms. These topics are discussed[20] in the lectures by Pottier.
- The perspectives of using muonic atoms to investigate the neutral current coupling of muons to u- and d-quarks are discussed[21] in the lecture by Missimer.
- Leptonic and semileptonic couplings for charged fermions can be tested from measurements of $e^+e^- \rightarrow \mu^+\mu^-$ and $e^+e^- \rightarrow q\bar{q}$ at the PETRA storage ring. The combined data on cross section and the asymmetry in the angular distribution are explained[22] by the standard theory with $\sin^2\theta_W = 0.24 \pm 0.06$.

6. MAJORANA NEUTRINOS

It is apparent that neutrinos are different from other known fermions. First, the masses of the observed neutrinos are, if any, much smaller than those of the charged leptons and quarks.

$$m_{\nu_e}/m < 10^{-4}, \ m_{\nu_\mu}/m_\mu < 1/200, \ m_{\nu_\tau}/m_\tau < 1/10 \tag{44}$$

Second, only left handed neutrinos have ever been observed. In fact, all present data on charged current weak interactions are consistent with (V-A) currents, so only the left handed components of neutrinos are produced in the processes. In contrast, quarks and charged leptons are known to have both right and left handed helicity states.

With these considerations in mind, we explore the possibility offered by Majorana fields to generate a mass term without the existence of right handed neutrino fields. Automatically, the Majorana mass term is lepton number violating, $\Delta L = 2$, leading to non vanishing probability for no neutrino double beta decay[23]. The propagators for Majoranas are discussed in section 6.1. In section 6.2 the implications[24] of the presence of several generations are studied. For Majorana neutrinos, CP violation is possible with only two generations. The phenomenon of "neutrino-antineutrino" oscillation appears in contrast with the Dirac case. Finally in section 6.3 we show the origin of atom mixing[25] and the perspectives to observe no neutrino double electron capture.

6.1 Majorana Field

For a fermion field, in the representation in which γ_5 is diagonal, it is possible to write a particular four-spinor solution with the following choice of components

$$\psi(x) = \begin{Bmatrix} i\sigma_2\xi^*(x) \\ \xi(x) \end{Bmatrix} \tag{45}$$

where

$$\begin{Bmatrix} 0 \\ \xi(x) \end{Bmatrix} = \tfrac{1}{2}(1 + \gamma_5) \; \psi(x) \equiv \psi_L(x) \tag{46}$$

The equation satisfied by the two-spinor is

$$(i\partial_t - i\vec{\sigma}\cdot\vec{\nabla}) \; \xi(x) + mi\sigma_2\xi^*(x) = 0 \tag{47}$$

its conjugate being the one corresponding to the other components $i\sigma_2\xi^*(x)$. In the case $m = 0$, equation (47) coincides with the Weyl equation. The solution (45) induced by $\xi(x)$ is a Majorana field.

One can work with four-component spinors with an additional restriction imposed by

$$\psi(x) = \eta\psi(x)^c = \eta C\bar{\psi}^T(x) \tag{48}$$

where C is the charge conjugation matrix

$$C\gamma^\mu C^{-1} = -\gamma^{\mu T}; \; C = -C^T = -C = -C^{-1} = C^* \tag{49}$$

and η is a phase. This leads to the following expansion

$$\psi(x) = \frac{1}{(2\pi)^3} \int \frac{d^3p}{2E(\vec{p})} \sum_\lambda \{u(\vec{p},\lambda)a(\vec{p},\lambda) \; e^{-ipx}$$

$$+ \eta u^c(\vec{p},\lambda)a(\vec{p},\lambda)e^{ipx}\} \tag{50}$$

where u and u^c are the four-spinors defined as for Dirac fermions.

For a Dirac fermion, $\psi_L(x)$ and $\psi_R(x)$ are independent. With both, it is possible for the fermion to acquire a Dirac mass. If, as assumed in spontaneous broken gauge theories, this mass is generated by the non vanishing vacuum expectation value of the Higgs field we can write a term in the Lagrangian

$$g_{\chi\bar{\psi}\psi} < 0 \mid \chi(x) \mid 0 > \bar{\psi}_R(x) \; \psi_L(x) \tag{51}$$

where the Higgs field $\chi(x)$ has zero lepton number and a weak isospin $\tfrac{1}{2}$. This is the scenario which is presented at the level of the standard $SU(2) \times U(1)$ gauge theory.

The question which arises for neutrinos is whether, with the absence of $\psi_R(x)$, it is possible to generate any mass at all. We have seen above that there are particular solutions, with only $\psi_L(x)$, which satisfy a Dirac equation with non vanishing mass. Therefore, we have an affirmative answer to our question. In fact, $\psi_R(x)$ is a left-handed component, so that $\bar\psi_R(x)$ could be replaced by $\psi_L^c(x)$. The only difference between this last spinor and $\psi_L(x)$ is in the Lorentz properties, but this can be arranged.

For charged leptons and quarks, which have non-zero electric charge and color, a mass term in the Lagrangian $q_L q_L$ or $\ell_L \ell_L$ is forbidden, because it is not a scalar under $SU(3)_c \times U(1)_{em}$. These are local gauge symmetries supposed to be exact, according with the present dogma. But neutrinos are colorless and have zero electric charge, so for them a mass term in the Lagrangian written as

$$m \ \nu_L^T(x) \ C \ \nu_L(x) = m \ \overline{\nu_L^c}(x) \ \nu_L(x) \tag{52}$$

seems quite legal. Equation (52) constitutes a Majorana mass term. The presence of the Majorana mass term implies lepton number violation in two units. This is not in contradiction with the dogma.

To generate a Majorana mass by spontaneous symmetry breaking, we cannot stay at the level of the standard theory with a weak doublet of scalars. In fact,

$$m = g_{\chi\nu\nu} \ < 0 \ |\chi(x)| \ 0 > \tag{53}$$

needs a $\chi(x)$ which is a member of a triplet in weak isospin. This appears naturally in grand unified theories as the ones based on the gauge group SO(10). At the level of the electro-weak theory, one needs an extension of the Higgs sector.

With the expansion of the Majorana field as given by equation (50) we can build the following two point Green functions. The Dirac propagator corresponds to

$$S(p) \equiv -i \int d^4x \ < 0 \ |T(\psi(x)\bar\psi(0))| \ 0 > e^{ipx}$$

$$= \frac{\not p + m}{p^2 - m^2 + i\epsilon} \tag{54}$$

as for Dirac fields. However due to the simultaneous presence of both $a(\vec p, \lambda)$ and $a(\vec p, \lambda)$ in the Majorana field $\psi(x)$, the propagator

$$\tilde S(p) \equiv -i \int d^4x \ e^{ipx} > 0 \ |T(\psi(x)\psi(0)^T)| \ 0 >$$

$$= -\eta \ \frac{\not p + m}{p^2 - m^2 + i\epsilon} \ C \tag{55}$$

is non vanishing. Equation (55) corresponds to the so-called "neutrino-antineutrino" propagation, which is a $\Delta L = 2$ term. When (55) is inserted between left-handed currents, the result is proportional to the neutrino mass.

6.2 Neutrino Oscillations

When several generations are considered, the mass parameter m in the Lagrangian is replaced by a matrix M. In the process of diagonalization, one gets from the weak current neutrino fields $(\nu_L)_i$ the definite mass neutrino fields $\psi(x)$ in the form

$$\psi(x) = V \nu_L(x) + V'C \bar{\nu}_L(x)^T \tag{56}$$

The non-trivial unitary transformation V induces the phenomenon of oscillations. The definite mass field (56) is a Majorana field, because it satisfies

$$\psi(x) = V V'^T \psi(x)^c \tag{57}$$

and $V V'^T$ is a diagonal matrix of phases. The matrix V is responsible, as for the Dirac case, of neutrino oscillations due to flavor violation, governed by the first type propagation with time.

The existence of the non-vanishing second type propagation (55) induces a new kind of neutrino oscillation. In fact, with double charge exchange, one can ask for the probability amplitude that a charged lepton ℓ_j^+ becomes a charged lepton ℓ_i^-. It is given by

$$\ell_j^+ \rightarrow \ell_i^- : $$

$$\tag{58}$$

$$X(\ell_j^+ \ell_i^-) = - \sum_K V_{ki}^* \frac{\eta_k m_k}{p^2 - m_k^2 + i\epsilon} V_{kj}^*$$

The result (58) is responsible of the set of new phenomena associated with $\Delta L = 2$ processes. If two generations are coupled, only a real parameter in V is necessary and a relative phase $e^{i\alpha}$ between the two neutrino fields. The amplitude for $\beta\beta$-decay without neutrinos would be proportional to

$$m_1 \cos^2\theta + e^{i\alpha} m_2 \sin^2\theta \tag{59}$$

where $m_{1,2}$ are the masses of neutrinos. For $\mu^+ \rightarrow e^-$ conversion, however, one gets an amplitude proportional to

$$(m_1 - e^{i\alpha} m_2) \sin\theta \tag{60}$$

Values of $\alpha \neq 0$, π imply a signal of CP violation in the leptonic sector which, contrary to the quark sector, is possible even with only two generations. The value $\alpha = 0(\pi)$ corresponds to equal (opposite) CP-eigenvalues of the two neutrino species.

The phenomenon of $\Delta L = 2\nu-\bar{\nu}$ oscillations[26] can be discussed as follows. Assume that at time $t = 0$ an antineutrino (using conventional wisdom) is produced from the charged lepton ℓ_j^+. We ask for the probability that at time $t > 0$ it behaves as a neutrino capable of producing a charged lepton ℓ_j^-. The time evolution is given by

$$P(e^+ \overset{t}{\to} e^-) = \frac{m_1 m_2}{E^2} \{ \frac{m_1}{m_2} \cos^4\theta + \frac{m_2}{m_1} \sin^4\theta$$

$$+ \tfrac{1}{2}\sin^2 2\theta \cos[(E_1-E_2)t + \alpha] \} \qquad (61)$$

$$P(e^+ \overset{t}{\to} \mu^-) = \frac{m_1 m_2}{E^2} \frac{1}{4} \sin^2 2\theta \{ \frac{m_1}{m_2} + \frac{m_2}{m_1} - 2 \cos[(E_1-E_2)t + \alpha] \}$$

where the CP-violating phase appears as a phase shift in the oscillatory term with time.

6.3 Atom Mixing

We have seen that the $\Delta L = 2$ neutrino mixing has the implication of non-vanishing no neutrino double beta decay of nuclei

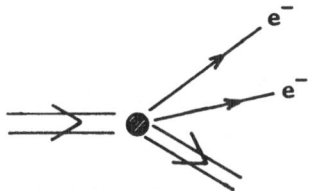

Analogously, the no neutrino single electron capture (NONSEC) in atoms[27] becomes accessible

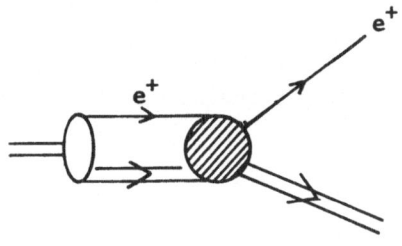

We are going to discuss with some detail the case of double electron capture[28](NONDEC):

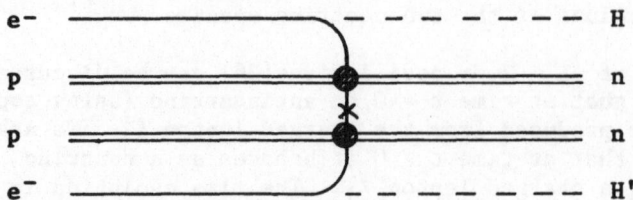

associated with a two nucleon mechanism $e^-e^- pp \to nn$, H and H' being the vacancies in the daughter atom of the two captured electrons. From the point of view of the atom, the process $(Z,A) \to (Z-2, A)^{H,H'}$ is a virtual mixing of the parent atom with the daughter atom with two electron holes. The process becomes real as the daughter atom de-excites $(Z-2,A)^{H,H'} \to (Z-2, A)+ ---$. To summarize, we have

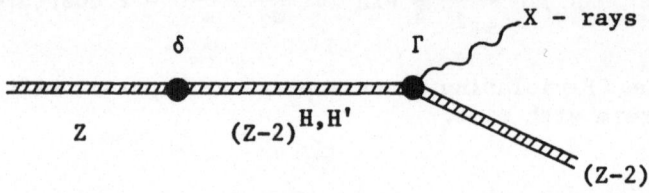

which can be discussed as a two states problem, where δ is the non diagonal term in the atomic mass matrix

$$M = \begin{pmatrix} m + \tfrac{1}{2}\Delta & \delta \\ \delta & m - \tfrac{1}{2}\Delta - i\,\dfrac{\Gamma}{2} \end{pmatrix} \;;\quad \begin{array}{l} m = \dfrac{m_1+m_2}{2} \\[2mm] \Delta = m_1 - m_2 \end{array} \qquad (62)$$

This implies that the stationary states of matter, the ones with definite mass and life time, are not the atoms but linear combinations of Z and (Z-2) atoms!

In the limit $\delta \ll \Delta$, Γ the eigenvalues of the mass matrix M are given by

$$\lambda_1 \simeq m + \tfrac{1}{2}\Delta - i\,\frac{\delta^2}{\Delta^2+\Gamma^2/4}\,\frac{\Gamma}{2}$$

$$\lambda_2 \simeq m - \tfrac{1}{2}\Delta - i\,\frac{\Gamma}{2} \qquad (63)$$

so that the originally stable atom (in the absence of δ) has actually a NONDEC decay rate

$$1/\tau \simeq \frac{\delta^2}{\Delta^2+\Gamma^2/4}\,\Gamma \qquad (64)$$

The strategy suggested by equation (64) is the one saying that the most interesting cases are those for which Δ is closest to zero. The ideal situation would correspond to the condition $\Delta \sim \Gamma$, but Δ is a difference in nuclear energy levels, typically MeV, and Γ is the atomic X-ray decay rate, typically eV. So we are asking for a "monumental" coincidence[29]. From present knowledge of atomic masses, one finds some cases with $\Delta = 0 \pm$ few keV, but the errors are too big.

These limitations preclude the search for atom oscillations with time. Any realistic time interval for an experiment satisfies $t \gg \Gamma^{-1} \sim 10^{-15}$ sec, implying that the daugther atom disappears much before. Furthermore, $t \gg \Delta^{-1} \sim 10^{-18}$ sec, so the oscillation is washed out. Only asymptotic times are available, for which the Z atom has a (Z-2) component with a probability $\delta^2/(\Delta^2+\Gamma^2/4)$. This leads to the NONDEC decay rate τ^{-1} given in equation (64).

To estimate the non diagonal mass δ, one has to proceed with the calculation of the diagram

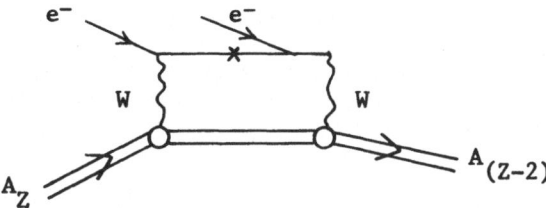

This calculation has been made[25] in some cases of good degeneracy, as it is the case for the mixing $^{112}_{50}\text{Sn} \leftrightarrow ^{112}_{48}\text{Cd}$ $(1.87)^{(1s)-2}$. One gets

$$\delta \simeq 2 \ 10^{-22} \text{eV} \ \frac{\langle m_\nu \rangle}{30 \text{eV}} \qquad (65)$$

which is about sixteen orders of magnitude smaller than the $K_L - K_S$ mass difference.

If, for the sake of the discussion, we assume $\delta \sim 100$ eV, then the NONDEC and TWONDEC lifetimes are comparable, of the order 10^{23} years. The two decays have distinct signatures:

NONDEC $\rightarrow$ 2 X-rays + γ -rays cascade
TWONDEC $\rightarrow$ 2 X-rays

and 50 decays of each type would take place per year and per kilogram of ^{112}Sn ($\sim$100 Kg of natural tin).

We have discussed the richness of new phenomena if neutrinos are described by Majorana fields. There is a new kind of lepton number violating $\Delta L = 2$ neutrino–antineutrino oscillation, with probabilities proportional to neutrino masses. Because of the restriction imposed to Majorana fields, CP violation appears possible for the leptonic sector even with only two generations. In the $\Delta L = 2$ neutrino oscillation, the CP violating phase manifests itself as a phase shift in the oscillatory term with time.

Induced by the second type propagation for Majorana neutrinos, there is a mixing of atoms, leading to non stability of otherwise stable atoms. The decay rate for corresponding process of no neutrino double electron capture is, consequently, resonance enhanced. For the best found cases, perhaps not the best existing cases, one finds a lifetime $\tau = 10^{23}$ years $\left(\frac{\Delta \ x30}{m_\nu x100}\right)^2$.

Acknowledgements

The author thanks the Organizers of the School for the kind invitation extended to him, as well as for the proper atmosphere created around the participants in the School. Many of the ideas, or the way to present them, discussed here have been a consequence of the collaboration of the author with F. J. Botella, A. De Rújula, C. Jarlskog, P. Pascual and A. Pich. I would like to acknowledge to all of them their contribution.

REFERENCES

1. C. N. Yang and R. L. Mills, Phys.Rev., 86:191 (1984).
2. S. L. Glashow, Nucl.Phys., 22:579 (1961).
3. S. Weinberg, Phys.Rev.Letters, 19:1264 (1967).
4. A. Salam, in: "Elementary Particle Physics," N. Svartholm, ed., Stockholm (1968).
5. M. Baker and S. L. Glashow, Phys.Rev., 128:2462 (1962).
6. J. Goldstone, A. Salam and S. Weinberg, Phys.Rev., 127:965 (1962).
7. P. W. Higgs, Phys.Rev., 145:1156 (1966); T. W. B. Kibble, Phys.Rev., 155:1554 (1967).
8. U A 1 Collaboration, G. Arnison, Phys.Letters, 122B:103 (1983); U A 2 Collaboration, G. Banner, Phys.Letters, 122B:476 (1983).
9. J. Kim, Rev.Mod.Phys., 53:211 (1981); I. Liede and M. Roos, Nucl.Phys., B167:397 (1980).
10. J. C. Pati and A. Salam, Phys.Rev., D10:275 (1974).
11. M. Kobayashi and K. Maskawa, Prog.th.Phys., 49:652 (1973).
12. S. L. Glashow, J. Illiopoulos and L. Maiani, Phys.Rev., D2:1285 (1970).
13. M. A. Beg, Phys.Rev.Letters, 22:1252 (1977).
14. P. W. Hung and J. J. Sakurai, Phys.Letters, 69B:323 (1977).

15. See the Lectures by W. Fetscher and H. Burkard.
16. L. Ph. Roesch, <u>Phys.Rev.Letters</u>, 46:1507 (1981).
17. F. J. Botella and J. Bernabéu, <u>Nucl.Phys.</u>, A414:456 (1984).
18. G. Barbiellini, "ν-N Structure of Weak Interaction," CERN report (1981).
19. M. Jonker, <u>Phys.Letters</u>, 102B:67 (1981).
20. See the lectures by L. Pottier.
21. See the lecture by J. Missimer.
22. F. Barreiro, Proceedings of the Winter Meeting on Fundamental Physics, Santillana del Mar, Spain (1984).
23. H. Primakoff and S. P. Rosen, <u>Phys.Rev.</u>, 184:1925 (1969).
24. J. Bernabéu and P. Pascual, CERN Th 3393 (1982), to appear in <u>Nucl.Phys.</u>, B.
25. J. Bernabéu, A. De Rujula and C. Jarlskog, <u>Nucl.Phys.</u>, B223:15 (1983).
26. J. Schechter and J. W. F. Valle, <u>Phys.Rev.</u>, D23:1666 (1981).
27. J. D. Vergados, CERN Th 3396 (1982).
28. H. M. Georgi, S. L. Glashow and S. Nussinov, <u>Nucl.Phys.</u>, B193:297 (1981).
29. R. G. Winter, <u>Phys.Rev.</u>, 100:142 (1955).

DETERMINATION OF THE MUON DECAY PARAMETERS

W. Fetscher

Institut für Mittelenergiephysik der ETH-Z
c/o SIN
CH-5234 Villigen

The investigation of the purely leptonic decay $\mu^+ \to e^+ \nu_e \bar{\nu}_\mu$ of the muon is of fundamental interest since it allows to put constraints on theories of weak interactions. To improve our present-day knowledge about charged weak interactions it is necessary both to measure the decay parameters with high precision <u>and</u> to make full use of the mutual interdependence of the decay parameters yielding additional constraints.

The decay is here described by the most general four-fermion interaction without derivatives introduced by Kinoshita and Sirlin[1] in charge retention form. This is most appropriate for data obtained without detecting the neutrini. The effective leptonic Hamiltonian density is given by[2]

$$= \sum_i \{ C_i (\bar{e} \Gamma_i \mu)(\bar{\nu}_\mu \Gamma^i \nu_e)$$

$$+ C_i' (\bar{e} \Gamma_i \mu)(\bar{\nu}_\mu \Gamma^i \gamma_5 \nu_e) + h.c.\} \quad,$$

where Γ_i stands for the scalar, vector, tensor, axial vector and pseudoscalar operators, respectively.

The C_i and C_i' are complex. We can set one of the constants, for example C_V, to be real, and thus obtain the parameter set $\{P_1\} = \{C_S, C_V, \ldots, C_A', C_P'\}$ which amounts to 19 independent parameters. If the two neutrini are not detected, ten parameters can be determined. Then it is more convenient to use the parameter set $\{P_2\}$ proposed by Derenzo[3] and Scheck[2]:

$$\{P_2\} = \{g_S, g_P, g_V, g_A, g_T, \phi_{PS}, \phi_{VA}, \phi_{TT}, \psi_{PS}, \psi_{VA}\} ,$$

which are defined by

$$g_i = \sqrt{|C_i|^2 + |C_i'|^2} ,$$

$$\cos \phi_{ij} = - \frac{\mathrm{Re}(C_i C_j'^* + C_i' C_j^*)}{g_i g_j} ,$$

and

$$\sin \psi_{ij} = - \frac{\mathrm{Im}(C_i C_j'^* + C_i' C_j^*)}{g_i g_j} .$$

Experimental results favor the V-A form of charged leptonic weak interactions. It is characterized in our parameter set $\{P_2\}$ by

$$g_V = g_A , \qquad \cos \phi_{VA} = 1$$

$$g_S = g_P = g_T = 0 \quad \text{and} \quad \sin \psi_{VA} = 0 ,$$

while ϕ_{PS}, ϕ_{TT} and ψ_{PS} remain undefined.

Previous experiments, however, allow large deviations from the pure V-A type of interaction (Derenzo[3]):

$$g_S \leqq 0.33 \, g_V$$

$$g_P \leqq 0.33 \, g_V$$

$$g_T \leqq 0.28 \, g_T$$

$$0.76 \, g_V \leqq g_A \leqq 1.20 \, g_V .$$

The measurement of the polarization vector (P_L, P_{T_1}, P_{T_2}) (for definition of the components see Fig. 1) of the positron yields the five parameters $\{\xi', \alpha/A, \beta/A, \alpha'/A, \beta'/A\}$ with

$$\xi' = (2 g_S g_P \cos\phi_{PS} + 8 g_V g_A \cos\phi_{VA} - 6 g_T^2 \cos\phi_{TT})/A.$$

$$\alpha/A = (g_S^2 - g_P^2)/A$$

$$\beta/A = (g_V^2 - g_A^2)/A$$

$$\alpha'/A = - 2 g_S g_P \sin\psi_{PS}/A$$

$$\beta'/A = - 2 g_V g_A \sin\psi_{VA}/A \qquad \text{and}$$

$$A = (g_S^2 + g_P^2) + 4 (g_V^2 + g_A^2) + 6 g_T^2 \qquad .$$

The well known low energy parameter η is equal to $\eta = (\alpha - 2 \beta)/A$. α/A and β/A signal the unequality of the strength of P and S interactions and of V and A interactions, respectively, while α'/A and β'/A are sensitive to the violation of T-invariance.

V-A predicts

$$P_L \lesssim 1, \quad <P_{T_1}> \approx - 0.003, \quad <P_{T_2}> = 0 \quad ,$$

where the brackets indicate the energy average of the corresponding quantities.

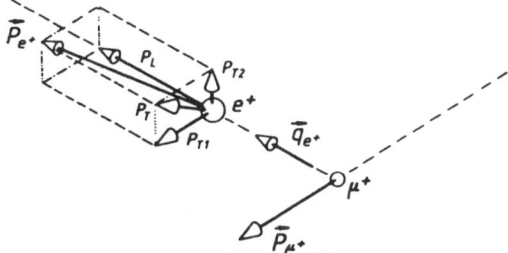

Figure 1. Definition of the components of the positron polarization, $\vec{P}_{e^+}$ with respect to a special reference system. P_L is along the positron's line of flight $\vec{q}_{e^+}$, P_{T_1} lies in the plane formed by the muon polarization $\vec{P}_{\mu^+}$ and by $\vec{q}_{e^+}$, P_{T_2} is perpendicular to that plane.

The two transverse polarization components P_{T_1} and P_{T_2} have been measured as a function of energy[4]. The principle of the experiment is described in Fig. 2. The energy dependence of P_{T_1} and P_{T_2} is shown in Fig. 3 (preliminary results including all data). Their energy average is

$$\langle P_{T_1}\rangle = (16 \pm 23) \times 10^{-3} \quad \text{and}$$

$$\langle P_{T_2}\rangle = (7 \pm 23) \times 10^{-3} \quad .$$

Both values are in good agreement with V-A predictions.

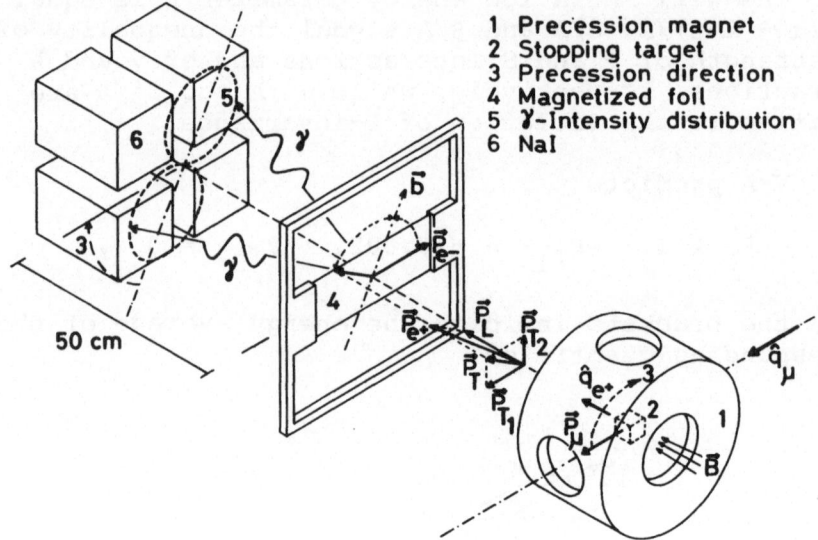

1 Precession magnet
2 Stopping target
3 Precession direction
4 Magnetized foil
5 γ-Intensity distribution
6 NaI

50 cm

Figure 2. Principle of the measurement of a possible transverse positron polarization in the decay of polarized muons. The muons from the SIN muon channel stop in a target where they decay. The positron polarization is measured by probing the pattern of the intensity distribution of the quanta from annihilation in flight with the polarized electrons in a magnetized iron foil. By coherent precession of the original muon spins the pattern is swept across pairs of sodium iodide counters.

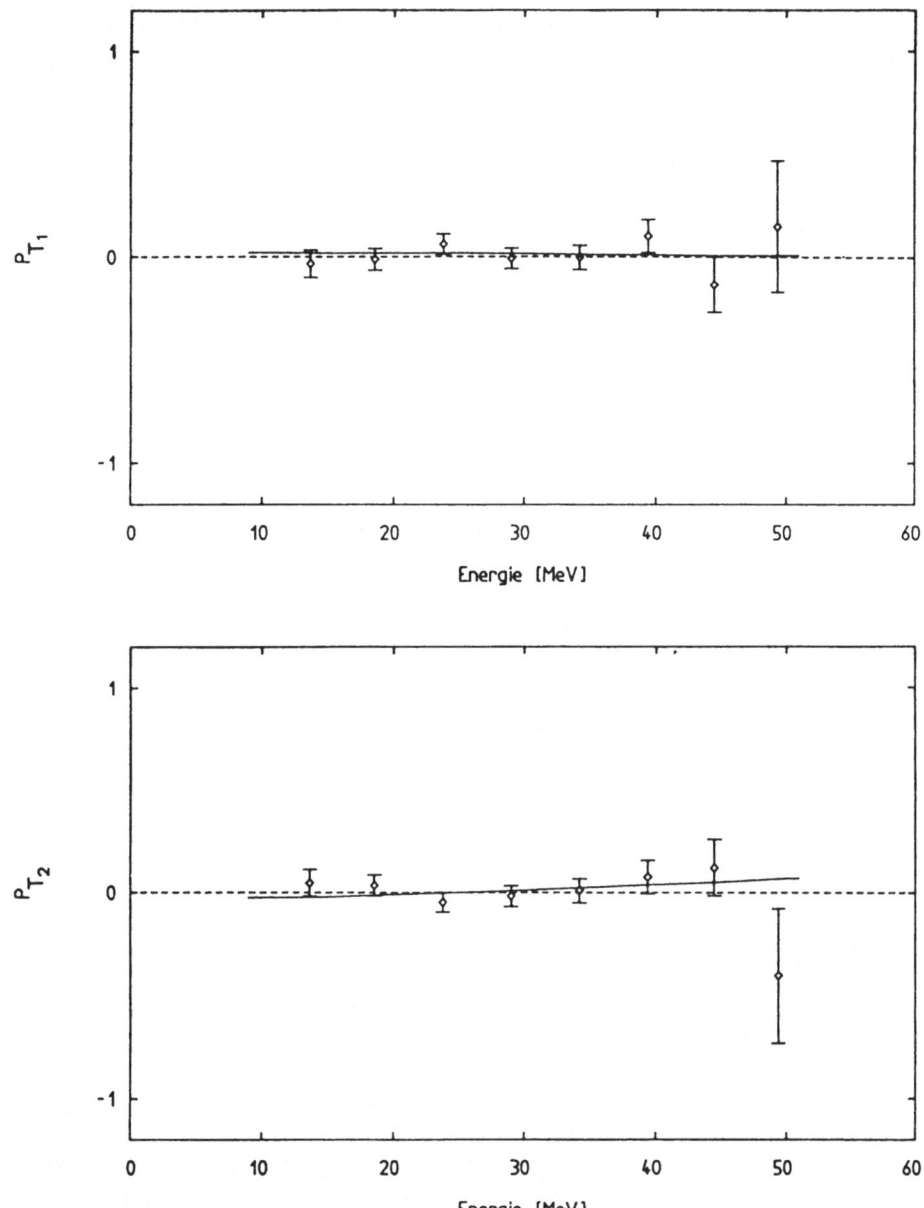

Figure 3. The transverse polarization components P_{T_1} and P_{T_2} as obtained from a direct analysis of the data and from an analysis for α/A, β/A, α'/A and β'/A (solid curves) as a function of the e^+ total energy. The errors include both statistical and systematic contributions.

Table 1. Initial results for α/A, β/A, α'/A and β'/A. The errors include both statistical and systematic contributions. (1) General analysis, (2) Analysis assuming an interaction of the form V-(1+ε)A, where ε is a complex number.

	$10^3 \frac{\alpha}{A}$	$10^3 \frac{\beta}{A}$	$10^3 \frac{\alpha'}{A}$	$10^3 \frac{\beta'}{A}$
(1)	114 ± 107	− 38 ± 37	− 115 ± 104	29 ± 37
(2)	0	− 2 ± 17	0	− 7 ± 16
	$10^3 \, \eta$			
(1)	190 ± 176			
(2)	4 ± 33			

The analysis of the energy dependence (of a sub-sample of the data) yields the values for α/A, β/A, α'/A and β'/A in Table 1[4]. Row (1) shows the results of a data analysis without restricting the types of interaction. For the results of row (2) an interaction of the form V-(1+ε)A was assumed[2], where ε is a complex number.

The limits obtained for ε are compared in Fig. 4 with previous measurements obtained for η [3] and ξ [5].

Fig. 5 shows a model independent diagram for α/A and α'/A (P, S-interaction) and for β/A and β'/A (V, A-interaction). The outer circles display the mathematical boundaries of the variables. The shaded circles show the 90% confidence limits obtained by this measurement.

The following two quantities can be calculated directly by using the approximate normalization

$$A \approx 4 \, (g_V^2 + g_A^2)$$

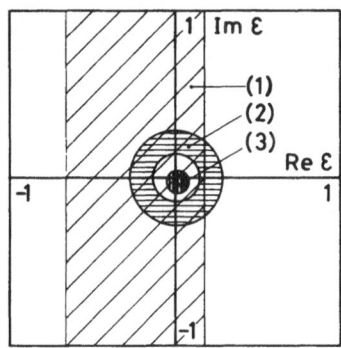

Figure 4. The complex ε plane. The shaded regions show, in the V-(1+ε)A type of interaction in the charge retention form, the limitations placed on ε by the measured values of (1) η = - 0.12 $\pm$ 0.21[3], (2) ξ = 0.975 $\pm$ 0.015[5], and (3), from this experiment, α/A = 0.0, β/A = - 0.002 $\pm$ 0.017, α'/A = 0.0, and β'/A = - 0.007 $\pm$ 0.016. Here 2η = Re ε, $2(1-\xi)$ = $|\varepsilon|^2$, and $4(-\beta+i\beta')/A$ = ε. The boundaries on ε correspond to 68% confidence limits.

through

$$g_A/g_V = \sqrt{\frac{1-4\ \beta/A}{1+4\ \beta/A}} \qquad \text{and}$$

$$\sin\psi_{VA} = \frac{-4\ \beta'/A}{\sqrt{1-(4\ \beta/A)^2}}$$

With the data of Table 1, row (1) we obtain

$$1.00\ g_V \lesseqgtr g_A \lesseqgtr 1.36\ g_V$$

and
$$\sin\psi_{VA} = -0.116 \pm 0.148 \qquad .$$

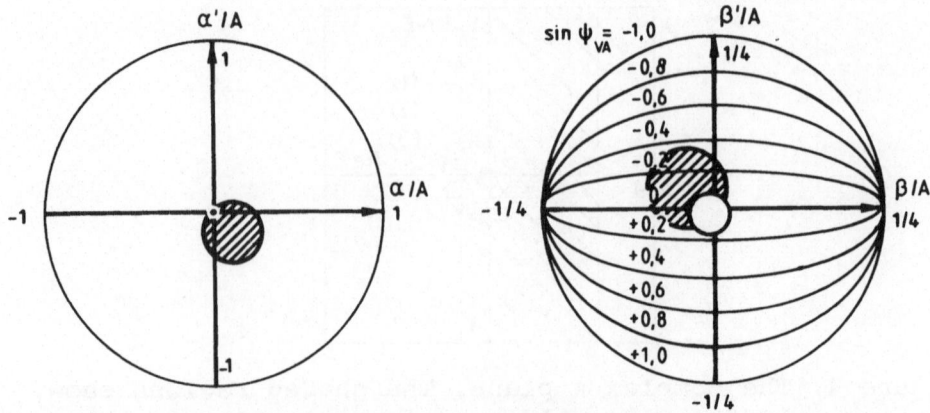

Figure 5. Model independent diagram for α/A and α'/A
(P, S-interaction) and for β/A and β'/A
(V, A-interaction). The shaded circles show
the 90% confidence limits obtained by this
measurement. By using the constraints given
by the other muon decay parameters the areas
in the diagrams are further reduced (to the
dot in the center of the left diagram and to
the unshaded circle in the right diagram).

We can obtain better limits for these quantities
by combining the result of our measurement of the
longitudinal polarization $P_L = \xi'$ of the positron[7] with
previous measurements of other decay parameters (ρ[3],
δ[8] and $\xi\frac{\delta}{\rho}$[6]). This yields an upper bound for $a = g_S^2 + g_P^2$
and thus for α/A and α'/A, since $|\alpha| \leqq a$, $|\alpha'| \leqq a$
(see dot in the center of the corresponding diagram in
Fig. 5). This method, together with our new P_L value, is
described elsewhere[9].

With these improved limits for $|\alpha|$ and $|\alpha'|$ we can
derive better limits for β and β'. The correlation
coefficient between α and β and between α' and β',
respectively, is equal to - 0.895. Small values of α
and α' thus imply small values of β and β'. It is shown
that β and β' acquire essentially the values of row (2)
in Table 1 (corresponding to $\alpha = \alpha' \equiv 0$, but without
making this assumption. This yields the new and better
limits for g_A/g_V and for the T-invariance violating
quantity $\sin\psi_{VA}$ in Table 2. All the results agree well
with the standard V-A theory.

Table 2. Summary of results.

	V-A	This experiment	Previous experiments
g_A/g_V	1	(0.94, 1.08)	(0.76, 1.20) (Derenzo 69)
$\sin\psi_{VA}$	0	$+ 0.028 \pm 0.064$	None
$10^3 \, \eta$	0	4 ± 33	$- 120 \pm 210$ (Derenzo 69)

References

1. T. Kinoshita and A. Sirlin, Phys. Rev. 108:844 (1957).
2. F. Scheck, Phys. Rep. 44:187 (1978).
3. S.E. Derenzo, Phys. Rev. 181:1854 (1969).
4. F. Corriveau, J. Egger, W. Fetscher, H.-J. Gerber, K.F. Johnson, H. Kaspar, H.J. Mahler, M. Salzmann and F. Scheck, Phys. Lett. 129B:260 (1983).
5. V.V. Akhmanov, I.I. Gurevich, Yu.P. Dobretsov, L.A. Makar'ina, A.P. Mishakova, B.A. Nikol'skii, B.V. Sokolov, L.V. Surkova, and V.D. Shestakov, Sov. J. Nucl. Phys. 6:230 (1968).
6. J. Carr, G. Gidal, B. Gobbi, A. Jodidio, C.J. Oram, K.A. Shinsky, H.M. Steiner, D.P. Stoker, M. Strovink, and R.D. Tripp, Phys. Rev. Lett. 51:627 (1983).
7. F. Corriveau, J. Egger, W. Fetscher, H.-J. Gerber, K.F. Johnson, H.J. Mahler, M. Salzmann, H. Kaspar, and F. Scheck, Phys. Rev. D24:2004 (1981).
8. D. Fryberger, Phys. Rev. 166:1379 (1968).
9. H. Burkard, F. Corriveau, J. Egger, W. Fetscher, H.-J. Gerber, K.F. Johnson, H. Kaspar, H.J. Mahler, M. Salzmann, and F. Scheck, submitted to Phys. Lett. B (1984).

MEASUREMENT OF THE ξ-PARAMETER IN μ-DECAY

H. Burkard

Institut für Mittelenergiephysik der ETH-Z
c/o SIN
CH-5234 Villigen / Switzerland

1. INTRODUCTION

In a continuing program for improving the knowledge of electron's observables in muon decay, we propose to make a precise measurement of the integral asymmetry parameter $P_\mu \xi$[1]. Our objective is to reach an accuracy of 1×10^{-3} with an intermediate step of 2×10^{-3}. To do so we require muons of known high transverse polarization and no depolarization of these muons before they decay. To solve this problem a novel approach will be used which utilizes both pion and muon decay in flight. To measure $P_\mu \xi$ we will use a decay-electron-detector sensitive to the emission of electrons parallel and antiparallel to P_μ and to the full range of electron energies.

2. MOTIVATION

The angular distribution of the electrons from the decay of polarized muons is given by

$$
\frac{d^2\Gamma}{dx\ d\cos\phi} \sim (x^2 - x^3)\ (1 + \frac{1}{3}\ P_\mu \xi\ \cos\phi)
$$

$$
+ \frac{2}{9}\rho\ (4x^3 - 3x^2)\ (1 + P_\mu \xi \frac{\delta}{\rho}\ \cos\phi)\ , \qquad (2.1)
$$

where terms of the electron mass with respect to the muon mass have been neglected, ϕ is the angle between the muon polarization P_μ and the electron momentum, x is the reduced electron energy. The "Michelparameters" ξ, δ and ρ are functions of the different coupling constants which appear in a general four fermion Hamiltonian[2]. For the maximum electron energy (x = 1) one gets the following angular distribution:

$$\frac{d\Gamma}{d\cos\phi} \sim 1 + P_\mu \xi \frac{\delta}{\rho} \cos\phi \qquad . \qquad (2.2)$$

To derive $P_\mu \xi$ alone we integrate (2.1) over all energies x. This yields

$$\frac{d\Gamma}{d\cos\phi} \sim 1 + \frac{1}{3} P_\mu \xi \cos\phi \qquad . \qquad (2.3)$$

A precise knowledge of $P_\mu \xi$ is of fundamental importance in testing theories of charged leptonic weak interaction. The present value of $P_\mu \xi = 0.975 \pm 0.015$[3] is not sufficient to have an implication for present-day theories.

Beg et al.[4] have proposed a left-right symmetric model where charged currents are mediated by left- and right-handed gauge bosons W_L and W_R. The masseigenstates of these bosons are then given by

$$W_1 = W_L \cos\zeta - W_R \sin\zeta$$

$$W_2 = W_L \sin\zeta + W_R \cos\zeta$$

with masses m_1, m_2. A 1% measurement of $P_\mu \xi$ would set the following limits on the mixing angle ζ and $m_2 = m(W_R)$ (see also Fig. 1):

$$m(W_R) > 530 \frac{GeV}{c^2}$$

and $- 0.023 < \zeta < 0.020$ at 90% C.L.

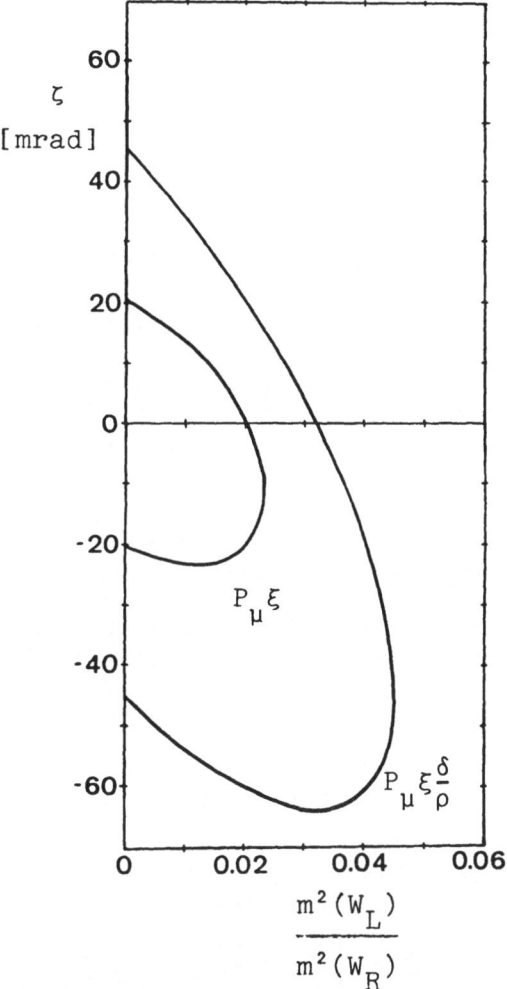

Figure 1. 90%-confidence limits on the $W_{L,R}$ mass squared ratio and mixing angle ζ describing possible right-handed charged currents. Shown are the allowed regions for $P_\mu \xi \frac{\delta}{\rho}$ 5 and for our proposed measurement of $P_\mu \xi$ 1 at a precision of 1×10^{-3}.

The present limits set by $P_\mu \xi \frac{\delta}{\rho}$ [5] are:

$$m(W_R) > 380 \; \frac{GeV}{c^2}$$

and $- \; 0.065 < \zeta < 0.046$ at 90% C.L.

3. METHOD

We propose to measure $P_\mu \xi$ by using pion decay in flight to produce transverse polarized muons[6] and to measure the azimuthal distribution from the decay of these muons also in flight. The whole experimental set-up is shown in Fig. 2.

3.1. Production of Muons with High Transverse Polarization

For the evalution of ξ from the muon decay asymmetry, the knowledge of the muon polarization vector is needed to the same precision. This vector lies in the plane formed by the pion and muon line of flight. The transverse polarization ζ_T depends on the muon emission angle θ_μ. Assuming the neutrino helicity h_ν in π-decay to be $|h_\nu| = 1$, the transverse polarization is given by

$$\zeta_T = \frac{\sin\theta_\mu}{\sin\theta_o} \; . \tag{3.1}$$

For muons emitted under their maximum laboratory angle θ_o ("Jacobian"-peak) the polarization vector is exactly perpendicular to the laboratory line of flight (Fig. 3). Since our apparatus accepts only muons less than 20 mrad away from the maximum muon emission angle θ_o, the muon sample is highly transversely polarized (average greater than 0.97). The polarization will be determined by measuring the muon laboratory line of flight for each event by the low mass drift chamber K_1 (Fig. 2). In practice the pion direction and the true muon trajectory are known with limited accuracy only. However Monte-Carlo-calculations have shown that we are able to reconstruct the transverse polarization to better than 1% if the pion beam's directional spread and the muon's multiple scattering are limited to totally $\pm$ 8 mrad.

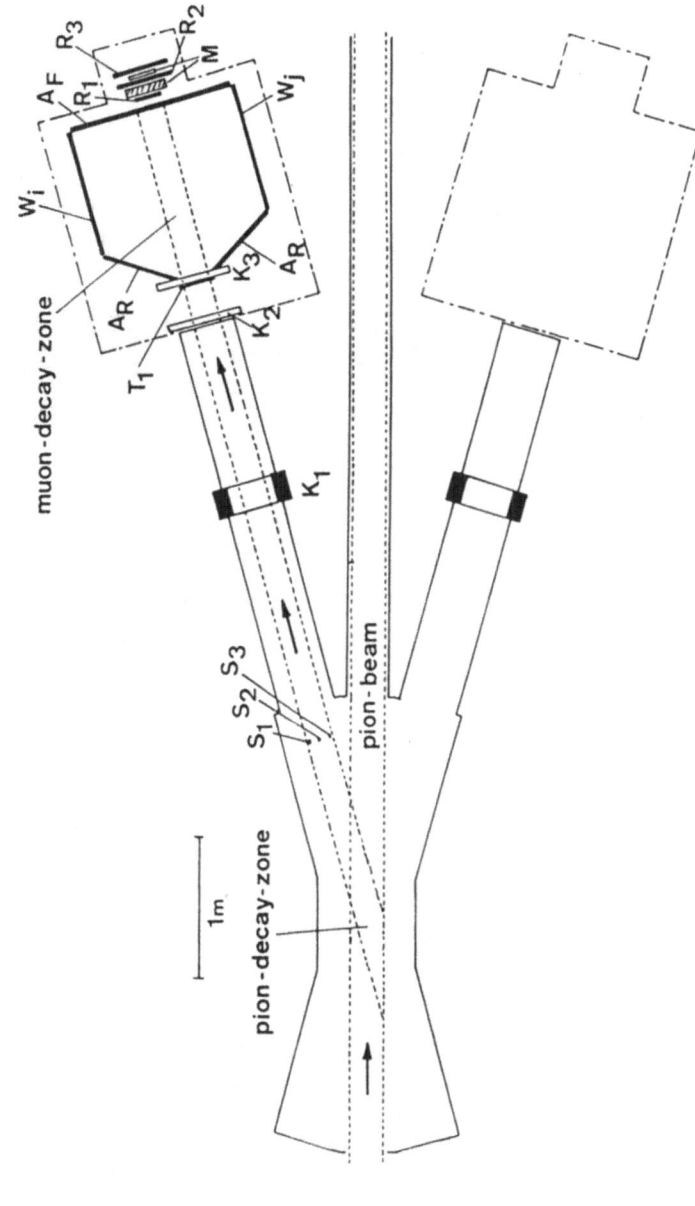

Figure 2. The experimental set-up. K_1 is a longitudinally operated drift chamber, K_2 and K_3 are conventional MWPC's, S_1–S_3, T_1, R_1–R_3, W_i, W_j, A_R, A_F are plastic scintillator counters.

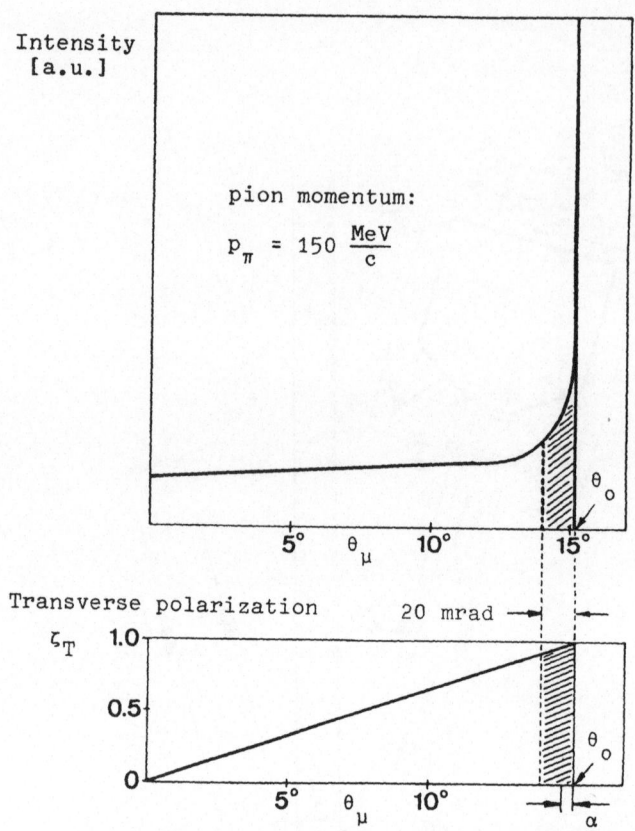

Figure 3. Principle of determination of the muon
polarization. The muon polarization is
determined by measuring the angle α
between the muon line of flight θ_μ and
the maximum angle of the sample θ_o.

The drift chamber K_1 is continually calibrated by muons
traversing one of the three thin counters S_1 to S_3 and
MWPC K_2. In addition, a stop telescope behind the
electron asymmetry detector, consisting of three
plastic scintillator counters R_1 to R_3 and moderator M,

is tuned to the muons at $\theta_\mu = \theta_0$. This prepares a sample of muons with average transverse polarization greater than 0.999 and serves to continuously calibrate the Jacobian-peak in the chamber K_1.

3.2. Detection of the Electron in μ-Decay

The electron asymmetry detector consists of two MWPC's K_2 and K_3 to measure the trajectory of the muons. A thin plastic scintillator T_1 is used in coincidence with the electron counters. The space between K_3 and A_F defines the decay volume. Muons and electrons from muon decay at small angles will be rejected by the anti-counter A_F. Only electrons with a polar emission angle greater than 30^0 and originating from the decay volume will be detected by two plastic scintillator walls W_i and W_j each consisting of seven long scintillators. The remaining solid angle will be covered by additional anticounters, since it is necessary that each decay electron is detected in order to reduce accidental coincidences. The wall counters W_i and W_j are each 5 mm thick and serve as energy loss counters with low thres-hold for the decay electrons. Thus we achieve the inte-gration over energy which enables us to measure exclu-sively the $P_\mu\xi$ term in the decay distribution.
To minimize the effects of possible misalignments between the symmetry axis of the electron detector and the chambers we will symmetrize our electron detector in the following way:

a) We will rotate the electron detector (exchange the two electron counter walls W_i and W_j)

b) We will move the whole detector to the other side of the π-beam. This is equivalent to a change of the transverse muon polarization.

By calculating the expected asymmetry and comparing with the measured one, determined from the four different electron detector positions, we get the value of $P_\mu\xi$.

4. REFERENCES

1. H. Burkard, W. Fetscher, H.-J. Gerber, K.F. Johnson,
   M. Salzmann, F. Scheck, SIN Proposal: "Measurement
   of the ξ-Parameter in μ Decay".
2. F. Scheck, Phys. Rep. 44:187 (1978).
3. V.V. Akhmanov, I.I. Gurevich, Yu.P. Dobretsov,
   L.A. Makar'ina, A.P. Mishakova, B.A. Nikol'skii,
   B.V. Sokolov, L.V. Surkova, and V.D. Shestakov,
   Sov. J. Nucl. Phys. 6:230 (1968).
4. M.A.B. Bég, R.V. Budny, R. Mohapatra, and A. Sirlin,
   Phys. Rev. Lett. 38:1252 (1977).
5. J. Carr, G. Gidal, B. Gobbi, A. Jodidio, C.J. Oram,
   K.A. Shinsky, H.M. Steiner, D.P. Stoker,
   M. Strovink, and R.D. Tripp, Phys. Rev. Lett.
   51:627 (1983).
6. M. Bardon, D. Berley, and L.M. Lederman, Phys. Rev.
   Lett. 2:56 (1959).

MUON CAPTURE IN HYDROGEN

J. Martino

Service de Physique Nucléaire - Haute Energie

CEN Saclay, 91191 Gif-sur-Yvette Cedex, France

Muon capture on the proton is a privileged tool to study the hadronic weak currents at low energy, especially in what concerns the pseudo-scalar form factor. In this lecture I will mainly speak about the recent results[1] obtained by the Saclay-CERN-Bologna (SCB) collaboration on the muon capture rate at rest in liquid hydrogen. In the first talk I shall first briefly remind the theoretical description of the capture process, then describe the experimental difficulties and procedures and finally present the results obtained. In the second talk I shall compare them to other capture experiments on the proton and present the informations that are provided by muon capture in hydrogen, especially for the PCAC hypothesis, the second class currents and the μ-e universality. Finally I shall present the possible muon capture experiments on the proton that could still be undertaken in order to improve our present knowledge of the subject.

1. THEORETICAL DESCRIPTION OF THE CAPTURE ON THE FREE PROTON

Muon capture on a free proton is a fundamental weak process :

$$\mu^- + p \rightarrow n + \nu_\mu \tag{1}$$

that occurs after the formation of a μp *atom*, where the weak interaction has time to manifest its low strength. The major advantage of studying muon capture on the free proton is to avoid any nuclear structure complications, and hence to reach fundamental coupling constants. The capture process is described by the well known V-A current-current hamiltonian leading to the following

43

capture rate[2] :

$$\lambda_c(\mu^- + p \rightarrow n + \nu_\mu) = 2\pi \times \left[\frac{G_F \cos\theta_c}{\sqrt{2}}\right]^2 \times \frac{\left[\alpha\ m'_\mu\right]^3}{\pi} \times (1 - 2\ m'_\mu.\alpha.R_p)$$

$$\times \int \left|\left[\overline{u}_{\nu_\mu}(p_{\nu_\mu}).\gamma_\mu.(1+\gamma_5).u_\mu{}^-(p_\mu{}^-)\right]\right. \tag{2}$$

$$\left.\times\left[V_\alpha + A_\alpha\right]\right|^2 \frac{E_{\nu_\mu}^2/(1 + E_{\nu_\mu}/E_n)}{(2\pi)^3}\, d\check{p}_{\nu_\mu} \times \left[1 + \frac{\alpha}{\pi}\Delta_{rad}(\mu^- + p \rightarrow n + \nu_\mu)\right]$$

where G_F is the Fermi coupling constant =
(1.16632 ± 0.00002) × 10^{-5} GeV^{-2} [ref.[3]]
θ_c is the Cabibbo angle $\cos(\theta_c)$ = 0.9737 ± 0.0025 [ref.[4]]
m'_μ is the reduced mass of the μp atom
α is the fine structure constant
R_p is the charge radius of the proton = 0.862 ± 0.012 fm
[ref.[5]]
Δ_{rad} is the radiative correction = 2.95 ± 0.15 [ref.[6]]
E_{ν_μ} is the energy of the emitted neutrino = 99.16 MeV
E_n is the energy of the emitted neutron = m_n + 5.22 MeV
$\check{p}_{\nu_\mu}$ is the unit vector in the neutrino direction.

The hadronic weak currents V_α and A_α (Vector and Axial respectively) are parametrised by (in the most general way insuring Lorentz
invariance) :

$$V_\alpha = \overline{u}_n\left[\gamma_\alpha.F_V(q^2) + \sigma_{\alpha\beta}.q_\beta.F_M(q^2) - iq_\alpha.F_S(q^2)\right].u_p$$

$$A_\alpha = \overline{u}_n.\gamma_5\left[\gamma_\alpha.F_A(q^2) + \sigma_{\alpha\beta}.q_\beta.F_T(q^2) - iq_\alpha.F_P(q^2)\right].u_p \tag{3}$$

where the form factors F_K only depend on the value of the momentum
quadritransfer $q^2 = -0.2514$ fm^{-2}.

The CVC hypothesis[7] states that V_α and the electromagnetic
current belong to the same isospin multiplet, so that :

$$F_V(q^2) = F_D^p(q^2) - F_D^n(q^2) = 0.9748 \pm 0.0009 \; \left[\text{ref.}^8\right]$$

$$(m_n + m_p).F_M(q^2) = F_P^p(q^2) - F_P^n(q^2) = 3.5875 \pm 0.0152 \; \left[\text{ref.}^8\right] \tag{4}$$

$$F_S(q^2) = 0$$

where F_D^i (F_P^i) are the Dirac (Pauli) electromagnetic form factors and m_i the mass of the nucleon i (i = p or n).

The axial form factor F_A is given at $q^2 = 0$, assuming μ-e universality, by the free neutron β-decay rate[9] :

$$F_A(0) = 1.254 \pm 0.007 \tag{5}$$

and its q^2-dependance is measured by neutrino scattering on proton and deuton,[10] giving :

$$F_A(q^2) = \frac{F_A(0)}{\left[1 + \dfrac{q^2}{M_A^2}\right]^2} = 1.226 \pm 0.007 \tag{6}$$

with

$$M_A = 930 \pm 30 \text{ MeV.} \tag{7}$$

Finally the today well accepted hypothesis of absence of second class current[11] leads to :

$$F_T(q^2) = 0. \tag{8}$$

At the end we are left – in this puzzle of coupling constants – with only one unknown : *the "induced" pseudo-scalar coupling constant F_P*, and hence the measure of the muon capture rate by the proton is a direct determination of F_P. In fact muon capture is the only way to reach F_P because (i) in β-decay the quadri-transfer is zero, so that the influence of F_P is suppressed (see eq.(3) where F_P is multiplied by q_α) ; and (ii) in high energy neutrino scattering the PCAC hypothesis (see section 6) implies that F_P is a rather unimportant correction.

After all calculations, the two possible capture rates on the proton, parametrised by $m_\mu.F_P(q^2) = g_p$, are :

$$\lambda_S = 804.92 + 16.17 \; g_p + 0.0812 \; g_p^2 \; s^{-1}$$

$$\lambda_T = 5.32 - 0.08 \; g_p + 0.0812 \; g_p^2 \; s^{-1}$$

$$(9)$$

These two capture rates refer to the two possible hyperfine spins of the μp atom ground state (S = 1 = Triplet state ; S = 0 = Singlet state). For the canonical value of g_p given by the PCAC hypothesis (see section 6), $g_p = -8.3$, we get :

$$\lambda_S = 676.3 \; s^{-1}$$

$$\lambda_T = 11.6 \; s^{-1}$$

$$(10)$$

2. EXPERIMENTAL PROBLEMS OF A CAPTURE RATE MEASURE

Before presenting the experimental set-up and results I want to summarize the *"classical"* difficulties generally encountered in measuring muon capture rate in hydrogen :

1) It is a *rare* process in competition with the muon desintegration :

$$\mu^- \to e^- + \bar{\nu}_e + \nu_\mu \qquad (11)$$

which occurs with a rate $\lambda_o = 455160 \pm 8 \; s^{-1}$ (see section 4), which is about 1000 times more frequent than the Singlet capture rate. On a nucleus the capture rate, roughly proportional to Z^4 [ref.[12]], can become more important than the decay rate, and its measure is then much more easy. But complications arising due to nuclear structure effects cancel this advantage in order to determine g_p unambiguously.

2) Only *neutral* reaction products are emitted, so that the capture on the proton is experimentally hard to detect.

3) As we saw previously, the sensitivity to the *initial* spin state is extremely high, so that this one has to be known unambiguously.

4) A very high sensitivity to *impurities* because of high transfer rates can very quickly distort the measurement.

The former points 3 and 4 need some more explainations concerning the "chemical story" of a negative muon in hydrogen (see Fig. 1) :

Point 3 : initially the μp atom reaches very quickly (in some pico-seconds) its ground state (1s) in a statistical mixture of Singlet

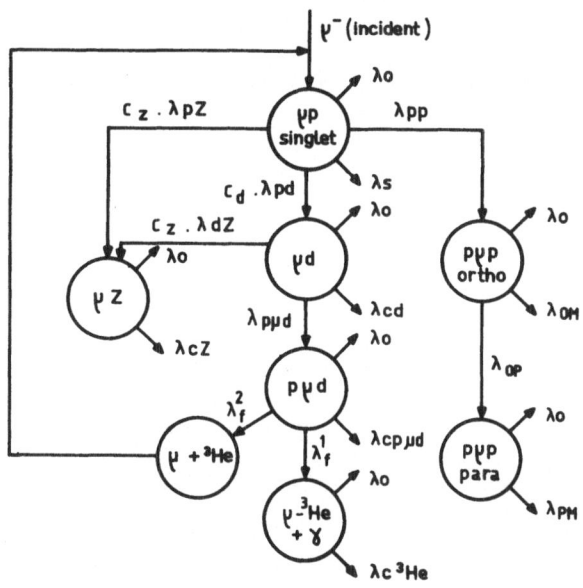

Fig. 1.   Muon chemistry in (liquid) hydrogen.

and Triplet hyperfine levels. If the hydrogen density is high enough (10 atm gas) the collision of the μp atom with other protons irreversibly populates the Singlet state in less then 15 ns. All these points have been checked experimentaly.[13] If the density increases further, then $p\mu p$ molecules are formed. The $p\mu p$ molecule has two bound states (Ortho and Para) [ref.[14]] and is only built in its upper level (Ortho). This assumption, although not experimentally proved, is quite safe according to the type of transition involved ([ref.[14]] : E1 to the Ortho-state and E0 to the Para-state). The rate of formation of the $p\mu p$ molecule at liquid hydrogen density has been measured[13] and is equal to $\lambda_{pp}=(2.3 \pm 0.2).10^6$ s^{-1}, so that the muon remains, on an average, about 400 ns in the atomic (1s) state before building a molecule. The cascade to the ground Para-state has always been until now *neglected*. Its upper limit is experimentaly[15] 10 % of the decay rate λ_0, in agreement with simple theoretical estimation.[14] But the different spin dependance of these two states[14] leads to two different capture rates :

$$\lambda_{OM} = 2 \gamma_o \left[\frac{3}{4} \lambda_S + \frac{1}{4} \lambda_T \right]$$
$$\lambda_{PM} = 2 \gamma_p \left[\frac{1}{4} \lambda_S + \frac{3}{4} \lambda_T \right] \tag{12}$$

where γ_o and γ_p are molecular muon-proton overlap coefficients[16] :

$$2 \gamma_o = 1.009 \pm 0.001$$
$$2 \gamma_p = 1.143 \pm 0.001 \tag{13}$$

Thus for the canonical values (10) of λ_S and λ_T we get :

$$\lambda_{OM} = 514.7 \text{ s}^{-1}$$
$$\lambda_{PM} = 203.2 \text{ s}^{-1} \tag{14}$$

and hence an evolution of the capture rate appears if such a transition occurs during the measure. Practically, experiments attending to reach a high level of precision in liquid hydrogen have to take the eventuallity of such a transition into account, that is to *measure* its rate λ_{op}, since a theoretical calculation cannot avoid uncertainties due to atomic physics assumptions (as we will detail in section 5).

Point 4 : The μp atom is a small neutral object (1/200 of the hydrogen atom size) that can very easily cross the electronic clouds to approach other nuclei. Then the *irreversible* transfer :

$$\mu p + Z \rightarrow \mu Z + p \tag{15}$$

occurs with a very high probability (about 10^6 time greater than the decay rate,[17]) leading to the formation of a parasite muon source for which the capture rate is no more the one to measure.

On the other hand, the *charged* pμp molecule is kept away from other nuclei and does not induce any contamination in μZ atoms. This is one of the major advantages, as we will see, of the measure in liquid hydrogen.

3. EXPERIMENTAL STATUS BEFORE THE SCB EXPERIMENT

The different experimental results obtained before our measurement (see Table 1, experiments 1 to 6, analyzed with $\lambda_{op} = 0.$) were obtained by *counting* the absolute amount of capture neutrons. A major difficulty is then to recognize these neutrons (from other neutron sources and among a high γ-ray background, as we will see later) and to know the detection efficiency of the neutron counters (through sophisticated Monte-Carlo simulations). These two points mainly explain the limited precision achieved (about 10 %).

The analysis of the results in term of g_p (neglecting the Ortho-Para transition for liquid data) show a discrepency between the various experimental conditions. The very first experiments, with Bubble Chamber (BC) technique are hard to analyse critically,

Table 1

Experimental values of muon capture rate in hydrogen. Deduced values of the pseudo-scalar coupling constant g_p with and without Ortho-Para transition

	Experiment		Conditions			Published Capture Rate (s^{-1})	$m_\mu \cdot F_P = g_p$	
							$\lambda_{op} = 0.\ s^{-1}$	$\lambda_{op} = 1$ 4.6 ± 1.36 $10^4\ s^{-1}$
1	Chicago	[ref.^{35}]	liq.	BC	n	428 ± 85	-19.4 ± 8.9	-17.4 ± 9.0
2	CERN-Bologna	[ref.^{36}]	liq.	BC	n	450 ± 50	-17.1 ± 5.0	-15.2 ± 5.1
3	Columbia I	[ref.^{37}]	liq.	SC	n	515 ± 85	-8.5 ± 7.8	-5.9 ± 8.0
4	Columbia II	[ref.^{38}]	liq.	SC	n	464 ± 42	-13.2 ± 4.2	-10.7 ± 4.3
						average $3\div4$	-12.1 ± 3.7	-9.6 ± 3.8
5	CERN-Bologna	[ref.^{39}]	gas	SC	n	651 ± 57	-10.0 ± 3.9	-10.0 ± 3.9
6	Dubna	[ref.^{40}]	gas	SC	n	686 ± 88	-7.6 ± 5.9	-7.6 ± 5.9
						average $5\div6$	-9.3 ± 3.3	-9.3 ± 3.3
						average $3\div6$	-10.5 ± 2.4	-9.4 ± 2.5
7	Saclay	[réf.^{1}]	liq.	SC	e	see text	-16.5 ± 1.9	-9.2 ± 2.7
						average $3\div7$	-14.2 ± 1.5	-9.3 ± 1.8
						χ^2 (5 points)	4.5	0.4

BC : Bubble Chamber ; SC : Scintillation Counter.
n : neutrons detected : absolute counting ; e : electrons detected : lifetime method.

because no final publication has never appeared. Nevertheless π^--contamination of the incoming μ^--beam, impurities in the bubble chamber hydrogen and track recognition lead to uncertainties difficult to estimate. Therefore in all following discussions I will forget these data. The experiments labeled Liquid and Gas (hydrogen target) were performed with electronic counter technique : liquid scintillators were used to recognize and count the neutrons (as we shall detail in section 5). The comparison between Liquid and Gas results shows a slight disagreement among the values of g_p. The question is then to know if this effect is not resulting from the neglected Ortho-Para transition.

These considerations brought our group to propose a new measurement in liquid hydrogen, with a precision at the level of 4 % and taking into account the Ortho-Para transition. These goals obliged us i) to apply a new technique to measure the capture rate, avoiding neutron counting uncertainties, and ii) for the first time to measure precisely the Ortho-Para transition rate.

4. SCB EXPERIMENT : CAPTURE RATE MEASURE BY THE LIFETIME METHOD

The measure of the capture rate is performed by comparison of the μ^+ to the μ^- lifetime. In a given target (here liquid hydrogen) μ^+ can only disappear by their decay (11), so that :

$$\tau_{\mu^+} = \frac{1}{\lambda_o} \tag{16}$$

whereas μ^- have two escape channels : decay and capture, giving :

$$\tau_{\mu^-} = \frac{1}{\lambda_o + \lambda_c} \tag{17}$$

where λ_c is the capture rate to measure.

This lifetime method is particularly adapted to the Saclay Linac *pulsed* beam structure : the muons are stopped in the target during a 3 µs beam burst (repetition rate of 3000 Hz) and the decay *electrons* of the surviving muons are detected *after* the burst by telescopes of scintillators. Their time distribution gives an accurate measurement of the muon lifetime because of two favorable effects : the low background level and the absence of any regeneration of the muon "radioactive" source during the decay study, between two bursts. In such conditions, systematic errors are smaller than 10^{-5} and the capture rate can be measured with a 4 % uncertainty, as soon as the required statistical precisions on the lifetimes are achieved ($\Delta\tau/\tau = 1/\sqrt{N}$, N being the number of decay events in the measured time distributions ; thus, 4 % on λ_c requires about 3×10^{-5} on τ_μ, that is 10^9 detected electrons (and positrons)).

Moreover there is no limitation on the number of muons that can be present at the same time in the target, since no correlation is requested between the muon and its decay electron. In fact such a measure is the study, after the beam pulse, of an only decaying "radioactive" muon source built during the burst. The number of muons present in the target only influences the intensity of the decaying source, but in no way its lifetime. Thus, of course, the use of a muon-stop telescope is avoided.

The liquid hydrogen is contained in a spherical copper vessel (Fig. 2). The muons enter the target through a lead collimator and a copper degrader. More than 99 % of the muons are stopped in hydrogen. Outside hydrogen muons can only stop in high Z elements like copper or lead where they are rapidly captured (capture rate $\alpha\ Z^4$) and hence do not contribute inside the measuring gate which starts only 1.5 μs after the burst.

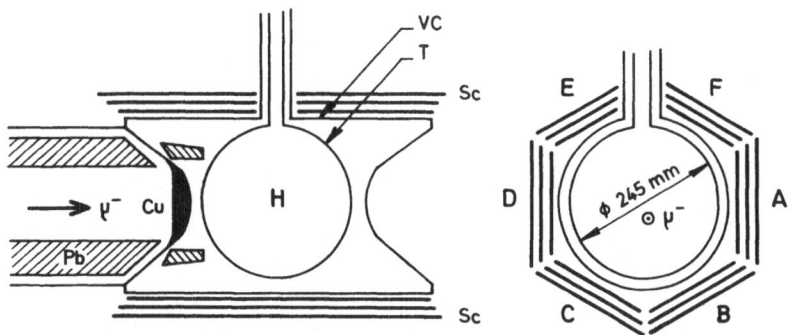

Fig. 2. Experimental set up : target and electron detectors. Sc : Scintillators ; T : Target ; VC = Vacuum Chamber.

The high purity required to avoid distortion from the transfer processes is obtained in a, now, traditional way by using a *"protium"* gas (contaminated by 2.7 ± 0.1 ppm of deuterium) passing through a palladium filter and then a heat exchanger. The efficiency of the palladium purifier is measured by applying on the input side a neon or nitrogen pressure up to 10 bars. The corresponding rate of pressure rise in the target is then less than 10^{-4} torr/hour, ensuring an impurity leaking of less than 10^{-9} in concentration during the filling operations with the protium gas. Before each run, the target is evacuated by a cryogenic pump down to 2×10^{-7} torr and baked for several days, until the outgassing is found to be less than 10^{-4} torr/day at room temperature. The much lower outgassing one expects at the liquid hydrogen temperature corresponds to a contamination well below 10^{-9}. Finally, because of the pμp molecule formation allowing transfer only in the first 400 ns of the muon presence in the target, the parasite contamination in μZ atoms is quickly quenched and disappears faster than the μp population, so that during the measuring gate the remaining distortion on the capture rate due to impurities is less than 0.5%.

The decay electrons are detected in 6 independent telescopes of 3 plastic counters surrounding the target. The total energy threshold (6 MeV) and the threefold coincidence insures a sufficient rejection of the neutral delayed events (mostly γ-rays from thermal neutron capture). The measuring gate opens 6 independent clocks supplied by a common quartz oscillator of 500 MHz (with a stability better than 10^{-5}). The first event occuring in a telescope closes the corresponding clock. If two or more electrons occur in the same gate for a given telescope, this "multiple" events are rejected.

The rate of "single" events is expected to follow the exponential law :

$$R(t) = R_o.exp(-t/\tau_\mu) \tag{18}$$

As shown in ref.[1] the observed law is rather :

$$R'(t) = R(t)*(1 + A.R(t)/R_o) + B.R_o \tag{19}$$

The distortion term A is due to the finite time resolution (15 ns) of the electronics circuits, which reduces the "multiple" events rejection efficiency. This effect is proportional to the event rate in the telescope and is *measured* by increasing the nominal rate by a factor 4 to 5 for the τ_μ+ measurement where we use a π^+ beam. For the nominal rate between 0.06 and 0.1 event per gate and telescope, A was found of the order of 5×10^{-4} with a statistical uncertainty less than 10^{-5}, inducing an error less than 0.5 % on the final value of the capture rate.

The distortion term B mainly represents the cosmic ray background which has no time dependence. At the level of 10^{-5} there is a time dependent component (period of 160 μs) due to thermal neutrons following the beam burst. These backgrounds are *measured* from the distribution of delayed events between two machine bursts (see Fig. 3). By stopping μ⁻ in the copper walls of the empty target one checks the behaviour of the background in the μ-decay region (as shown on Fig. 4).

The lifetimes τ_μ+ and τ_μ- are obtained respectively from the 12.5×10^8 decay positrons and the 13.0×10^8 decay electrons time distributions after background substraction and extrapolation to ideal zero acquisition rate. The results obtained are displayed on Fig. 5, showing a very high stability. Fig. 6 shows the distribution of all elementary results (per telescope and run), in excellent agreement with the expected gaussian distribution. Summing all data we get :

$$\tau_{\mu^+} \text{ (SCB)} = 2197.078 \pm 0.073 \text{ ns}$$

$$\tau_{\mu^-} \text{ (SCB)} = 2194.908 \pm 0.067 \text{ ns}$$

(20)

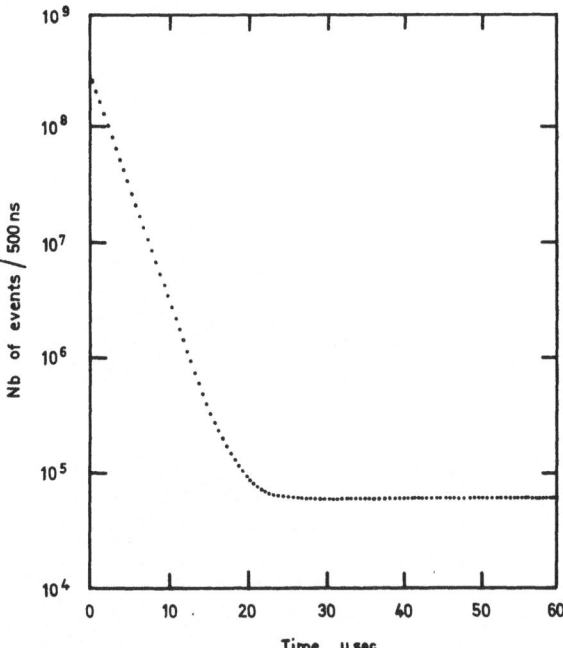

Fig. 3. Decay electron time distribution for the threefold coincidence events inside the measuring gate.

The comparison of our value τ_{μ^+}(SCB) with the Previous World Average value[18] :

$$\tau_{\mu^+} \text{ (PWA)} = 2197.109 \pm 0.076 \text{ ns} \qquad (21)$$

and with the recently published TRIUMF result[19] :

$$\tau_{\mu^+} \text{ (TRI)} = 2196.95 \pm 0.06 \text{ ns} \qquad (21)$$

shows an excellent agreement and leads to a New World Average value of τ_{μ^+} :

$$\tau_{\mu^+} \text{ (NWA)} = 2197.031 \pm 0.040 \text{ ns} \qquad (22)$$

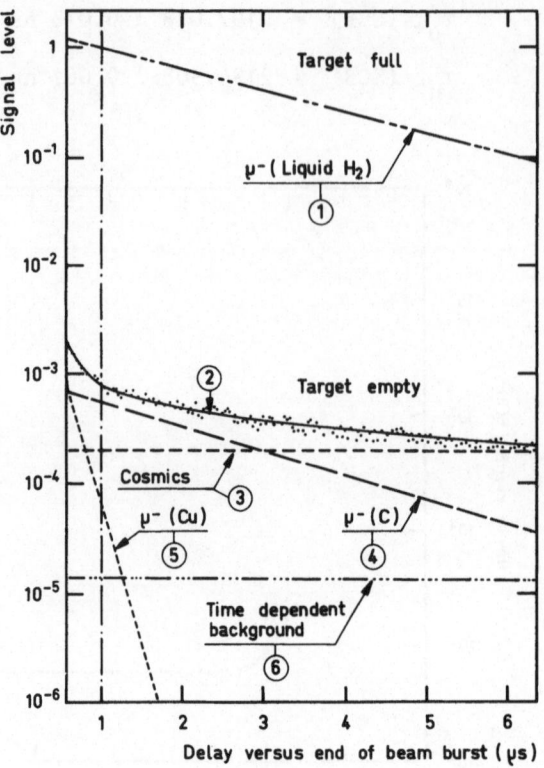

Fig. 4. Time distribution of events, target empty, starting at the
end of the beam burst. The muons are stopped in the copper
walls. The different components are : (1) signal level
with full target ; (2) signal level measured with empty
target, that can be decomposed as follows : (3) cosmics ;
(4) μ^- reaching the scintillators ; (5) μ^- stopping in the
copper walls ; (6) time dependant background (see text).

The capture rate obtained from $\tau_\mu+$(NWA) and $\tau_\mu-$(SCB) is corrected
by $- 21 \pm 7$ s^{-1} for the small deuterium contamination (this effect
was *measured* at two different deuterium concentrations). A second
slight correction of $+ 12$ s^{-1} due to the μ^- atomic binding by a
proton[20] is also applied. Finally we get :

$$\lambda_c = 431 \pm 18 \text{ s}^{-1} \tag{24}$$

In the error quoted above for λ_c, the statistical and systematical
uncertainties are in the (quadratic) proportion of 2 to 1. As it
is not too difficult to reduce the systematical error at the percent
level, an experiment, that could also, in a reasonable time,
achieve the same statistical precision, could be done with the same
technique.

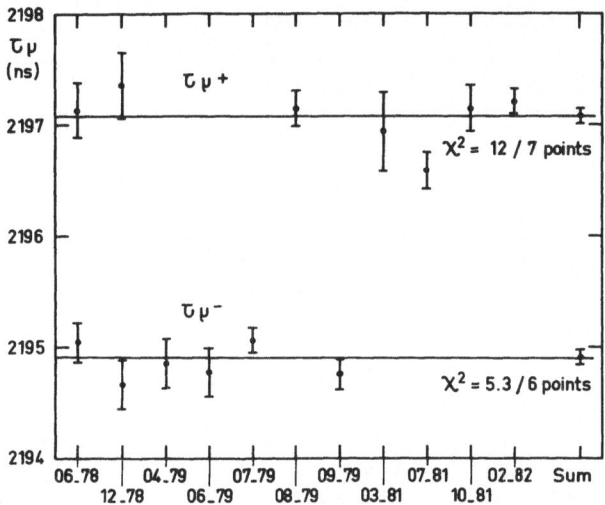

Fig. 5. Values of τ_{μ^+} and τ_{μ^-} obtained from the different runs.

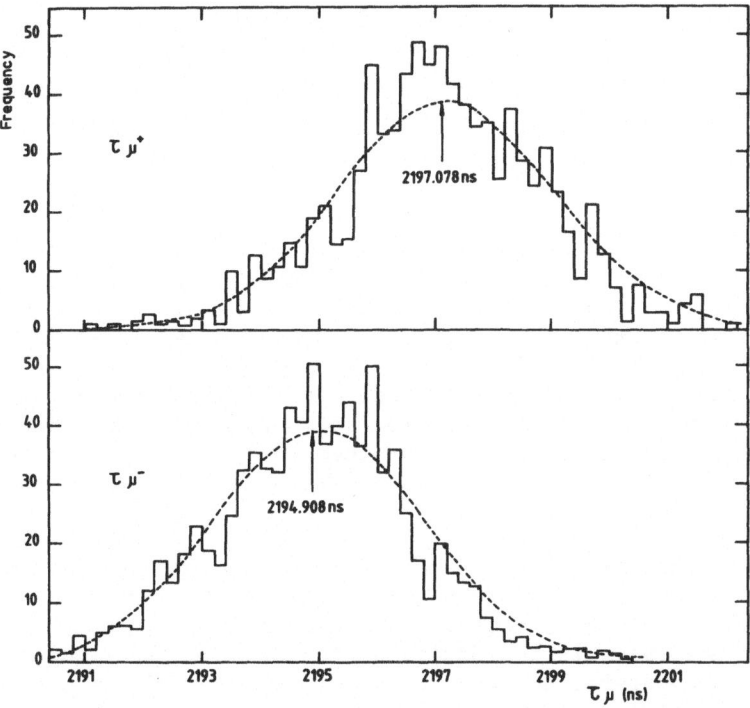

Fig. 6. Experimental distributions of all the elementary lifetime results.

5. SCB EXPERIMENT : ORTHO-PARA TRANSITION

As we already pointed out, in order to interpret λ_c in term of the fundamental couplings of the weak interaction, we have to know precisely the initial state of the $p\mu p$ molecule. A classical method to investigate the effect of a transition in the initial state in muon capture is to study the time distribution of the capture products.[21] If $N_\mu(t)$ is the number of living muons at the time t, the rates of the decay electrons and capture neutrons are :

$$e(t) = \lambda_o * N_\mu(t)$$
$$n(t) = \lambda_c * N_\mu(t)$$
(25)

An Ortho-Para transition results in a time dependence of the capture rate λ_c, and hence the mean time of the neutron distribution compared to that of the electrons is :

$$\tau_n = \tau_e \left[1 - \frac{\lambda_{op}}{\lambda_o} \cdot \frac{\lambda_{OM} - \lambda_{PM}}{\lambda_{OM}} \right]$$
(26)

if the Ortho-Para transition rate is small compared to λ_o. In case of a non zero value of λ_{op}, the measured capture rate (24) has to be corrected in the following way in order to get λ_{OM} :

$$\lambda_{OM} = \lambda_c \left[1 + K \frac{\lambda_{op}}{\lambda_o} \frac{\lambda_{OM} - \lambda_{PM}}{\lambda_{OM}} \right]$$
(27)

where K is *dependent of the experimental conditions* (pulse width, measuring gate delay, lifetime method or neutron counting technique, ...). For the SCB experiment one has K = 2.76 ± 0.05. It is worth noting that the correction to be applied on λ_c is, at our level of precision, independent of λ_{OM} and λ_{PM}, because the neutron mean time (26) exactly includes the right combination of λ_{OM} and λ_{PM} needed for the capture rate correction. In order to match the experimental precision obtained on λ_c (24), λ_{op} (if of the order of 0.1 λ_o) must be measured with 30 % precision and hence τ_n at the level of 10^{-2}.

The measure of the Ortho-Para transition rate is a measurement of the capture neutron *time distribution*. Hence *no knowledge* of any detection efficiency is needed and no limitation on the final precision is introduced by any Monte-Carlo simulation of the neutron counters. But the technical problems of capture neutron identification remains : i) the 10^3 much more frequent decays provide a huge Bremsstrahlung γ-rays source that have to be discriminated

from the neutrons ; ii) the capture neutrons only represent a fraction (about 70 %) of the total amount of detected neutrons :
"neutrons-like' events from cosmical or photonuclear origin must be
recognized and substracted.

The capture neutrons (energy = 5.2 MeV) are detected in a set
of 6 cylindrical (∅ = 15 cm, L = 10 cm) liquid scintillators N
(NE213 ; Nuclear Enterprise). A sheet of plastic scintillators E
(5 mm thick) surrounds the target in order to veto or detect the
μ^--decay electrons. A necessary condition for an event to be defined
as a neutron is the requirement $N\overline{E}$.

About 30 % fo the μ^--decay electrons, radiating in the target
walls, are not vetoed by E. The ratio of these $N\overline{E}$ Bremsstrahlung
photons to the detected capture neutrons is about 3000. Therefore
the rejection of these photons is quite essential. Two techniques
were employed, using the shape of the scintillators light response.
First (PSD1) the prompt and delayed components of the light pulse
are compared in two 100 ns wide gates with 0 and 40 ns delay (see
Fig. 7), respectively. Second (PSD2) a discrimination is made from
the risetime of the integrated pulse (ORTEC circuit). Both technique
have a rejection factor better than 10^5 against photons, for isolated events. However, any distortion in the pulse tails due to pile-
up of small energy photons reduces the rejection factor. The resulting "fake" neutrons have a mean lifetime of $\tau_e/2$. Such a type
of pile-up events are even much more frequent in the NE events
(these events are mainly due to μ^--decay electrons and therefore
are expected to have the electron lifetime τ_e), and will affect
the observed lifetimes.

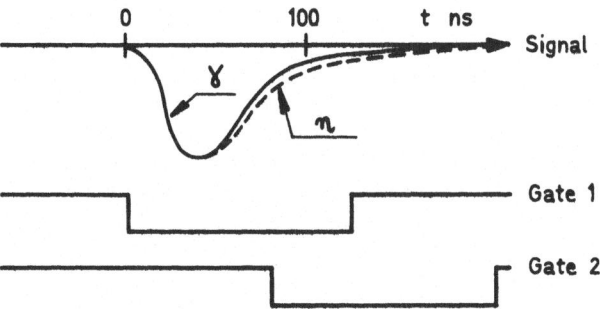

Fig. 7.   Neutron-gamma discrimination method PSD1 : signal and
gates.

Experimentally we found that the PSD1 and PSD2 responses are
different for pile-up events allowing their almost total discrimination in a PSD1-PSD2 plane, as shown on Fig. 8. The residual
pile-up events in region 1 of Fig. 8 ($N\overline{E}$ capture neutrons) are

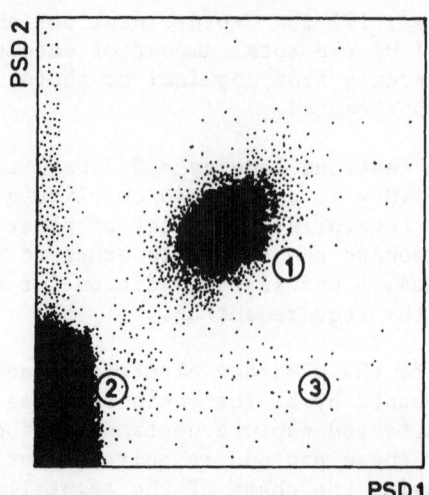

Fig. 8. Distribution of the $N\overline{E}$ events in the PSD1-PSD2 plane. The neutrons (1) are in the central region, the gamma-ray (2) in the lower left corner, and the pile-up events (3) are scattered over all the plane.

obtained from the number of pile-up events in region 3 : the coefficient of extrapolation is chosen in order to restore, with the same treatment, the expected lifetime τ_e, for the NE events.

About 45000 neutron-like events $N\overline{E}$ have been taken for a threshold of 1 MeV on the recoil proton energy. After these events have been isolated by only accepting a small region of the PSD1-PSD2 plane, the following background components are still to be substracted : i) Neutrons due to cosmic rays, *measured* in the late part of the observation gate. ii) Residual pile-up events, as explained before. iii) Fast delayed neutrons, following the beam burst, *measured* in runs with empty target. Fig. 9 shows the behaviour of these neutrons. They have to be measured very carefully, since contrary to the other neutron components neither their time dependence nor their yield is known. iv) *Photo-neutrons* (or photo-protons), generated by photonuclear reactions mainly inside the neutron counters. After substraction of the neutron components i) to iii), the remaining NE and high energy $N\overline{E}$ (more than 5.2 MeV) events are photo-neutrons (with the electron lifetime τ_e). It is then possible to substract the low energy $N\overline{E}$ photo-neutron component (less than 5.2 MeV, in the capture neutron region) by *assuming* the same energy spectrum for NE and $N\overline{E}$ photo-neutrons. This assumption has been checked by increasing the target wall radiation length to enhance photo-neutron yield. The two methods give the same photo-neutron level, which is displayed on Fig. 10 for the $N\overline{E}$

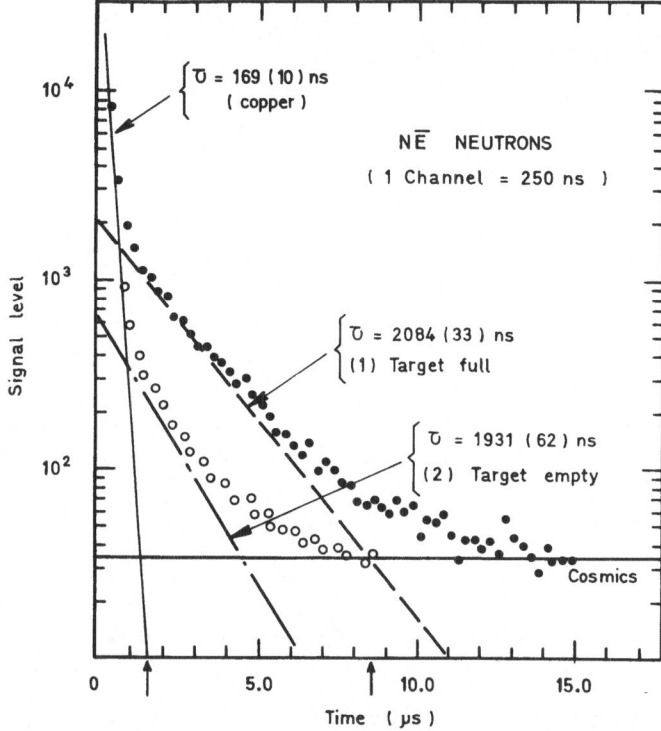

Fig. 9.   Time spectrum of the $\overline{NE}$ neutrons : (1) target full and
(2) target empty. The arrows indicate the window kept
for the analysis.

events. Finally the comparison of the high energetic $\overline{NE}$ photo-
neutrons energy spectrum and lifetime with the expected contributions of impurity events (measured by increasing the nitrogen and
neon contamination) allows to set an experimental limit of ± 30 ns
for the impurity distortion on τ_n, that confirms the value of
± 10 ns deduced from the outgassing measures (described in section 4).

As we are not interested in the absolute yield of capture
neutrons, the optimal cuts may be applied on the various parameters in order to minimize the backgrounds : energy cut between 3
and 5 MeV on the recoil proton energy (see Fig. 10) and window on
the first 7 µs of the time distribution (see Fig. 9). The mean
lifetime obtained for the capture neutron time distribution is
finally (for 28280 neutrons) :

$$\tau_n = 2078 \pm 28 \text{ ns} \tag{28}$$

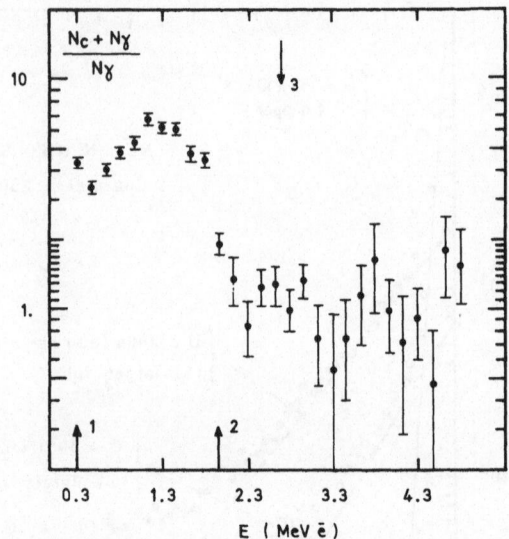

Fig. 10. Capture and $\overline{NE}$ photo-neutrons energy spectra. Arrows 1
and 2 define the capture neutron region retained for
analysis. Arrow 3 show the lower bound retained for
photo-neutrons.

which is quite different from the lifetime τ_e (see value (20)) and
leads to (see Fig. 11) :

$$\lambda_{op} = (4.61 \pm 1.35) \times 10^4 \text{ s}^{-1} \qquad (29)$$

While our measure was in progress, a theoretical calculation of
λ_{op} has been undertaken by the Dubna group[16] that has shown the
existence of mecanisms allowing for an Ortho-Para transition. But
some theoretical inputs, especially in what concerns the molecular
complexes that describe the initial state of the $p\mu p$ molecule, are
hard to checked experimentally, and hence limit the precision of
such a calculation (see Table 2). The final result obtained :

$$(\lambda_{op})^{th} = (7.1 \pm 1.2) \times 10^4 \text{ s}^{-1} \qquad (30)$$

is in good agreement with the experimental value (29) which will
be usued in the following analysis.

From eqs.(26) and (27) and values (20) and (28) one gets the
experimental value of λ_{OM} :

$$\lambda_{OM} = 505 \pm 29 \text{ s}^{-1} \qquad (31)$$

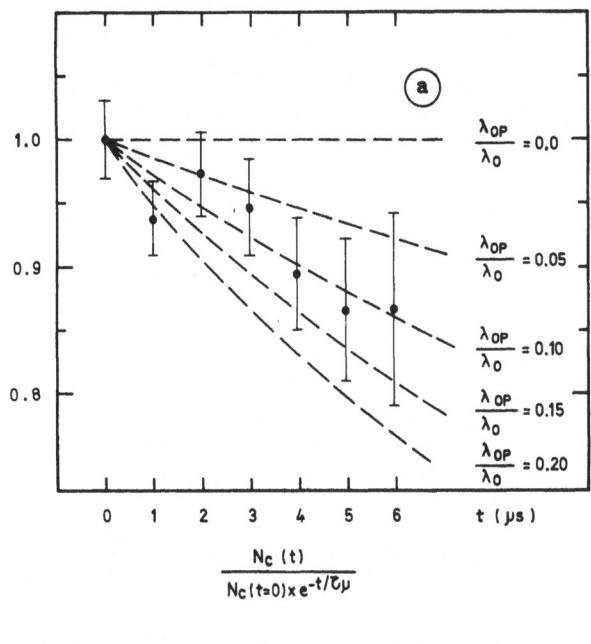

$$\frac{N_C(t)}{N_C(t=0) \times e^{-t/\tau_\mu}}$$

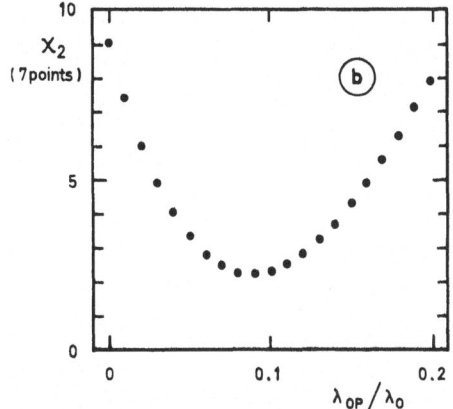

Fig. 11.   (a) Time distribution of the capture neutrons. The number of events in each time interval has been multiplied by $\exp(t/\tau_e)$ to remove the μ^--decay time dependence.
(b) Variation of the χ^2 for the 7 points of Fig. 11a.

The correction due to the Ortho-Para transition ($= 74 \pm 22$ s^{-1}) is then quite non negligible. The error on this correction can be quadratically decomposed in 2 main terms : i) ± 17 s^{-1} which takes into account the statistic on the capture neutrons and ii) ± 13 s^{-1} which represents the uncertainty due to the fast delayed neutrons. At our level of precision the contribution of the photo-neutrons is not the most important one.

Table 2

Variation of the calculated Ortho-Para transition rate according to the different molecular complexes formed by the Ortho-pμp system. The major uncertainty of the theoretical calculation arises from the weights that are to be given to each population (see ref.[16]).

Complex	$[(p\mu p)pe]^+$	$[(p\mu p)e]$	$[(p\mu p)2p2e]^+$	$[(p\mu p)p2e]$
$\lambda_{op}(10^4 \text{ s}^{-1})$	4.3	6.3	7.4	9.5

6. EXTRACTION OF THE PSEUDO-SCALAR COUPLING CONSTANT g_p AND PCAC HYPOTHESIS

Table 1 displays the standard analysis of μ^--capture data on the proton leading to the pseudo-scalar coupling constant, in the theoretical frame of section 1, that is assuming i) the CVC hypothesis defining the Vector current, ii) the μ-e universality giving $F_A(q^2)$ and iii) the absence of second class currents eliminating F_T. The results for g_p are obtained first neglecting the Ortho-Para transition ($\lambda_{op} = 0.$) and secondly taking it into account, that is applying a correction of the type of eq.(27) with K varying with each experimental conditions.[22]

We, at first, notice that the agreement between Liquid and Gas data is *much better* when we account for the Ortho-Para transition. We also immediately see that the lifetime method is much more sensitive to the Ortho-Para transition than the neutron counting method, so that a new measure of λ_{op} with an improved precision (about 10 %) would already accurate the information provided by the SCB measure.

The new value of g_p obtained from the SCB experiment is :

$$g_p = -9.2 \pm 2.7 \tag{32}$$

The combined value of g_p from all hydrogen measures (except the very fist bubble chamber experiments) can now be calculated because of the high level of compatibility between the results, as soon as the Ortho-Para correction is introduced (see Table 1). This average value of g_p is :

$$g_p = -9.3 \pm 1.8 \tag{33}$$

The theoretical frame to calculate g_p is the PCAC hypothesis [ref.[23]]. It starts from the leptonic pion decay, that can only occur via the Axial current, leading to :

$$< 0|\partial_\alpha A_\alpha|\pi^+ > = f_\pi m_\pi^3 < 0|\phi_\pi|\pi^+ > \tag{34}$$

where f_π is the pion decay constant ($= 0.94 \pm 0.01$ [ref.[24]]), m_π the pion mass and ϕ_π the pion field. Hence the divergence of the Axial current has the pion field quantum numbers. The PCAC hypothesis is then the following simple assumption :

$$\partial_\alpha A_\alpha = f_\pi m_\pi^3 \phi_\pi \tag{35}$$

In the chiral limit where $m_\pi = 0$, PCAC restores the Axial current conservation, as expected. The PCAC hypothesis has a lot of consequences (low energy theorems) that are very well reproduced by experiments with low energy pion.[25]

The PCAC hypothesis allows to calculate the divergence of the Axial current between nucleon states :

$$< n|\partial_\alpha A_\alpha|p > = F_A(q^2) - \frac{q^2}{2M_N} F_p(q^2) = < n|f_\pi m_\pi^3 \phi_\pi|p >$$

$$= f_\pi f_{\pi pn}(q^2) \frac{1}{1 - \frac{q^2}{m_\pi^2}} \tag{36}$$

where $f_{\pi pn}$ is the strong interaction pion-nucleon coupling constant for pion with quadri-momentum q^2 and nucleons on mass shell. Real pions scattering on nucleons gives[26] :

$$f_{\pi pn}(m_\pi^2) = 1.41 \pm 0.01 \tag{37}$$

As a function of q^2, $f_{\pi pn}$ is generally parametrized by :

$$r(q) = \frac{f_{\pi pn}(q^2)}{f_{\pi pn}(0)} = \frac{\Lambda^2}{\Lambda^2 - q^2} \tag{38}$$

with $\Lambda = 1250 \pm 250$ MeV [ref.[27]].

For μ^--capture on the proton we get :

$$\left[g_p \right]_{PCAC} = \left[m_\mu \, F_p(q^2) \right]_{PCAC} = - \, m_\mu \, \frac{m_n + m_p}{m_\pi^2} \, \frac{F_A(q^2)}{1 - \dfrac{q^2}{m_\pi^2}} \, (1 + \varepsilon) \qquad (39)$$

with

$$\varepsilon = - \frac{m_\pi^2}{q^2} \left[1 - \frac{r(q^2)}{F_A(q^2)/F_A(0)} \right] \qquad (40)$$

A simple study of the above expression (39) shows that the value of $g_p(q^2)$ reaches a maximum at about $q^2 = - m_\pi^2$, that is very close to the value of the μ-capture quadri-transfert equal to $-0.88 \, m_\mu^2$. This confirms the privileged position of muon capture to measure g_p.

Numerically we finally obtain with the parametrisation (38) for $r(q^2)$:

$$(g_p)_{PCAC} = - \, 8.1 \pm 0.1 \qquad (41)$$

where the quoted error arises from the numerical uncertainties and not from the validity of the different hypothesis. Another calculation by Primakoff[2] assuming a slightly different behaviour of $r(q^2)$ gives :

$$(g_p)_{PCAC} = - \, 8.5 \pm 0.1 \qquad (42)$$

The comparison of values (41) and (42) gives an estimation of the uncertainty that has to be affected to the theoretical calculation. Therefore the value we will retain for the following discussion will be :

$$(g_p)_{PCAC} = - \, 8.3 \pm 0.2 \qquad (43)$$

Thus the experimental results (32) and (33) for g_p are in good agreement with PCAC theoretical prediction. At this point it is also interesting to compare the hydrogen result for g_p to the ones obtained from measurements of the nuclear polarizations of ^{12}B after muon capture on ^{12}C. Roesch et al. found[28] :

$$g_p = - \, 11.1 \pm 2.1 \qquad (44)$$

and a new measure by the Tokyo group gives[29] :

$$g_p = - \, 11.8 \pm 3.1 \qquad (45)$$

The overall agreement with the hydrogen results is correct, but the deviation with the PCAC prediction (43) is slightly bigger than for hydrogen.

7. BEYOND PCAC : SECOND CLASS CURRENTS AND MUON-ELECTRON UNIVERSALITY

Another way of analyzing data is to *assume the PCAC hypothesis* (which is, as we saw previously, well reproduced by ^{12}C capture experiments and which implications like for "low energy theorems" give very reliable results) and to test other crucial assumptions like the presence of second class currents and muon-electron universality.

First, if for the hydrogen results the Ortho-Para transition had been neglected, as was done in the past, one would get a difference with PCAC that could be interpreted as an evidence for second class current. The pressence of second class current, indeed, corrects in a very simple way the muon capture information by only adding up the second class term to g_p :

$$g_p = m_\mu . F_p + 2M . F_T \qquad (46)$$

M being the nucleon mass. Thus, the comparison of the experimental and PCAC values of g_p allows to set a limit on the second class current term :

$$2M . F_T = -1.0 \pm 1.8 \qquad (47)$$

which shows that F_T is compatible with zero when consistency with PCAC is required. This agrees with the limit obtained by Baker et al. in high energy quasi-elastic neutrino scattering on deuterium [ref.[30]] :

$$|2M . F_T| < 2. \qquad (48)$$

and with the analysis of Morita et al.[31] on β-ray angular distributions in the A=12 systems, giving :

$$2M . F_T = 0.5 \pm 1.1 \qquad (49)$$

In a recent paper[32] Holstein demonstrates that muon capture provides the strictest present limit for a second class contribution to the Vector current. If we assume that the whole second class effect (47) is due to F_S (see eq.(3)) rather than to F_T, we get :

$$2M.F_S = 1.3 \pm 2.3 \qquad\qquad (50)$$

Another way of analyzing data is to assume both PCAC and the absence of second class current hypothesis and then to test the muon-electron universality in muon capture, that is to fix g_p to its value (43) and to leave $F_A(0)$ free. In this way we find (taking into account the Ortho-Para transition) :

$$F_A^\mu(0) = -1.237 \pm 0.031 \qquad\qquad (51)$$

where the suffix μ indicates that this value is related to the muon and not to the electron like for β-decay. Again the error given on (51) does not take into account the possible systematic uncertainty due to the theoretical assumption of the PCAC expression (39).

Value (51) agrees very well with the one obtained from neutron β-decay (see formula (5)), thus giving an important experimental confirmation of muon-electron universality in two low energy processes : muon capture by the proton and neutron β-decay, at a level of about 3 %. Again, if in the future an experimental evidence for second class current is found with a value $|2M.F_T| > 2.$, the above conclusion for $F_A^\mu(0)$ would be changed.

8. FUTURE OF MUON CAPTURE ON THE PROTON

A look at Table 1 and formula (9) shows that it would be very interesting to measure the Singlet capture rate in gaseous hydrogen at the same level of precision achieved in liquid hydrogen (with the SCB lifetime technique), that is about ± 15 s^{-1}. As no Ortho-Para correction is then needed, the pseudo-scalar coupling constant g_p could be determined with an accuracy of ± 1, which would be a sensible improvement. The major difficulties to overcome in gaseous hydrogen are : i) the muon beam has to be stopped in a thin target, because of the very low density of gaseous hydrogen ; ii) the absence of $p\mu p$ molecules formation implies that, though the density is low, the transfer to impurities are as much (if not more) dangerous as in a liquid target, because they are allowed during the whole life of the muon ; iii) the thermal motion of the μp atom can bring it in the vicinity of the target walls, with which transfer can also occur. This third point is surely the most difficult to overcome.

For the measure in liquid hydrogen we have already seen that a first step is to improve the Ortho-Para transition rate, which, for the SCB experiment, almost reduces the final precision by a factor 2. The same technique of measuring the capture neutrons time distribution can be usued, but the backgrounds have to be reduced (see end of section 5). A way to do this could be to have a good shielding against cosmical and beam neutrons and to eliminate photo-neutrons. At the present time the exact nature of photo-neutrons is not clear : if they were photo-protons produced in the detector (as thought from photonuclear cross sections and energy spectra) it should be possible to eliminate them by confining the decay electrons in a region where they could not radiate (with a high magnetic field solenoid) to avoid any Bremsstrahlung. Another way to eliminate photo-protons could be to require a double scattering of the capture neutron : such a device, by its directionality, could also help to reduce the backgrounds not coming from the target. Moreover a better neutron-gamma discrimination could also be achieved with new sampling techniques (transient recorder with flash ADC). The background reduction would give a cleaner time distribution, so that the Ortho-Para transition could be observed in a longer time window (let say 10 to 15 µs instead of the 7 µs of the SCB experiment), which would again lead to a better precision.

A muon capture observable which has never been measured is the Triplet capture rate. Its measure would not increase the precision on g_p (unless if a very high precision could be achieved), but would be a very interesting test of the V-A structure of the weak currents. Unfortunately its direct measure is impossible because, even in the most favourable conditions (a very low density target or the Para level of the $p\mu p$ molecule), we have a statistical mixture of Tripplet and Singlet states, where the proportion of Singlet capture rate is still too high.

An indirect way to reach the ratio $x = \lambda_T/\lambda_S$ is to measure more precisely the capture neutrons time distribution in liquid hydrogen, taking advantage of the modulation introduced by the Ortho-Para transition. As a fact, the expression for this time distribution (neglecting the capture rates in front of λ_o and λ_{op}) is :

$$\frac{dN_n}{dt}(t) = N e^{-\lambda_o t} \left[1 + K(x) e^{-\lambda_{op} t} \right] \qquad (52)$$

with

$$K = \frac{\lambda_{OM} - \lambda_{PM}}{\lambda_{PM}} \simeq 2 - 8.x \qquad (53)$$

Hence a three parameter fit (normalisation N, K and λ_{op}) of this distribution allows to reach x. For instance, to measure x with an accuracy of 50 %, we should have to detect 3×10^8 capture neutrons (without background), and hence to work near a very intense beam like high energy (30 GeV) proton machine. The other experimental difficulties are the same as the one encountered previously for the improved measure of λ_{op}.

Another direction to reach g_p is to measure the rate of radiative muon capture on the *proton* (RMCP) :

$$\mu^- + p \rightarrow n + \nu_\mu + \gamma \qquad (54)$$

Unfortunately this rate is very low (maximal evaluation of 0.1 s^{-1}) [ref.[33]]. Hence such an experiment needs a large solid angle detector for γ-rays with a good electron induced photons identification. Moreover it is an absolute yield of RMCP γ-rays that has to be measured, so that the uncertainty in the detection efficiency directly influences the final precision. A proposal to study RMCP in a gaseous high pressure and liquid hydrogen target has been submitted at TRIUMF by the TPC collaboration.[34] Measurement of RMCP has never been done before because of its extremely low rate. It presents the advantage over RMC on nuclei of being free of nuclear model uncertainties. Calculations[33] also show that it is three times more sensitive to g_p than ordinary muon capture. Although its rate is 10^{-4} that of ordinary muon capture, Primakoff[2] has argued that it is considerably more sensitive to the existence of a second class current term than ordinary muon capture.

9. CONCLUSION

A major advantage of muon capture in hydrogen is to provide informations, especially for the Axial form factors of the weak hadronic current, that are free from any nuclear structure complication. As these form factors are fundamental parameters of the theory, their measure is quite essential. To improve the present knowledge on these parameters, we have seen that the future experiments will be very difficult : i) either we will have to achieved high levels of precision, and hence to be very carefull in the experimental procedure ; ii) or we will have to enter the region of very rare processes where again the experimental techniques are at the frontier of our possibilities. Nevertheless I think that such experiments are unavoidable in order to definitely clarify the fundamental informations provided by muon capture on the proton. In such a future, muon capture on nuclei would not suffer any more from the lack of knowledge on the elementary coupling constants, and could become a very usefull tool to study nuclear problems.

REFERENCES

1. i)   G. Bardin et al., Nucl. Phys. $\underline{A352}$: 365 (1981) ;
   ii) G. Bardin et al., Phys. Lett. $\overline{104B}$: 320 (1981) ;
   iii) G. Bardin et al., Phys. Lett. to be published (1984) ;
   iv) see also in G. Bardin, These Doctorat d'Etat, University
        of Paris-Sud, n°2647 (1982) ;
   v)   and in J. Martino, These Doctorat d'Etat, University of
        Paris-Sud, n°2567 (1982) ;
   vi) a detailed description of the target is given in M. Adam
        et al., Nucl. Instr. Meth. $\underline{177}$: 305 (1980).
2. We follow closely the theoretical exposition of· H. Primakoff,
   in Muon Physics II (Academic Press, Inc., New York, 1975), p.3,
   and use the same symbols.
3. See in ref. 1 (iii) and (v) and in ref. 2.
4. See in ref. 1 (v) and E.R. Shrock et al., Phys. Rev. Lett. $\underline{41}$:
   1692 (1978).
5. G.G. Simon et al., Nucl. Phys. $\underline{A333}$: 381 (1980).
6. M.R. Goldman, Nucl. Phys. $\underline{B49}$: $\overline{621}$ (1972).
7. R.P. Feynman et al., Phys. Rev. $\underline{109}$: 193 (1958) ;
   S.S. Gershstein et al., Zh. Eksp. Teor. Fiz. $\underline{29}$: 698 (1955)
   (Sov. Phys. JETP $\underline{2}$: 76 (1957).
8. See in ref. 5 for the proton, and for the neutron in V.E.
   Krohn et al., Phys. Rev. D8:1305 (1973) ; L. Koester et al.,
   Phys. Rev. Lett. 36:1021 $\overline{(1976)}$; G.G. Simon et al., Z.
   Naturforsch. $\underline{35a}$: $\overline{1}$ (1979).
9. Particle Data Group, Review of Particle Properties, Rev. Mod.
   Phys. $\underline{52}$:
10. i) N. Armenise et al., Nucl. Phys. $\underline{C16}$: 397 (1977) ;
    ii) N.J. Baker et al., Phys. Rev. $\underline{D23}$: 2499 (1981).
11. S. Weinberg, Phys. Rev. $\underline{112}$: 1375 $\overline{(1958)}$.
12. H. Primakoff, Rev. Mod. Phys.· $\underline{31}$: 802 (1959).
13. See i) E. Zavattini, in Muon Physics II (Academic Press, Inc.,
            New York, 1975) p. 219 ;
       ii) S.S. Gershtein et al., in Muon Physics III (Academic
            Press, Inc., New York, 1975) p. 142.
14. S. Weinberg, Phys. Rev. Lett. $\underline{4}$: 585 (1960) ; see also in
    ref. 1 (iv) and in ref. 16.
15. See in ref. 13 (i).
16. D.D. Bakalov et al., Nucl. Phys. $\underline{A384}$: 302 (1982).
17. See in ref. 13 (ii).
18. This is the combination of the results quoted in ref. 9 and in
    G. Bardin et al., Phys. Lett. $\underline{B79}$: 52 (1978).
19. K.L. Giovanetti et al., Phys. Rev. $\underline{D29}$: 343 (1984).
20. H. Uberall, Phys. Rev. $\underline{119}$: 365 $\overline{(1960)}$.
21. R.Winston, Phys. Rev. $1\underline{29}$: 2766 (1963).
22. See in ref. 1 (iv).
23. See in ref. 2 and in ref. 1 (v).

24. See in ref. 2.
25. See for instance (i) S.L. Adler, Ann. Phys. $\underline{50}$: 189 (1968) and
    ii) M. Chemtob et al., Nucl. Phys. $\underline{A163}$: 1 (1971).
26. D.V. Bugg et al., Phys. Lett. $\underline{B44}$: 248 (1973) ;
    M. McGregor et al., Phys. Rev. 182: 1714 (1969).
27. K. Holinde, Phys. Rep. $\underline{68}$ (3) (1981).
28. L.P. Roesch et al., Phys. Rev. Lett. $\underline{46}$: 1507 (1981).
29. Y. Kuno et al., to be published ; see also in M. Fukui et al.,
    Phys. Lett. $\underline{B132}$: 255 (1983).
30. See in ref. 10 (ii).
31. M. Morita et al., Phys. Lett. $\underline{B73}$: 17 (1978).
32. B.R. Holstein, Phys. Rev. $\underline{C29}$:623 (1984).
33. See in ref. 34, table 1.
34. G. Azuelos et al., TRIUMF Research Proposal.
35. R. Hildebrand, Phys. Rev. Lett. $\underline{8}$: 34 (1962) ; J.H. Doede et
    al., quoted by C. Rubbia in Proc. Intern. Conf. on Fundamental
    Aspects of Weak Interactions, Brookhaven (1963).
36. E. Bertolini et al., Proc. Conf. High Energy Physics, CERN
    (1962) ; S. Focardi et al., quoted by C. Rubbia in Proc. Intern.
    Conf. on Fundamental Aspects of Weak Interactions, Brookhaven
    (1963).
37. E.J. Bleser et al., Phys. Rev. Lett. $\underline{8}$: 288 (1962).
38. J.E. Rothberg et al., Phys. Rev. $\underline{132}$: 2664 (1963).
39. A. Alberigi Quaranta et al., Phys. Rev. $\underline{177}$: 2118 (1969).
40. V.M. Bystritskii et al., Sov. Phys. JETP $\underline{39}$: 19 (1974).

MUON CAPTURE IN DEUTERIUM

M. Piccinini

Istituto Nazionale de Fisica Nucleare, Sezione de Bologna
and Dipartimento di Fisica dell'Università di Bologna
Bologna, Italy

I. INTRODUCTION

The deuteron represents the simplest bound state of nucleons existing in nature. From the study of the weak nuclear capture process

$$\mu^- + d \rightarrow n + n + \nu_\mu \tag{1}$$

one can then hope to get information in two main directions. On one side, if the hadronic form factors in the weak Hamiltonian describing reaction (1) are known, then the deuteron wave function, the final state interaction between the produced neutrons and the meson exchange currents effects can be investigated[1-3]. On the other hand, assuming this information from other sources, one can determine the above-mentioned form factors obtaining similar or complementary information with respect to those obtained from the simplest nuclear muon capture by protons. In particular, due to the presence of two neutrons in the final state, the Pauli exclusion principle effect gives an opportunity of observing an almost pure Gamow-Teller coupling[4].

Moreover, it must be noted that the operators responsible for the initiating processes of the stellar nucleosynthesis

$$p + p \rightarrow d + e^+ + \nu \tag{2a}$$

$$p + e^- + p \rightarrow d + \nu \tag{2b}$$

are the same as for the muon absorption in the deuteron (equation (1)), therefore a good knowledge of this reaction allows to make accurate predictions on the rates of solar neutrino production[2].

71

For these reasons process (1) has been the object of theoretical and experimental investigation, in parallel to the ones performed on the fundamental weak process[5].

$$\mu^- + p \rightarrow n + \nu_\mu \tag{3}$$

II. THEORETICAL PREDICTIONS

The calculation of the nuclear muon capture rate λc for process (1) has been faced by many authors, mainly in the frame of the impulse approximation (where it is assumed that the total momentum transferred is delivered to either one of the nucleons without subsequent exchange of mesons). Here we will outline only the main theoretical elements needed for the calculation:

a) The current-current Hamiltonian (in the framework of the V-A weak interactions theory)[6]

$$H_w = \frac{1}{\sqrt{2}} G \cos\theta_c \, J_\alpha^1 \, J_\alpha^h, \text{ where}$$

$G = (1.16634\pm0.00002) \times 10^{-5}$ GeV^{-2} is the Fermi coupling constant, $\cos\theta_c = 0.9730\pm0.0024$ is the cosinus of the Cabibbo angle[7]

$$\langle \nu_\mu | J_\alpha^1 | \mu^- \rangle = \bar{u}_\nu \gamma_\alpha (1+\gamma_5) u_\mu$$

$$\langle n | J_\alpha^h | p \rangle = V_\alpha + A_\alpha$$

$$V_\alpha = \bar{u}_n \{ \gamma_\alpha F_V(q^2) + \sigma_{\alpha\beta} q_\beta F_M(q^2) \} u_p$$

$$A_\alpha = \bar{u}_n \gamma_5 \{ \gamma_\alpha F_A(q^2) - i q_\alpha F_P(q^2) \} u_p$$

$$\sigma_{\alpha\beta} = \frac{1}{2i} (\gamma_\alpha \gamma_\beta - \gamma_\beta \gamma_\alpha)$$

(second class currents are neglected).

b) The deuteron wave function, calculated on the basis of a wide range of nuclear potentials (hard-core[8], Reid soft-core[1,2], Hamada-Johnston[2], etc.) and with different levels of approximately (mainly concerning the contribution of the D-state part of the deuteron).

c) The final state wave function, where different calculation techniques were employed to take into account the final state interaction in a more or less refined way[1,9-11].

d) A correction due to meson exchange currents, necessary to perform calculations with an accuracy better than 10%[2-3].

In this case too, as in process (3), the calculations foresee a strong hyperfine effect for the values of λc relative to capture reactions starting from the doublet (λc^D) or from the quartet (λc^Q) state of the μd system[5].

In Table 1 the main theoretical results on the value of λc^D are reported[1-13]. After a careful examination of these calculations one can see that, starting from the same set of updated values for the hadronic form factors, and including all the above mentioned elements (a) to d)) the following theoretical value can be assumed for λc^D:

$$\lambda\text{c}^D(\text{Th}) \approx (410 \pm 20) \ \text{s}^{-1} \ , \tag{4}$$

where the 5% uncertainty is mainly due to the different nuclear potentials assumed. On the other side the theoretical predictions for λc^Q range between 7 and 15 s^{-1}. In equation (4), deuteron structure effects (i.e. the influence on λc^D of variations of the deuteron wave function at small distances) as well as off-shell effects in the final state are not included[1,10,11].

III. EXPERIMENTS

Experiments Based on the Detection of Capture Neutrons

With respect to the hydrogen case (equation (3)), measurements of λc based on the "classical" method of detecting one of the final neutrons of reaction (1) have to face two additional difficulties: i) the emitted neutrons are not monoenergetic (actually they have a mean energy of 1.5 MeV[8,9]), ii) in pure deuterium the formation of dμd

Table 1. Comparison of Experimental and Theoretical Values of λc^D (Capture Rate Concerning μd Systems in the Doublet State of Spin).

Year	Exp(s^{-1})	Th(s^{-1})
1958		250 [4]
1965	365±96 [14]	334 [8]
1972		313 [12]
1973	445±60 [15]	
1974		377 [1]
1975		387 [13]
1976		405 [2]
1979		413 [3]
1984	480±31*	

*LT measurement (see text)

molecules (even if fairly slow in low-density targets), leads to a prompt fusion reaction with a 2.5 MeV neutron as a possible final product, perturbating the capture neutrons measurement (see Figure 1). Two experiments have been done in the past by means of the neutron detection technique (see Table 1); in both measurements the target was made of a mixture of hydrogen and deuterium, in order to overcome the fusion neutrons problem.

The Columbia experiment[14] was carried out using a liquid hydrogen target contaminated by 0.32% deuterium; in these conditions the dμd formation is negligible, but the formation of pμd molecules is important. These muonic molecules can fuse following the reaction

$$p\mu d \rightarrow \mu^3He + \gamma \tag{5}$$

(see Figure 1) with the production of μ^3He systems, and then the neutron rate in this experiment is a combination of neutrons coming from muon capture in hydrogen, deuterium and ^3He. The analysis of their time distribution allows to extract the muon capture rate relative to the doublet state of the μd system (λ^D_c). The particular analysis performed limits the precision of the result (~25%) and it contains the assumption that the formation of pμd molecules proceeds starting from μd atoms in the statistical mixture of hyperfine structure states.

The second experiment[15] (Bologna-CERN collaboration) applied the same method to avoid the dμd formation, but in this case the target was a gaseous one with 5% deuterium. In these conditions the isotopic transfer reaction

$$\mu p + d \rightarrow \mu d + p \tag{6}$$

remains fast despite the low density of the target but the pμd formation is negligible, allowing the neutron rate to be produced mainly by muon capture within μd systems. The obtained value of λ^D_c is compatible with the updated theoretical predictions for λ^D_c (see Table 1 and equation (4)).

As one can see, a correct interpretation of these results depends on the knowledge of the hyperfine population of the μd atoms in the hydrogen-deuterium mixture and then on the quartet to doublet transition rates λ^{QD}_d and λ^{QD}_p governed by the elastic scattering processes

$$\mu d(s_i) + d \rightarrow \mu d(s_f) + d \text{ , and} \tag{7a}$$

$$\mu d(s_i) + p \rightarrow \mu d(s_f) + p \tag{7b}$$

respectively (here s_i (s_f) represents the spin state of the μd atom, quartet or doublet, in all possible combinations). It must be re-

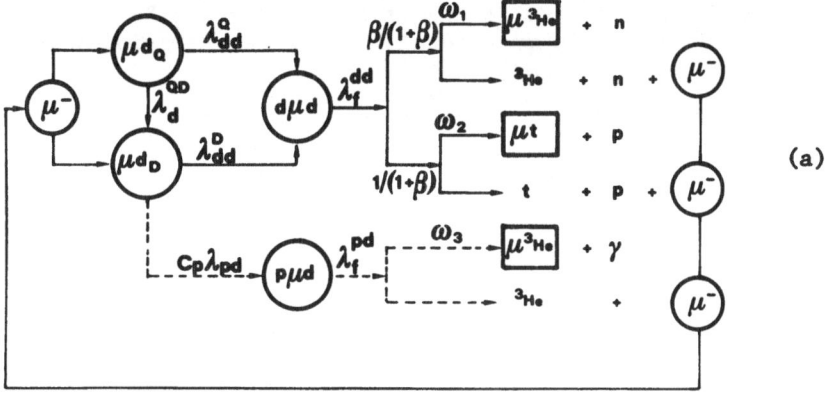

(a)

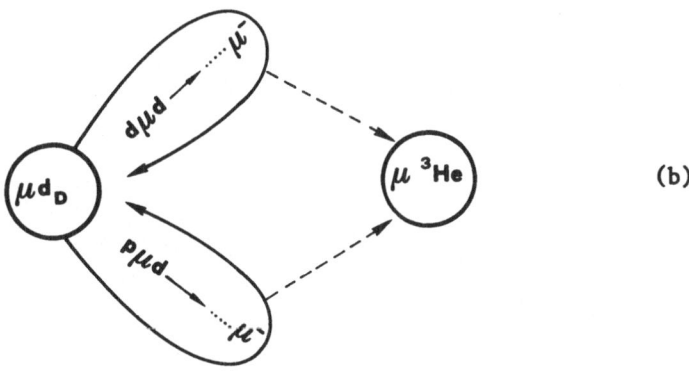

(b)

Fig. 1. a) Simplified scheme of the μ-molecular processes which
take place when negative muons are stopped in a deuterium
target contaminated by a concentration c_p of hydrogen
(the solid lines represent the processes which are relevant
for the LT measurement). The symbols are defined in the
text. b) Symbolic representation of the main populations
contributing to the value of $\lambda\overset{*}{c}$.

called here that in a hydrogen target with a little amount of deuterium, μd systems are mainly formed through reaction (6), with a
kinetic energy of 45 eV[15]. Given that the cross sections of processes (7) are functions of energy[16], one can understand that the
knowledge of λ_d^{QD} and λ_p^{QD} needs systematic investigations by varying
density and relative concentration conditions of the mixture. Recent
results on this subject have been obtained by a Vienna-SIN group in
the frame of a measurement whose main goal is the extraction of the
neutron-neutron scattering length from the study of the correlation

Table 2. Up-to-date Values of the Parameters Relevant to the Main mu-molecular Processes Pertaining to the LT Measurement

Parameter	Value
λ_d^{QD}	$(4.26 \pm 0.17) \times 10^7 \ s^{-1}$ [18]
λ_{dd}^D	$(4.5 \div 6.3) \times 10^4 \ s^{-1}$ [18,20]
$\lambda_{dd}^Q / \lambda_{dd}^D$	79.5 ± 8.0 [18]
λ_f^{dd}	$10^9 \ s^{-1}$ [21]
β	1.46 [22]
ω_1	$0.11 \div 0.15$ [23]
ω_2	0.01 [21]
λ_{pd}	$(5.80 \pm 0.30) \times 10^6 \ s^{-1}$ [14]
λ_f^{pd}	$(3.05 \pm 0.10) \times 10^5 \ s^{-1}$ [14]
ω_3	0.84 ± 0.04 [14]

between the final neutrons of reaction (1)[17]. They give

$$\lambda_d^{QD} = (37.0 \pm 1.5) \times 10^6 \ s^{-1} , \qquad (8a)$$

a value obtained in gaseous deuterium at 34 K[18], and

$$\lambda_p^{QD} > 4.5 \times 10^6 \ s^{-1} \qquad (8b)$$

obtained in liquid hydrogen with deuterium concentrations ranging from 0.1% to 22%[19] (values normalized to liquid hydrogen density).

Given the uncertainty remaining on the value of λ_p^{QD} (equation 8b)) we are induced to interpret the Columbia result as an upper limit for λ_c^D, and the Bo-CERN one like a lower limit for the same quantity: taking the two-standard-deviation range one gets

$$325 \ s^{-1} < \lambda_c^D < 557 \ s^{-1} \qquad (9)$$

The Lifetime Technique (LT) Experiment

The above mentioned difficulties brought our group[20] to undertake a measurement of process (1) (that is running at the 600 MeV Saclay Linear Accelerator (ALS)) with a different experimental approach: in order to have a well defined µ-atomic system as initial state and to avoid the problems concerning the neutrons detection, the lifetime τ_μ- of negative muons stopped in liquid deuterium is measured to a very high accuracy (10^{-5}) from the delayed time distribution of their decay electrons, and it is compared to the lifetime τ_μ+ of positive muons to get an average capture rate:

$$\lambda_C^* = (1/\tau_\mu-) - (1/\tau_\mu+) . \tag{10}$$

When negative muons are stopped in a ultrapure liquid deuterium target, they undergo several processes before decaying or being captured through reaction (1). These processes can be summarized as follows (see Figure 1 and Table 2):

i) Deuterium muonic atoms µd are formed, initially in a statistical mixture of doublet $(µd)_D$ and quartet $(µd)_Q$ total spin state. At the density and temperature of liquid deuterium, however, the transition rate λ_d^{QD} from the quartet to the (lower-lying) doublet state is very high[18] with respect to the muon decay rate (see equation (8a)) and such a process is irreversible[16].

ii) The primary µd atoms form dµd molecular ions, according to the reaction[20]

$$µd + d \rightarrow dµd \tag{11}$$

at a rate λ_{dd}^Q or λ_{dd}^D, depending on the spin state of the starting µd system.

iii) The dµd systems immediately undergo either fusion reaction[21]

$$dµd \rightarrow n + {}^3He + µ^- \tag{12a}$$

$$dµd \rightarrow p + t + µ^- \tag{12b}$$

with branching ratio β[22] (the 2.44 MeV neutrons released in process (12a) overlap to the spectrum of the capture neutrons from process (1)).

iv) The muons from reactions (12a and b) which do not stick to the ^{3}He and t nuclei (about 80% of the fusing dµd's; the sticking probabilities are ω_1 and ω_2 respectively[23]) are still available to form µd atoms, and re-enter the cycle.

v) Even the ultrapure deuterium is usually contaminated by a certain concentration c_p of hydrogen, so that the primary µd systems can also form pµd ions that subsequently can undergo the fusion process of equation (5), giving $µ^3He$ systems or recycled muons.

This apparently complicated story turns out to be fairly simple if one calculates the population of the various muonic systems sketched in Figure 1 assuming the rate values reported in Table 2. In fact from this calculation one gets that, starting from about 1 µs after its stopping in the target, a negative muon spends about 97% of its life in the $(\mu d)_D$ state and 3% in the μ^3He state. All the other systems have a sufficiently short lifetime to be neglected. One can then conclude that, in this case, the muon nuclear capture reactions occurring in ultrapure deuterium proceed essentially from the $(\mu d)_D$ state.

According to the above conclusion, the value λ_c^* (equation (10)) can be written as

$$\lambda_c^* = \lambda_c^D + \partial\lambda_{d\mu d} + \partial\lambda_{p\mu d} \tag{13}$$

where $\partial\lambda_{d\mu d}$ and $\partial\lambda_{p\mu d}$ represent two corrections due to that fraction of muons that stick to the ^{3}He nuclei, following respectively processes (12a) and (5) (see Figure 1b).

The data taken until now were carried out in two different target conditions,' in which the ultrapure liquid deuterium was contaminated respectively by a hydrogen concentration $c_p = (0.180\pm 0.015) \times 10^{-2}$ and $c_p = (1.14\pm0.10) \times 10^{-2}$. This was done to provide the possibility of extrapolating the results obtained to the ideal condition of zero-hydrogen concentration, obtaining thereby a result which does not depend on the formation of $p\mu d$ molecular ions ($\partial\lambda_{p\mu d} = 0$). On the other hand, the correction $\partial\lambda_{d\mu d}$ can be split into two terms ($\partial\lambda_{d\mu d} = \partial\lambda_s + \partial\lambda_f$) as follows: i) $\partial\lambda_s$, representing the biggest part of the correction, takes into account the slow formation of μ^3He systems from the $(\mu d)_D$ states and it depends on the product $\alpha_s \lambda_{dd}^D$ (where $\alpha_s \simeq (\beta/(1+\beta))\omega_1$) and on the muon capture rate $\lambda_{c,He}$ within the μ^3He atoms. ii) $\partial\lambda_f$ describes the effect due to the μ^3He systems promptly formed (before opening the observation gate) via the $(\mu d)_Q$ state, and it depends on λ_d^{QD}, $\alpha_s \lambda_{dd}^Q$ and $\lambda_{c,He}$.

The term $\alpha_s \lambda_{dd}^D$ can be measured by exploiting the fusion reaction (12a): it occurs in fact that the time distribution of the neutrons from this reaction is proportional to the population of $(\mu d)_D$ alone:

$$dn/dt \propto \exp\{-(\lambda_o + \lambda_s)t\} \tag{14}$$

where $\lambda_o = (1/\tau\mu^+)$ represents the muon decay rate and $\lambda_s \simeq \alpha_s \lambda_{dd}^D + $ (hydrogen contribution). Then the analysis of the lifetime of these neutrons (τ_n) allows to obtain the required product (once the above mentioned extrapolation procedure has eliminated the hydrogen contribution). For this reason the lifetime of the neutrons released following the stopping of muons in the ultrapure deuterium target was also observed independently in the present measurements, turning in a source of useful information the main perturbating effect of the "classic" neutron detecting experiments. The fact that the parameters observed are lifetimes makes the present analysis independent on the evaluation of any absolute detection efficiency.

78

The same lifetime technique was already used by the Saclay-CERN-Bologna collaboration to measure the muon nuclear capture in liquid hydrogen[25] and the ortho-to-para transition rate[26] within the $p\mu p$ molecular ion. We will not describe here the experimental apparatus[27], the details of the performed analysis and the background sources which are common to both experiments, and have been described by Martino during his lectures[28]. Concerning the fusion neutron runs, a preliminar analysis has been published[20]. Here we will just recall that the favorable signal-to-noise ratio reduces the corrections due to the different backgrounds to a few 10^{-4} to τ_n. Two secondary effects which are peculiar to the deuterium measurement will be discussed later.

The results obtained by the analysis of the data taken until now are reported in Table 3. (For the exact calculation of $\partial\lambda d\mu d$, the ratio $\lambda_{dd}^Q/\lambda_{dd}^Q$ and the value of λ_d^{QD} were taken from reference 18 and the value of $\lambda c,He$ was obtained from reference 24). Correcting λ_c^* for the values of $\partial\lambda_s$ and $\partial\lambda_f$ and adding a final correction of 12 s^{-1} to account for the relativistic effect on the muon decay rate due to the mu-atomic bond, one gets:

$$\lambda_c^D = (487 \pm 38) \text{ s}^{-1} . \tag{15}$$

The result of equation (15) on λ_c^D was obtained with the minimum of assumptions other than the informations given by the LT experiment.

On the other side, during the course of the experiment additional information appeared in literature concerning some of the parameters needed to obtain $\partial\lambda d\mu d$: in particular, the sticking parameter ωl (see Figure 1a and Table 2) has been accurately measured[23]. Given that the analysis of the time distribution of the fusion neutrons (equation (12a)) gives results in perfect agreement with the corresponding values of Table 2 [20], one can express and calculate $\partial\lambda d\mu d$ directly as a function of these independent measurements, obtaining the value

$$\partial\lambda d\mu d = - (64 \pm 15) \text{ s}^{-1} . \tag{16}$$

Table 3. Lifetimes, Capture Rate and Corrections as Observed in the LT Measurement

Lifetimes (ns)	Rates (s^{-1})
$\tau_{\mu^-} = 2194.464 \pm 0.130$ $\tau_{\mu^+} = 2197.078 \pm 0.073$	$\lambda_c^* = 542 \pm 31$
$\tau_n = 2172.8 \pm 8.4$	$\partial\lambda_s = -54 \pm 22$ $\partial\lambda_f = -13 \pm 5$

Additionally, if one performs the calculations assuming as τ the world average (updated for the recent Saclay[29] and TRIUMF[30] results):

$$\tau_\mu{}^+ = (2197.031 \pm 0.040) \text{ ns} \tag{17}$$

the preliminary value for λ_c^D becomes:

$$\lambda_c^D = (480 \pm 31) \text{ s}^{-1} . \tag{18}$$

The result (18) is compared to the previous experimental results and to the theoretical predictions in Table 1.

Before concluding this section let us consider two secondary effects, peculiar to the deuterium measurement, which represent the price to be paid to deal with the particularly simple muonic system from which the capture process proceed (a μd atom in the doublet state of total spin):

Impurities. The measurement in liquid deuterium is particularly sensitive to the presence of impurity atoms within the target, because of the high probability with which the muon may be transferred to extraneous nuclei. (Typical transfer rates are in range 10^{10} to 10^{11} s^{-1}, and capture rates in high Z nuclei – for instance noble gases – are 10^4 to 10^6 s^{-1})[32]. In the present experimental conditions, nevertheless, it was concluded that (i) the careful pumping of the target and of the filling circuit, (ii) the filling procedure, consisting in passing the deuterium gas to be liquified through a palladium filter (whose efficiency ensured a transparency to impurities smaller than 10^{-7}) and a heat exchanger[27], and (iii) the low level of impurities in the original gas (about 20 ppm of nitrogen), ensured during the measurements a level of impurities smaller than 10^{-10}, despite the residual outgassing. (As an example, if one assumes that a concentration c= 10^{-9} of extraneous element would be present within the liquid deuterium as entirely solved nitrogen, one calculates that the corresponding systematic error on λ_c^* would be about 5 s^{-1}). During auxiliary measurements, moreover, the target was filled without using the palladium filter, and a rate λc$\approx$8000 s^{-1} was measured in these conditions while the total capture rate expected for totally dissolved 20 ppm of nitrogen is c$\lambda\approx$54000 s^{-1}. The above considerations make us confident on a negligible effect of impurities during our measurements.

Wall effect. The muons stopped near the copper walls of the target can indeed be transferred from the μd atoms to Cu atoms, where the nuclear capture occurs at a much higher rate. Also the muons recycled after a fusion process (see equations (12)) can have a sufficiently high energy to reach the target walls. To test this effect special measurements were performed, where the stopping region of the muon beam inside the target was varied by changing the momentum of the beam in order to enhance the number of muons stopped near

the target walls. (We recall here that in normal operating conditions more than 99% of the muons are stopped in the liquid deuterium). No significant effect of such a transfer could be evidenced
out, even in non-optimal muon stopping conditions.

IV. CONCLUSIONS

Our result (18) and the theoretical value of λ_c^D (equation (4))
show a discrepancy of about two standard deviations. Even if result
(18) is a preliminary one, this difference must stimulate to look for
possible explanations.

The most recent calculations on λ_c^D have paid a particular attention to the effects of non trivial off-shell behaviors of nucleon-
nucleon potentials[1,11], as well as to the contribution to λ_c^D due to
meson exchange currents[2,3]. Concerning the former topic, Sotona
and Truhlik[1] have investigated some deuteron structure effects and
off-shell effects in the $^1S_0$ nn final state. They calculate λ_c^D with
different deuteron and final state wave functions, obtained by generating phase-shift equivalent potentials from the Reid soft-core
potential[33], by means of particular unitary rank-one transformations[34]. (These transformations are chosen in such a way as to
provide the same two-body on-shell scattering amplitude and to obtain
the deuteron properties either unchanged or differing only slightly
from the untransformed case). In this way they find a maximum vari-
ation of $\sim 8\%$ for λ_c^D due to deuteron structure effects, and of $\sim 30\%$
due to off-shell effects in the final state, but always in the direc-
tion of lower values with respect to equation (4). In any case, the
possibility of finding unitary transformations (of higher rank, or
characterized by different form factors) corresponding to off-shell
effects leading to an enhancement of λ_c^D is not excluded by these
authors. Similar conclusions have been obtained by Ho-Kim et al.[11],
by adopting an approach independent from particular potentials.

As regards meson exchange currents, their calculation has been
faced by Dautry et al.[2] in the one-boson-exchange approximation, by
a method based on current algebra, PCAC and vector dominance[35], and
by Ivanov and Truhlik[3] in the same approximation, exploiting the
hard-pion method[36]. In both calculations an enhancement of $\sim 6.9\%$
is obtained for the impulse-approximation value of λ_c^D. (This correction is included in the corresponding values of Table 1, as well
as in equation (4)). The LT result on λ_c^D suggests a further effort
in the study of these topics.

Concerning the weak-interactions aspect of the problem, the
situation is represented in Figure 2 where the result of equation
(18) has been compared to the theoretical value of equation (4),
in a by now usual way[14,15]. (The regions defined by these two
values in a g_p^μ v/s g_A^μ plane are represented, where $g_p^\mu = m_\mu F_p(q^2)$

and $g_A^\mu = F_A(q^2)$ – see Section II. The hadronic form factors are taken at $q^2 \simeq m_\mu^2$, m being the muon mass). By the same procedure followed in the case of the experiment in liquid hydrogen[26,28] (see also the lectures of Martino), one can extract from equation (18) a value of g_P^μ:

$$g_P^\mu = -(1.4 \pm 3.8) \tag{19}$$

to be compared with the value $g_P^\mu = -(9\pm2)$ obtained from the hydrogen experiments (see the quoted lectures). Alternatively, a value for $F_A(0)$ can be obtained (assuming $g_P^\mu = -8.1$ from the PCAC hypothesis):

$$F_A(0) = -(1.373 \pm 0.066) \tag{20}$$

to be compared with $F_A(0) = -(1.242\pm0.035)$ one gets from the experiment on muon capture in hydrogen.

As one can see, assuming the nuclear aspects of the problem as well known the agreement between the information on the hadronic form factors coming from reactions (1) and (3) is not good. In particular, the conclusions concerning second class currents or μ-e universality one would get from equations (19) and (20) respectively, would be rather different from those obtained from the hydrogen experiments.

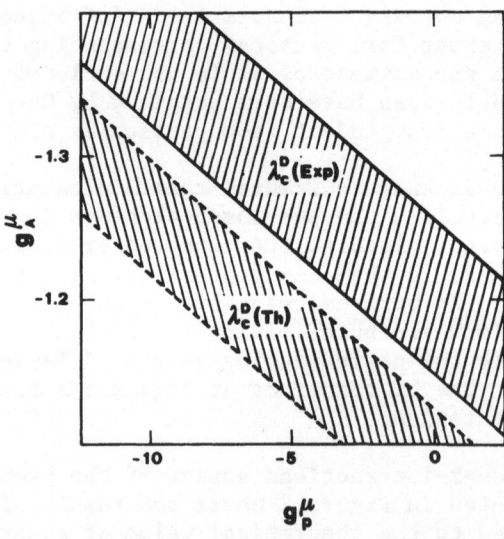

Fig. 2. Comparison of the experimental value of λ_c^D obtained in the LT measurement (equation (18)) with the theoretical value (equation (4) on the text).

This confirms that further experimental efforts are needed:

a) To measure better the fundamental process (3), both in the gaseous and in the liquid phase. In the first case the aim would be to increase the accuracy on g_P^μ. (We recall here that the present g_P^μ value is obtained from the analysis of two "gas" and three "liquid" experiments, and that the interpretation of the last ones is strongly dependent on the value of the rate λop relative to the ortho-to-para transition in the pμp ion). In the second case (liquid phase), the aim would be to determine with better accuracy λop and to explore the study of new hyperfine states of the μp atom, as already stressed by Martino.

b) To exploit thoroughly the lifetime method, to obtain precise results on the study of process (1). In particular, a measurement with a gaseous ultrapure deuterium target could represent an ulterior source of information on the structure of the weak interactions Hamiltonian. In this case in fact the μd atoms initially formed in a statistical mixture of total spin state remain in this state for a longer time if compared to the μp atoms case, because of the following main reason: the thermal energy $\varepsilon \cong 0.04$ eV (at room temperature) of the μd system is comparable with the hyperfine splitting of its fundamental state $\Delta E_{\mu d} = 0.049$ eV. (Opposite to the case of the μp atom where ε must be compared to $\Delta E_{\mu p} = 0.183$ eV, and the triplet state of the μp system has a short lifetime – see the lecture of Massa in this school). Then reaction (7a), responsible for the change of spin state of the muonic atom is active in both directions (s_i = quartet, s_f = doublet and vice versa). As a consequence, the possibility of measuring the capture rate relative to hyperfine states statistically populated ($\lambda_c^{st} = 1/3(\lambda_c^Q) + 2/3(\lambda_c^D)$) seems easier to realize than in the hydrogen case.

Acknowledgements

I wish to record my warm appreciation to Profs. A. Bertin and A. Vitale for useful criticism and suggestions. I want also to thank Prof. E. Zavattini for constant interest and fruitful discussions.

REFERENCES

1. M. Sotona dn E. Truhlik, Nucl.Phys., A229:471 (1974).
2. F. Dautry, M. Rho, and D. O. Riska, Nucl.Phys., A264:507 (1976).
3. E. Ivanov and E. Truhlik, Nucl.Phys., A316:451 (1979).
4. H. Uberall and L. Wolfenstein, N.Cim., 10:136 (1958).
5. E. Zavattini, in: "Muon Physics II," V. W. Huges and C. S. Wu, eds., Academic Press (1975).
6. H. Primakoff, Rev.of Mod.Phys., 31:802 (1959).
7. A. Bertin and A. Vitale, Pure leptonic weak processes in: "Fifty Years of Weak-Interactions Physics," A. Bertin, R. A. Ricci and A. Vitale, eds. (1984).

8. I. -T. Wang, Phys.Rev., B139:1539 (1965).
9. E. Truhlik, Nucl.Phys., B45:303 (1972).
10. M. Sotona and E. Truhlik, Phys.Lett., 43B:362 (1973).
11. Q. Ho-Kim, J. P. Lavine, and H. S. Picker, Phys.Rev., C13:1966 (1976).
12. P. Pascual, R. Tarrach, and F. Vidal, N.Cim., A12:241 (1972).
13. Nguyen Tien Nguyen, Nucl.Phys., A254:485 (1975).
14. I. -T. Wang, E. W. Anderson, E. J. Bleser, L. M. Lederman, S. L. Meyer, J. L. Rosen, and J. E. Rothberg, Phys.Rev., B139:1528 (1965).
15. A. Bertin, A. Vitale, A. Placci, and E. Zavattini, Phys.Rev., D8:3774 (1973).
16. S. Cohen, D. L. Judd, and R. J. Riddell, Phys.Rev., 119:397 (1960); L. I. Ponomarev, L. N. Somov, and M. I. Faifman, Yad.Fiz., 29:133 (1979). (Sov.J.Nucl.Phys., 29:67 (1979)).
17. W. H. Bertl, W. H. Breunlich, P. Kammel, W. J. Kossler, H. G. Mahler, C. Petitjean, L. A. Schaller, L. Schellenberg, and H. Zmeskal, SIN Jahresbericht C27 (1979).
18. P. Kammel, W. H. Breunlich, M. Cargnelli, H. G. Mahler, J. Zmeskal, W. H. Bertl, and C. Petitjean, Phys.Rev., A28:2611 (1983).
19. W. H. Bertl, W. H. Breunlich, P. Kammel, W. J. Kossler, H. G. Mahler, C. Petitjean, W. L. Reiter, L. A. Schaller, and L. Schellenberg, submitted to the Third Int. Conf. on Emerging Nuclear Energy Systems, Helsinki, Finland, June 6-9 (1983).
20. G. Bardin, J. Duclos, A. Magnon, J. Martino, A. Bertin, M. Capponi, M. Piccinini and A. Vitale, N.Cim.Lett., 36:79 (1983).
21. S. S. Gershtein, Yu. V. Petrov, L. I. Ponomarev, L. N. Somov, and M. P. Faifman, Zh.Eksp.Teor.Fiz., 78:2099 (1980). (Sov. Phys.JETP, 51:1053 (1980)).
22. B. P. Ad'yasevich, V. G. Antonenko, and V. N. Brasin, Yad.Fiz., 33:1167 (1981). (Sov.J.Nucl.Phys., 33:619 (1981)); L. N. Bogdanova, V. E. Markushin, V. S. Melezhik, and L. I. Ponomarev, Phys.Lett., B115:171 (1982).
23. D. V. Balin, A. A. Vorobyov, B. L. Gorskhov, A. I. Ilyin, E. M. Maev, A. A. Markov, V. I. Medvedev, V. V. Nebyubin, E. M. Oriscin, G. E. Petrov, L. B. Petrov, G. G. Semenchuk, Yu. V. Smirenin, and V. V. Sulimov, LINP preprint, 715 (1981); Gatchina Group, Submitted to the Third Int. Conf. on Emerging Nuclear Energy Systems, Helsinki, Finland, June 6-9 (1983).
24. O. A. Zaimidoroga, M. M. Kulyukin, B. Pontecorvo, S. M. Sulyaev, I. V. Falomkin, A. I. Filippov, V. M. Tsupko-Sitnikov, and Yu. A. Scherbakov, Phys.Lett., 6:100 (1963); L. B. Auerbach, R. J. Esterling, R. E. Hill, D. A. Jenkins, J. T. Lach, and N. H. Lipman, Phys.Rev., B138:127 (1965).
25. G. Bardin, J. Duclos, A. Magnon, J. Martino, A. Richter, E. Zavattini, A. Bertin, M. Piccinini, A. Vitale, and D. Measday, Nucl.Phys., A352:365 (1981).

26. G. Bardin, J. Duclos, A. Magnon, J. Martino, A. Richter, E. Zavattini, A. Bertin, M. Piccinini, and A. Vitale, Phys.Lett., B104:320 (1981).
27. M. Adam, G. Bardin, B. Coudou, J. Duclos, A. Godin, P. Leconte, A. Magnon, J. Martino, M. Maurier, A. Mougeot, and D. Roux, Nucl.Instr.and Meth., 177:305 (1980).
28. see also J. Martino, Thèse d'Etat, 26 Avril 1982, Orsay n. 2567 and G. Bardin, Thèse d'Etat, 28 Octobre 1982, Orsay n. 2647.
29. G. Bardin, J. Duclos, A. Magnon, J. Martino, E. Zavattini, A. Bertin, M. Capponi, M. Piccinini, and A. Vitale, Phys.Lett., B137:135 (1984).
30. K. L. Giovanetti, W. Dey, M. Eckhause, R. D. Hart, R. Hartmann, D. W. Hertzog, J. R. Kane, W. A. Orance, W. C. Phillips, R. T. Siegel, W. F. Vulcan, R. E. Welsh, and R. G. Winter, Phys.Rev., D29:343 (1984).
31. H. Uberall, Phys.Rev., 119:365 (1960).
32. S. S. Gershtein and L. I. Ponomarev, in: "Muon Physics III," V. W. Huges and C. S. Wu, eds., Academic Press (1975).
33. R. V. Reid, Ann.of Phys., 50:411 (1968).
34. J. P. Vary, Phys.Rev., C7:521 (1973); M. I. Haftel and F. Tabakin, Phys.Rev., C3:921 (1971).
35. M. Chemtob and M. Rho, Nucl.Phys., A163:1 (1971).
36. E. Ivanov and E. Truhlih, Nucl.Phys., A316:437 (1979).
37. A. Bertin, M. Capponi, I. Massa, M. Piccinini, G. Vannini, M. Poli, and A. Vitale, N.Cim., 72A:225 (1982).

PARITY VIOLATION IN ATOMS

Lionel Pottier

Laboratoire de Spectroscopie Hertzienne de l'E.N.S.

Paris, France

INTRODUCTION

The purpose of this lecture is to give in simple terms a con-
cise review of Parity Violation in Atoms. Stress is laid on guiding
principles rather than on technical details.

I. PARITY VIOLATION : GENERAL CONSIDERATIONS

Parity is said to be violated if the laws of physics are not
invariant under space reflection. Because they are invariant under
rotations, the decomposition of space reflection into a rotation
and a mirror reflection $(x, y, z) \rightarrow (-x, -y, z) \rightarrow (-x, -y, -z)$
shows that space reflection is physically equivalent to a familiar
mirror reflection. Therefore parity violation can be defined as a
contradiction between the laws of physics of the real world and
the laws of physics of the world seen in a mirror. In other terms,
parity is violated if the laws of physics are handed.

A direct consequence of the above considerations is the following recipe to search for parity violations : A suitable experiment is performed in a handed (say, left) configuration; let R_ℓ be
the measured result. In a mirror we see the same experiment performed in the configuration of opposite (say, right) handedness, with
a result R'_r which is the mirror image of R_ℓ (e.g. if R_ℓ is a mere
number, R'_r is the same number; if R_ℓ is a vector normal to the
mirror, R'_r is the opposite vector; etc...). Then we do perform
the experiment in its right configuration, and the result is R_r.
If $R_r \neq R'_r$, the laws of physics are different in the real world
and in the mirror world : so parity is violated. An alternative

recipe consists in performing an unhanded experiment, that coincides with its mirror image; if the result does <u>not</u> coincide with its mirror image, then parity is violated. All experiments reviewed below follow one of these two patterns.

In quantum mechanics, any observable physical quantity involves squared matrix elements. In the cases considered here, arrangement is made so that the relevant quantity is of the form $|A_{em} + A_W|^2$, where A_{em} and A_W are electromagnetic and weak matrix elements respectively. Parity violation shows itself in the opposite behavior of A_{em} and of a part of A_W under mirror reflection : one is odd, the other even . (Which one is which one depends on the considered physical quantity). Thus the squared matrix element takes on two different values $|A_{em} + A_W|^2$ and $|A_{em} - A_W|^2$ in the real world and in the mirror world; the difference lies in the <u>electroweak interference</u> term $\pm 2A_{em} A_W$. (For the sake of simplicity, A_{em} and A_W are here assumed to be both real, and the part of A_W which behaves like A_{em} under mirror reflection is neglected). The amount of parity violation is characterized by the normalized difference, called right-left asymmetry :

$$ = \frac{|A_{em} + A_W|^2 - |A_{em} - A_W|^2}{|A_{em} + A_W|^2 + |A_{em} - A_W|^2} \cong 2 \frac{A_W}{A_{em}} \quad (\text{since } A_W \ll A_{em}) \tag{1}$$

II. PARITY VIOLATIONS IN ATOMS : WHY ?

1. Some History

The way to the discovery of parity violation in atoms is marked by three milestones. The first one is the first discovery of parity violation in weak (charged current) interactions in 1956-57 ([1] - [4]). Yet, in the absence of any clue to possible parity violation <u>in atoms,</u> few works were devoted to this subject in the following years ([5] - [8]). After 10 years of slow ripening ([9] - [11]) came a second milestone event : the breakthrough of gauge theories ([12]), which led to the prediction of <u>weak neutral currents,</u> mediated by the neutral Z^O boson, and to the possible existence of <u>parity violation in atoms</u>. Yet orders of magnitudes in atoms looked desperately small. The third milestone was a more accurate estimation by M.A. BOUCHIAT and C. BOUCHIAT ([13], [14]), showing that the parity violating (PV) effects to be expected in an atom with Z protons are about Z^3 times larger than anticipated so far (Bouchiat-Bouchiat's Z^3 law). This result triggered the start of several experiments to search for PV in atoms. Further experiments started in the following years.

Here the reader is logically waiting for a fourth milestone date, that of the first observation of parity violation in an atom. Although such a first observation of course does exist ([15]), parity violation in atoms initially passed, as will appear below, through a period of contradictions. More than half a decade was necessary to really establish experimental consensus concerning the existence of PV in atoms with roughly the expected size, and to-date some minor quantitative contradictions still remain.

2. Orders of Magnitude : Quite Small ...

The origin of the difficulties is the huge mass of the Z^O ($M \cong 93$ GeV/c^2), resulting in the minute range of the weak interaction ($\lambda = \hbar / Mc \sim 2.10^{-18}$ m), a hundred million times smaller than the typical atomic size (Bohr radius $a_0 \sim .5 \times 10^{-10}$ m). The expected right-left asymmetry is given by eq. (1), where

$$A_w \cong g^2 / (M^2 c^2 + q^2),$$
$$A_{em} \cong e^2 / q^2.$$

Here q is the relevant momentum transfer; g^2 is a coupling constant for weak neutral current interactions. Electroweak unification implies $g^2 \cong e^2$; therefore the right-left asymmetry is

$$\mathcal{Q} = 2A_w / A_{em} \cong 2q^2 / (M^2 c^2 + q^2).$$

In atoms, a typical order of magnitude for the momentum transfer q is the momentum $mc\alpha$ (m = electron mass, $\alpha = 1/137$) of the electron on the 1st Bohr orbit of the hydrogen atom. This leads to

$$\mathcal{Q} \cong 2\alpha^2 (m^2 / M^2) \cong 3.10^{-15} \quad !$$

What are the possible remedies ?
. The most obvious one consists in increasing q, up to Mc if possible. This means doing high-energy rather than atomic physics experiments ! Historically this orientation was the first one that was adopted.
. Bouchiat-Bouchiat's more accurate estimation (§ II.1 and III.2) predicts $10^{-14} - 10^{-13} \times Z^3$ for $Z \gg 1$. A new remedy now consists in choosing atoms of large Z.
. A third remedy, which can be used together with the previous one, consists in reducing A_{em} by operating with a forbidden atomic (electromagnetic) transition. This apparent artifice has in fact the important advantage of reducing systematic effects, as we now explain :

Because the sought-for effects are so small, systematic effects are likely to reach, if one doesn't take care, a comparable or even larger size. In fact, the author's opinion is that systematic effects are the central problem in the practical realization of parity-violation experiments in atoms. This should be emphasized very strongly here, especially as excessive considerations on this unexciting subject will be hereafter avoided, in order not to bore the reader. Yet it should be pointed out that systematic effects, whose origin is by definition other than the effect of interest (here, the weak interaction), may originate in purely electromagnetic effects, e.g. due to imperfect reversal of the handedness of the experimental configuration. This is why a forbidden transition is a judicious choice as regards systematic effects.

3. Motivation for Searching for Parity Violation in Atoms

In the study of weak neutral currents, atomic physics experiments are interesting essentially because the informations which they provide are complementary to those obtained at high energy. One reason is that in high-energy experiments, the nuclei and nucleons are broken, so that the quarks act incoherently. On the contrary, in atomic physics the nuclei remain intact and the quarks act coherently. A second reason is the quite different range of momentum transfer values; typically, $q \sim 1 - 100$ GeV/c in high-energy experiments, while in atomic physics $q \sim 1 - 10$ MeV/c (in heavy atoms) or $q \sim 10$ keV/c (in hydrogen) [49].

The complementary character of high-energy and atomic physics experiments is illustrated by the Fayet model with 2 neutral bosons [16] : for suitable choices of the mass and coupling constant of the second neutral boson, this model coincides with the standard Weinberg-Salam (WS) model at high energies, but simultaneously predicts nearly anything (say, between −3 times and +3 times the WS predictions, including zero) in atomic physics. Another example, of great practical importance, is the fact that up to now the combinations of coupling constants for the electron-hadron interaction obtained from the two types of experiments are nearly orthogonal (§ IV.1).

III. A TOUCH OF THEORY

1. The Parity-Violating Hamiltonian

Preliminary remark : A naive argument, referring to the point-like range of the weak interaction and to the Coulomb repulsion between electrons, concludes that the weak $e^- - e^-$ interaction plays a negligible role in atoms. This argument is not valid in atoms of large Z, because the relativistic velocities of the electrons at the nucleus make the Coulomb repulsion ineffective. Yet the conclusion has been shown to remain valid ([14]). Consequently, in atoms only the weak <u>electron-nucleus</u> interaction need to be considered.

In the zero-range limit, the effective parity-violating electron hadron weak neutral current Hamiltonian is the sum of two terms :

$$\mathcal{H} = \mathcal{H}_{A_e, V_n} + \mathcal{H}_{V_e, A_n} \qquad (2\text{-a})$$

The justification of the subscripts ("A_e, V_n" for "Axial electronic, Vector nucleonic"; "V_e, A_n" for "Vector electronic, Axial nucleonic") is found in the location of the factors γ_5 in the explicit expressions :

$$\mathcal{H}_{A_e, V_n} = \frac{G_F}{\sqrt{2}} \left\{ c_u^{(1)} \int d^3 r \, (\overline{e}\gamma^\mu \gamma_5 e)(\overline{u}\gamma_\mu u) \; + \right.$$
$$\left. c_d^{(1)} \int d^3 r \, (\overline{e}\gamma^\mu \gamma_5 e)(\overline{d}\gamma_\mu d) \right\} , \qquad (2\text{-b})$$

$$\mathcal{H}_{V_e, A_n} = \frac{G_F}{\sqrt{2}} \left\{ c_u^{(2)} \int d^3 r \, (\overline{e}\gamma^\mu e)(\overline{u}\gamma_\mu \gamma_5 u) \; + \right.$$
$$\left. c_d^{(2)} \int d^3 r \, (\overline{e}\gamma^\mu e)(\overline{d}\gamma_\mu \gamma_5 d) \right\} . \qquad (2\text{-c})$$

Here e, u and d are the field operators of the electron and of the u and d quarks. Concerning the fundamental coupling constants $c_u^{(1)}$, $c_d^{(1)}$, $c_u^{(2)}$, $c_d^{(2)}$, each electroweak model (such as WS) makes a definite prediction. Yet they must be ultimately determined from experiment.

Although rigourous computations have to be relativistic, the underlying physics is more easily understood in the non-relativistic limit. $\mathcal{H}_{A_e, V_n}$ and $\mathcal{H}_{V_e, A_n}$ then become two parity-violating potentials $V^{(1)}$ and $V^{(2)}$ in the atomic Hamiltonian. Eq. (26) becomes

$$\mathcal{H}_{A_e, V_n} \to V^{(1)} = \frac{1}{2} Q_W \frac{G_F}{\sqrt{2}} \left[\vec{s} \cdot \frac{\vec{p}}{mc} \, \delta^3(\vec{r}) + \text{h.c.} \right], \quad (3)$$

where $\vec{s}$, $\vec{p}$ and m are the electron spin, momentum and mass. The parameter Q_W, which may be called the "<u>weak (vector) charge</u>" of the considered nucleus, is given by

$$-\frac{1}{2} Q_W = Z \left[2 \, c_u^{(1)} + c_d^{(1)} \right] + N \left[c_u^{(1)} + 2 c_d^{(1)} \right], \qquad (4)$$

where Z and N are the numbers of protons and neutrons in this nucleus. The weak vector charge is conserved, i.e. <u>additive</u> : the nuclear charge is the sum of the charges of the nucleons, which in turn are the sum of the charges of the quark constituents.

Expression (2-c) becomes

$$\mathcal{H}_{V_e,A_n} \curvearrowright V^{(2)} = A(Z,N) \left[\left(\vec{I}.\frac{\vec{p}}{mc} + 2\vec{I}.\vec{s} \times \frac{\vec{p}}{mc} \right) \delta^3(\vec{r}) + \text{h.c.} \right]$$

(5)

where $\vec{I}$ is the nuclear spin. The function $A(Z,N)$ results from summing the effects of all nucleons. While Q_W grows linearly with Z and N, the function $A(Z,N)$ remains of the order of unity, because axial coupling adds like spin, not like charge. Therefore

<u>in heavy atoms the vector nucleonic, axial electronic term is predominant,</u>
(6)

roughly by a factor Z or N. (In heavy atoms, N/Z is approximately constant and equal to 1.5). Consequently the V_e,A_n potential $V^{(2)}$ will be neglected hereafter (except in § IV.2 concerning hydrogen, Z = 1).

2. Bouchiat-Bouchiat's Z^3 Increase

The matrix elements of $V^{(1)}$ between atomic eigenstates have been computed by M.A. BOUCHIAT and C. BOUCHIAT ([13], [14]) in the central field approximation (where each electron moves in a central field representing the nucleus and the other electrons). The only non-vanishing elements turn out to be those connecting a $S^1/_2$ state (i.e. a state of orbital angular momentum $\ell = 0$) with a $P^1/_2$ state ($\ell = 1$, antiparallel to the electron spin s = 1/2). They are given by :

$$< n\ S^1/_2\ |V^{(1)}|n'\ P^1/_2 > =$$

$$\frac{3}{16\pi} \frac{i\hbar}{mc} \frac{G_F}{\sqrt{2}} Q_W \left[R_{nS^1/_2} \frac{d}{dr} R_{n'P^1/_2} \right]_{r=0}$$

(7)

where R is the radial wavefunction and n the principal quantum number. An estimate of $R_{nS^1/_2}$ (r=0) is given by the simple approximate formula of FERMI and SEGRÈ ([17] - [19]). This formula essentially involves the atomic number Z and the effective quantum number, i.e. the values of n to be inserted into the hydrogenic energy formula to match the real energy spectrum of the considered atom.

92

M.A. BOUCHIAT and C. BOUCHIAT have derived (^{14}) a generalized Fermi-Segré formula to estimate $\left| \, d^{\ell}R_{\ell}/dr^{\ell} \, \right|$ $_{r=0}$ for any ℓ in terms of the same quantities. Carrying their result into eq. (7) leads to

$$< n \, S^{1}/_{2} \, \left| V^{(1)} \right| n' \, P^{1}/_{2} > \; =$$

$$\frac{i \hbar}{4\pi mca_{o}^{4}} \; \frac{G_{F}}{\sqrt{2}} \; f \, K \, Q_{W} Z^{2} \tag{8}$$

Here f ($\cong$ a few units) is a function of the effective quantum number. Since the derivation is non-relativistic, a correcting factor K must finally be inserted; for Cs(Z=55), K $\cong$ 3; for Bi(Z=83), K $\cong$ 9.

An essential feature in (8) is the factor $Q_{W}Z^{2}$. Since N/Z is nearly constant in heavy atoms, eqs. (4) and (8) imply that

in heavy atoms the matrix elements of the dominant parity-violating potential $V^{(1)}$ grow approximately like Z^{3}. (Bouchiat-Bouchiat's Z^{3} law) (9)

Actually they grow even somewhat faster, since the relativistic correction K also grows. From the discussion of the end of § III.1, we similarly conclude that the matrix elements of $V^{(2)}$ grow only like Z^{2}.

IV. WHAT INFORMATIONS CAN ATOMIC PHYSICS YIELD ?

This question, implicit in § II.3, can now be answered in more precise terms.

1. Information from Heavy Atoms

In heavy atoms, parity violation is dominated by the axial-electronic, vector-nucleonic term. In the first order of perturbation, any parity-violation effect is proportional to a matrix element of $V^{(1)}$ (eq.(3)), i.e. to the nuclear weak charge Q_{W} times a purely atomic quantity. Consequently,

provided a reliable theoretical model of the considered atom is available, experiments in a heavy atom yield its nuclear weak vector charge Q_{W}. (10)

From the value of Q_{W} one extracts (eq.(4)) a value for the linear combination $(2Z+N) \, C_{u}^{(1)} + (Z+2N) \, C_{d}^{(1)}$ of the quark charges $C_{u}^{(1)}$ and $C_{d}^{(1)}$. Since N/Z is nearly a constant (1.4 - 1.5) in all heavy atoms, all of them yield approximately the same linear combination $7C_{u}^{(1)} + 8C_{d}^{(1)}$, nearly orthogonal to the combination $2C_{u}^{(1)} - C_{d}^{(1)}$

determined in high energy measurements [20]. Yet the need for a reliable atomic model obviously favors atoms with a single valence electron (i.e. alkali atoms), and especially the one of largest Z (cesium, Z = 55) in view of the Z^3 law. (Francium, unstable and spectroscopically almost unexplored, is not a good candidate).

2. Informations from Hydrogen (and Deuterium)

In hydrogen (Z = 1), the matrix elements of $\mathcal{H}_{A_e,V_n}$ and $\mathcal{H}_{V_e,A_n}$ are of comparable size. Those of $\mathcal{H}_{A_e,V_n}$ are proportional to the weak vector charge of the proton :

$$c_p^{(1)} = 2c_u^{(1)} + c_d^{(1)},$$

and those of $\mathcal{H}_{V_e,A_n}$ are proportional to the net axial coupling constant of the proton, which we shall note $c_p^{(2)}$.

[Note : Since weak axial coupling is not simply additive, the explicit expression of $c_p^{(2)}$ in terms of $c_u^{(2)}$ and $c_d^{(2)}$ generally requires a model of the proton. Yet in the case of the standard electroweak model, symmetry considerations connect $c_p^{(2)}$ with charged-current quantities known from β decay, and imply an explicit prediction for $c_p^{(2)}$ without reference to any model of the nucleon. The same is true for the axial coupling $c_n^{(2)}$ of the neutron employed just below].

In deuterium a weak axial coupling constant $c_n^{(2)}$ for the neutron is traditionnally introduced. The net axial coupling constant for the whole nucleus (deuteron) is approximately (yet not exactly) $c_p^{(2)} + c_n^{(2)}$, essentially because the deuteron is approximately in a triplet state (i.e. a state in which the proton and neutron spins are "parallel" and therefore "additive"). The vector coupling constant of the neutron is of course

$$c_n^{(1)} = c_u^{(1)} + 2c_d^{(1)}$$

Two features make experiments in hydrogen and deuterium extremely attractive :
  * The theoretical calculations, including higher-order corrections, can be performed with high accuracy. (The limitation lies in the hadronic models involved in some higher-order processes. In deuterium, a further limitation lies in the deuteron model).
  * Explicit computation of the relevant matrix elements show that the four coupling constants $c_p^{(1)}$, $c_n^{(1)}$, $c_p^{(2)}$ and $c_n^{(2)}$ can in principle be determined (by combining results from various atomic transitions, and also from hydrogen and deuterium).

However, experiments in hydrogen and deuterium can also be expected to be very difficult, since the predicted parity-violation effects are awfully small, for at least two reasons :
* Since Z = 1 there is no Z^3 enhancement.
* The standard model predictions for $C_p^{(1)}$, $C_p^{(2)}$ and $C_n^{(1)}$, proportional to $1-4 \sin^2\theta$, are very small ($\cong .05$) for the observed value $\sin^2\theta \sim .23$. Only $C_n^{(1)} = .5$ is large.

V. VIA WHICH PHYSICAL QUANTITY CAN PARITY VIOLATION BE OBSERVED IN ATOMS ?

To observe parity violation in atoms, the only practical possibility suggested up to now is to use an <u>off-diagonal matrix element of the electric dipole operator between two atomic states of "same" parity</u>.

In the absence of weak interaction, the atomic Hamiltonian is invariant under space reflection. Its eigenstates have a definite parity. Since the electric dipole operator $\vec{er}$ is odd under space reflection, its matrix elements (e.g. $E_1 = < a|ez|b >$) vanish between states of same parity. In presence of the parity-violating weak interaction (which can be treated as a perturbation), the states are slightly admixed with states of opposite parity. The new eigenstates $|\tilde{a} >$, $|\tilde{b} >$ are not strictly of definite parity, so that the matrix element $< \tilde{a}|ez|\tilde{b} > = E_1^{PV}$ is not strictly zero. It is called a "parity-violating, <u>transition</u> electric dipole" ("transition" means here "off-diagonal"; "E_1" is a traditional notation for transition electric dipole matrix elements). Diagonal (i.e. static) PV electric dipole matrix elements are forbidden since the Hamiltonian (including the weak interaction) is invariant under rotation and under time-reversal. (Rotational invariance implies $< \vec{er} > = k < \vec{j} >$; under time reversal the left side is conserved while the right side is reversed, so k must vanish).

Between two states of same parity the <u>magnetic</u> dipole operator $\vec{\mu}$ generally has non-vanishing matrix elements, e.g. $< a|\mu_z|b > = cM_1$. (With this traditional definition, M_1 has the dimension of an <u>electric</u> dipole, like E_1). Of course the minute perturbation of the eigenstates by the weak interaction does not appreciably change this, so that practically $< \tilde{a}|\mu_z|\tilde{b} > = < a|\mu_z|b > = cM_1$. Time reversal invariance implies ([13], [50]) that E_1^{PV} / M_1 is pure imaginary, therefore we take M_1 to be real and $E_1^{PV} = i \operatorname{Im} E_1^{PV}$ throughout. The ratio R = $\operatorname{Im} E_1^{PV} / M_1$ (always << 1) expresses how much smaller than the electromagnetic interaction the weak interaction is in the considered atomic transition. Reducing M_1 (by choosing a <u>forbidden</u> magnetic dipole transition) reduces systematic effects (§ II.2). Typical values of R in recent experiments are 10^{-3} (Tℓ), 10^{-4} (Cs), 10^{-7} (Bi, Pb).

VI. THE THREE TYPES OF EXPERIMENTS

Up to now, experiments to search for parity violation in atoms can be classed in three types. These are :

. Search for optical rotation near a permitted magnetic dipole transition of a heavy atom (Bi, Pb).
. Experiments in forbidden magnetic dipole transitions of heavy atoms (Tℓ, Cs). Parity violation has been observed via the parity-violating behavior of a spin polarization under reversal of the geometrical handedness of the experiment.
. Microwave experiments in hydrogen. The sought-for quantity is the dependence of a microwave transition rate on the handedness of the field configuration.

These three types of experiments will be presented in the three next sections. Contemporary experiments (achieved or in progress) are listed in table 1.

Table 1. Contemporary experiments to search for parity
violation in atoms

Type of experiment	Atom	Place	Ref.	
Optical rotation in permitted M_1 transition of heavy atoms	Bi	Oxford Seattle Novosibirsk Moscow	[22] [24] [15] [25]	
	Pb	Seattle	[26]	
Experiments in forbidden M_1 transitions of heavy atoms	Tℓ	Berkeley	[27]	Parity violation already reported
		Paris	[28]	
		Zürich Ann Arbor	[35] [36]	Parity violation not reported yet
μ-wave experiments in hydrogen	H	Ann Arbor Yale Seattle	[29] [30] [31]	

VII. ATOMIC OPTICAL ROTATION ([15], [21] − [26])

1. General

When a beam of linearly polarized light passes through a substance exhibiting optical rotation, the transmitted linear polarization makes an angle with the incident one, and the value of this
angle is the same for all directions of the incident polarization.
Optical rotation is common in substances possessing geometrical handedness (sugars, quartz, ...). In this case the two species of opposite handednesses exhibit opposite optical rotations : thus the mirror image of either case coïncides with the real other case, so that
parity is conserved. On the contrary, since atoms have no geometrical
handedness, if a vapor of atoms exhibits optical rotation, the mirror
image is a vapor of atoms identical with the real ones except they
have opposite optical rotation. Here the mirror image case contradicts reality, so parity is violated. The handedness is not in the
geometry of the atom, but in the physical laws of the weak interaction.

2. Permitted M_1 Transitions : Reasons and Consequences

In optical rotation induced by weak neutral currents, the rotation angle per unit length (for an atomic line of Lorentz shape) is
given by :

$$\Phi_{pv} = 2k_{max} \frac{\left(\dfrac{\omega - \omega_o}{\Gamma/2} \right)}{1 + \left(\dfrac{\omega - \omega_o}{\Gamma/2} \right)^2} \cdot Im \left(E_1^{pv} / M_1 \right) \qquad (11)$$

Here ω is the frequency of the light, ω_o is the center frequency of
the atomic line, Γ is the linewidth. The absorption coefficient
for unit length at line center, k_{max}, is proportional to M_1^2, so
that the rotation angle Φ_{pv} is proportional to M_1. Therefore one
must operate in permitted M_1 transitions, which lead in the most
favorable cases to rotations of $\sim 10^{-7}$ rad. Forbidden M_1 transitions
would lead to rotations much too small to be measured. This has two
important consequences :

. Good statistics is obtained in reasonable integration times.
(Example : fig. 1-c was recorded in 6 hours).
. But the problems of systematic effects are severe. Among the
four groups who study atomic optical rotation, three have obtained
later results in contradiction with their own earlier ones, which
they finally rejected. Still to-date, experimental results from

different groups do not completely agree yet. The most likely explanation of all this is the presence of initially underestimated systematic effects.

3. Common Features of the Atomic Optical Rotation Experiments

In all experiments the beam of a tunable laser is passed through the atomic vapor sandwiched between two slightly uncrossed polarizers. The transmitted intensity is

$$I = I_o (\Phi_{un} + \Phi_{pv} + \Phi_{res})^2 .$$

Here I_o is the incident intensity; Φ_{un} is the uncrossing angle; Φ_{pv} is the parity-violating optical rotation; Φ_{res} is a residual angle resulting from polarizer imperfections. For the best polarizers Φ_{res} is $\sim 3 \times 10^{-4}$ rad, while Φ_{pv} is typically 10^{-7} rad ! Reversing the uncrossing angle changes the transmitted intensity by

$$I(\Phi_{un}) - I(-\Phi_{un}) = 4I_o \Phi_{un} (\Phi_{pv} + \Phi_{res}) .$$

The sensitivy of this method is excellent : an angle $\Phi_{pv} + \Phi_{res}$ as small as 10^{-8} is easily detected. In practical experiments, variants are encountered (modulation instead of reversal of Φ_{un}, etc...).

To distinguish Φ_{pv} from the 1000 times larger Φ_{res}, the laser wavelength is swept (continuously or by discrete hops) across the atomic line profile (fig. 1). Since Φ_{res} is nearly independent of the wavelength, it is nearly removed. Further discrimination is achieved by comparing several hyperfine structure components of the line (including "null components" where the expected optical rotation is zero), or by removing the atomic vapor (by cooling, or by shifting the container aside).

A dc magnetic field component along the beam introduces handedness and induces an artificial optical rotation, called Faraday rotation. This effect, whose wavelength dependence is qualitatively similar to the PV optical rotation (fig. 1), is useful to identify the lines, tune the laser, and also estimate the optical thickness of the vapor for calibration purposes. However, Faraday rotation in stray magnetic fields can be a source of systematic effects. For all these reasons Faraday rotation must be understood and controlled. Figure 1 shows an example of set-up and signals.

4. Atomic Optical Rotation : Results

The results are summarized in table 2. The old figures in parentheses, now denied and rejected, are recalled to illustrate the problem of systematic effects, which are also the likely explanation for the dispersion in the experimental results concerning the 648 nm

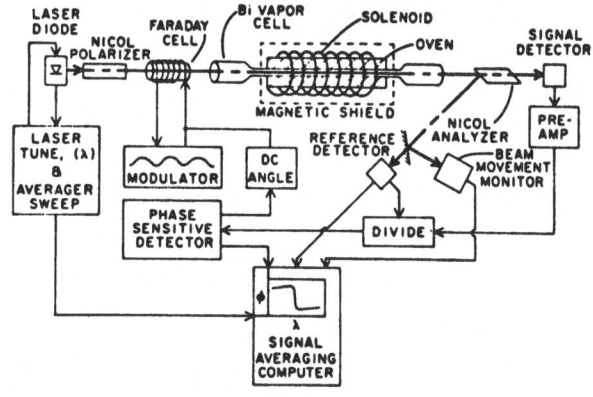

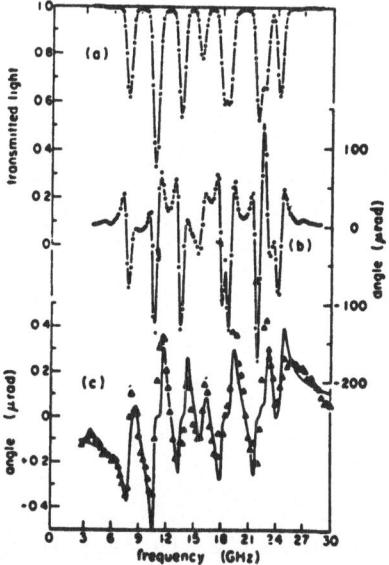

Fig. 1. Example of set-up and signals in an atomic optical
rotation experiment ([24]). The polarizer uncrossing
is replaced by Faraday rotation in water (Faraday
cell). The curves show absorption (a), Faraday
rotation (b) and parity-violating optical rotation
(c) in the 876 nm line of bismuth.

Table 2. Results $\left[\ 10^8 \times (\text{Im}\ E_1^{PV})\ /\ M_1\ \right]$ for parity-violating optical rotation in atoms. $\left[\ a\ \right]$: D.N. Stacey, private communication; * : new Moscow result, unpublished but known to be in agreement with the Oxford result -9.3 ± 1.5 $\left[\ a\ \right]$.

	Experiment		Theory (sin²θ = 0.23)	
Bi 648 nm	(+2.7 ± 4.7)	[21]	-13	[40]
	-8.8 ± 2.0	[22]		
	-9.3 ± 1.5	[a]	-17	[37]
			-17.8	[38]
	-20.2 ± 2.7	[15]		
			-11.1	[39]
	(-2.3 ± 1.3)	[25]		
	*	[a]		
Bi 876 nm			-8.3	[39]
	(-2.4 ± 1.4)	[23]		
	-10.4 ± 1.7	[24]		
			-11	[40]
			-13	[37]
Pb 1.28 μ	-9.5 ± 2.6	[26]	-13 ± 2	[26]

Bi line. The dispersion in the theoretical predictions for either Bi line is a consequence of the complicated structure of bismuth (5 electrons on the outmost shell).

These dispersions should not mask, however, the considerable technical advances (both in the experiments and in the atomic theoretical computations) accomplished in the course of these works. As table 2 shows, there is now no doubt at all that parity violating optical rotation is present in bismuth and lead with roughly the magnitude predicted by the standard electroweak model, an important result.

VIII. PARITY-VIOLATING ELECTRONIC POLARIZATION IN FORBIDDEN M_1 TRANSITIONS OF HEAVY ATOMS ([27], [28])

1. Definition

In these experiments, the sought-for quantity is an electronic polarization which is changed into the opposite of its mirror image when the handed experimental configuration is changed into its mirror image. (Since electronic polarization is an axial quantity, this parity-violating behavior coincides with the behavior of a vector in a situation where parity would be conserved).

2. Forbidden Magnetic Dipole Transitions : Advantages and Drawbacks

The advantage of choosing highly forbidden transitions is to reduce electromagnetic systematic effects (§ II.2). On the other hand, highly forbidden transitions imply up to now low counting rates, i.e. very long integration times (typically several weeks).

3. Common Features of the Experiments

One original feature of the PV experiments reported so far in forbidden M_1 transitions is the choice of an electroweak interference in which the electromagnetic partner is induced by an external dc electric field. The favoured direction of this field labels the signal with a particular symmetry which provides excellent discrimination against collisional or molecular effects, isotropic on the average : all background is eliminated. In addition, the field can be reversed (or modulated) to distinguish the interference term, and its size can be adjusted for optimum S/N ratio.

The forbidden transition is thus characterized by 3 transition amplitudes :

. the forbidden magnetic dipole M_1 ($\cong$ a few $\times\ 10^{-5}\ \mu_B/c$, while a normal M_1 is $\cong \mu_B/c$);

. the parity-violating electric dipole E_1^{pv} ($\cong 10^{-11} - 10^{-10}\ ea_0$, i.e. $10^{-4} - 10^{-3} \times M_1$; a "normal" atomic electric dipole is typically $\cong ea_0$);

. the electric dipole E_1^{ind} induced by the external electric field. This amplitude, proportional to the field, is in practical conditions the largest of the three ($\sim 10^5 - 10^6 \times E_1^{pv}$). Interference between E_1^{pv} and E_1^{ind} gives rise to the parity-violating polarization to be measured.

The basic experimental configuration (fig. 2) is conditioned by 3 parameters that can be separately reversed : the direction $\vec{\ell} = \pm z$ of the excitation laser; the sign $\xi = \pm 1$ of the helicity of the laser photons; the direction $\vec{E} = \pm y$ of the electric field. The

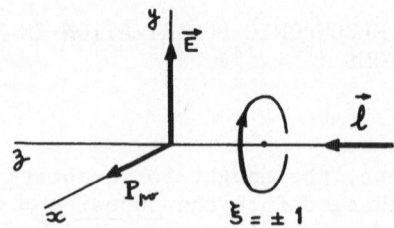

Fig. 2. Basic configuration for PV electronic polarization
experiments

parity-violating electronic polarization $\vec{P}^{PV}$ in the upper state of
the transition is along x. When the experimental configuration is
replaced by a second configuration which coincides with the image
of the first one in a mirror, the <u>parity-violating</u> component $\vec{P}^{PV}$ is
expected to be changed into the <u>opposite</u> of its image in this mirror.
A reflection in the yz plane as the mirror is performed on the experimental configuration by reversing ξ while conserving $\vec{l}$ and $\vec{E}$.
Under this reversal, $\vec{P}^{PV}$ is expected to be reversed. Similarly, the
zx mirror reflection is performed by reversing $\vec{E}$ and ξ simultaneously; it leaves $\vec{P}^{PV}$ unchanged. (Consequently, reversing $\vec{E}$ alone reverses ($\vec{P}^{PV}$). At last, the xy-mirror reflection is performed by reversing the laser direction $\vec{l}$ and the helicity ξ simultaneously; it
leaves $\vec{P}^{PV}$ unchanged.

Parity-conserving electronic polarization is of course also
present, and several orders of magnitude larger than the parity-
violating one. Yet their opposite behavior under mirror reflections
of the experimental configuration (together with nearly orthogonal
directions) is sufficient to allow clear distinction between them.

A more detailed description of the experiment in a forbidden
M_1 transition of Cs (the experiment in which the author has taken
part) is given in §§ X - XVI.

4. Results

The results concerning PV polarization in forbidden M_1 transitions are listed in table 3. In the case of cesium, comparison between theory and experiment is more significant than in optical rotation experiments, in view of smaller uncertainties both in the theory (because the atom is simpler) and in the experiment (because the
forbidden transition reduces systematic effects); two variants of
the experiment (§ XV) cross-check each other, resulting in still

Table 3. Results for PV electronic polarization in forbidden M_1 transitions of heavy atoms

Thallium : Im $(E_1{}^{PV}/M_1)$

exp. : $(1.40 \pm .35 \; {}^{+ \; .15}_{- \; .10}) \times 10^{-3}$ [27]

theor. : $\begin{cases} (1.05 \pm .35 \quad) \times 10^{-3} & \text{[41]} \\ (.82 \pm .12 \quad\;\;) \times 10^{-3} & \text{[43]} \end{cases}$

Cesium : Im $(E_1{}^{PV}/\beta)$, mV/cm   [28, 42]

exp., $\Delta F = 0$ $-1.34 \pm .22 \pm .11$

exp., $\Delta F = 1$ $-1.78 \pm .26 \pm .12$

exp., combined $-1.56 \pm .17 \pm .12$

theor. $-1.61 \pm .07 \pm \sim .2$

increased confidence. Within the quoted experimental and theoretical uncertainty, <u>the predictions of the standard electroweak model are clearly verified</u>.

IX. MICROWAVE EXPERIMENTS IN METASTABLE HYDROGEN [29 - 34]

These experiments, whose motivations have been presented in § IV.2, essentially consist in inducing a microwave transition with an oscillating rf field in presence of dc electric and magnetic fields. Reversal of the handedness of the field configuration is expected to cause a change in the transition rate, due to interference between the PV transition dipole $E_1{}^{PV}$ and some electromagnetic transition amplitude. All experiments in progress involve microwave transitions between sublevels of the metastable 2S state in an atomic beam. Sufficient frequency resolution is easily achieved to completely resolve the energy sublevel structure, resulting in the principle possibility of determining separately the four coupling constants $C_p^{(1)}$, $C_p^{(2)}$, $C_n^{(1)}$ and $C_n^{(2)}$.

All except one of these experiments operate at a magnetic field value where a P sublevel and a S sublevel cross one another. The original motivation was to increase the sensitivity by enhancing the PV admixture of the crossing S and P sublevels. In fact, near the crossing the rf power must be reduced in order to avoid unwanted state depletion via power-broadened adjoining rf transitions, and this second effect just compensates the first one $(^{32})$. Yet, sweeping the magnetic field across a level-crossing yields in principle characteristic dependence shapes useful to discriminate the PV signal. In view of the extreme difficulty of these experiments (§ IV.2), no parity result has been obtained in any group yet.

*****

The remaining sections contain a more detailed description (given orally in a seminar) of the experiment in cesium already mentioned in § VIII.

X. THE 6S-7S TRANSITION OF CESIUM

Among high Z atoms, cesium is the simplest case : only one valence electron orbiting around a fairly monolithic, rare-gas-like core. The $6S^{1/2}-7S^{1/2}$ transition was chosen as a highly forbidden transition whose lower state -the ground state- is spontaneously populated.

As explained in § VIII.3, the transition occurs via 3 amplitudes $M_1 \cong 4 \times 10^{-5} \mu_B/c$; $E_1^{PV} \cong 10^{-11}$ $ea_O \cong 10^{-4} \times M_1$; E_1^{ind} proportional to the dc electric field. Because of spin-orbit coupling, the induced dipole is not merely colinear with the field, but is of the form $\alpha \vec{E} + i\beta \vec{\sigma} \times \vec{E}$, where $\vec{\sigma}$ is the Pauli operator of the electronic spin. The coefficients α and β, respectively called the scalar and vector polarizabilities, are real; the i is here for time reversal invariance and β is approximately $\alpha/10$. In a field $E \cong 3$ V/cm αE is equal to M_1, but in real experimental conditions (E $\sim$ 100 - 600 V/cm), the inequalities αE, $\beta E \gg M_1 \gg E_1^{PV}$ hold. Finally the overall transition amplitude in a light-wave of direction $\vec{k}$ and polarization $\vec{\epsilon}$ is found to be

$$\left[\alpha \vec{E} + i\beta \vec{\sigma} \times \vec{E} - M_1 \vec{\sigma} \times \vec{k} + i(Im\ E_1^{PV}) \vec{\sigma} \right].\vec{\epsilon} \qquad (12)$$

The expressions of physically observable quantities concerning the atoms excited to the 7S state contain interferences between the 4 terms of (12). The PV signal (described in the next paragraph) involves either an αE - E_1^{PV} interference or a βE - E_1^{PV} interference, depending on the type of hyperfine structure component (respectively $\Delta F = 0$ or 1; F denotes the total angular momentum). In both cases the measurements are calibrated using an αE - βE interference (present only in $\Delta F = 0$ hfs components; in the $\Delta F = 1$ case the laser must be tuned back to a $\Delta F = 0$ component). The quantity actually measured turns out to be (Im E_1^{PV})/β in the $\Delta F = 0$ case, and (α/β)(Im E_1^{PV})/β in the $\Delta F = 1$ case. The two results can be compared since α/β is known accurately from separate measurements by several authors $(^{35},\ ^{36},\ ^{44})$.

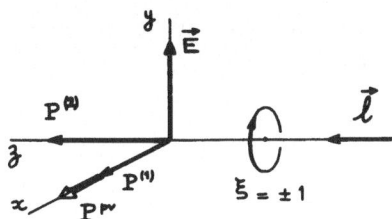

Fig. 3. Basic configuration for the cesium experiment.

XI. PRINCIPLE OF THE EXPERIMENT

The experimental configuration is shown in fig. 3. A laser beam, resonant for the forbidden transition, is directed along $\vec{\ell}$ = ±z; its polarization is in principle circular, right or left (helicity ξ = ±1). An external dc electric field $\vec{E}$ is applied along ±y. The handedness of this experimental configuration is defined by the three parameters $\vec{\ell}$ = ±z, ξ = ±1, $\vec{E}$ = ±y. A mirror reflection with the yz plane as the mirror is performed by simply reversing ξ; a zx-mirror reflection, by reversing $\vec{E}$ and ξ simultaneously; a xy-mirror reflection, by reversing $\vec{\ell}$ and ξ.

In this configuration, the atoms excited to the 7S state possess electronic polarization, which consists of 3 contributions :

. The most interesting one is the parity-violating contribution $\vec{P}^{pv}$, directed along x. It arises from $\alpha E - E_1^{pv}$ interference (in the $\Delta F = 0$ case) or from $\beta E - E_1^{pv}$ interference (in the $\Delta F = 1$ case). In a practical situation its typical magnitude is $2-3 \times 10^{-6}$.

. The parity-conserving component $\vec{P}^{(1)}$, also directed along x, arises from $\alpha E - M_1$ ($\Delta F = 0$) or $\beta E - M_1$ ($\Delta F = 1$) interference. Its typical magnitude is $3-5 \times 10^{-2}$.

. The parity-conserving component $\vec{P}^{(2)}$ is directed along z. In the $\Delta F = 0$ case it arises from $\alpha E - \beta E$ interference. Its magnitude ($\cong 8 \times 10^{-2}$), nearly independent of the field E, is accurately known. Therefore it is used as a standard to calibrate the electronic polarization measurements. In the $\Delta F = 1$ case the origin of $P^{(2)}$ is more complicated, and its size is reduced to $\cong 10^{-2}$. Calibration proceeds by tuning back to a $\Delta F = 0$ component and using again the standard just described.

Table 4. Even or odd behavior of $P^{(1)}$, $P^{(2)}$ and P^{pv} under reversal of ξ, $\vec{\ell}$ or $\vec{E}$

	ξ	$\vec{\ell}$	$\vec{E}$
P^{pv}	−	+	−
$P^{(1)}$	+	−	−
$P^{(2)}$	−	+	+

The principle of the experiments consists in replacing the experimental configuration by its mirror image and looking in the electronic polarization for a contribution that is changed into the _opposite_ of its mirror image. Since electronic polarization is axial, the behavior expected for $\vec{P}^{pv}$ coincides with that of a vector in a "normal" (i.e. parity-conserving) situation. Table 4 lists the even or odd character of $\vec{P}^{(1)}$, $\vec{P}^{(2)}$ and P^{pv} under reversal of each of the three parameters considered above. (Note : reversal of $\vec{E}$ $\equiv$ zx–mirror reflection × yz–mirror ref.). The particular behavior of $\vec{P}^{pv}$ shown in table 4 is precisely the feature on which the experimental discrimination is based.

XII. EXPERIMENTAL FEATURES

1. Measuring the Electronic Polarization in the 7S state

7S state atoms do not decay straight back to the ground state, since the transition is so forbidden. An atom decaying via the $6P\frac{1}{2}$ state (fig. 4) emits a photon ($\lambda = 1.36\ \mu$) to conserve energy. Conservation of angular momentum implies that the circular polarization of the fluorescence photons emitted in a given direction (e.g. x) is proportional to the electronic polarization component in the 7S state along this direction ($P^{(1)} + P^{pv}$).

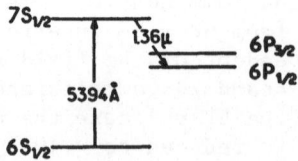

Fig. 4. Level scheme for the Cs experiment

2. Apparatus

Figure 5 shows a scheme of the set-up. The subset labeled "modulator" performs an operation equivalent to reversing the circular polarization ξ of the laser beam. Mirror M_2 reverses (approximately) the direction of the laser beam. These two functions are now described with more details.

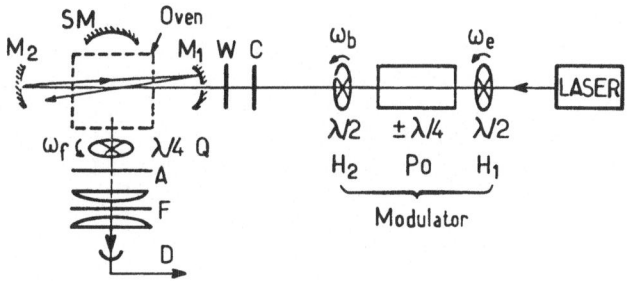

Fig. 5. Apparatus of the cesium experiment.
H_1, H_2 : rotating half-wave plates; Po :
Pockels cell; C : birefringence compensator;
W : entrance window of the sample tube;
M_1, M_2 : multipass mirrors; SM : spherical
mirror for improved fluorescence light collection;
Q : rotating quarterwave plate; A : analyzer;
F : interference filter; D : detector

For the right-left interchange to be clean, when the circular polarization is reversed, the intensity of the beam should remain unchanged, as well as its linear polarization (which is in principle zero). Existing polarization modulators produce spurious changes of these parameters at a typical level of several percent –but 1 percent is enough to simulate the parity violation signal ! We have developed a new type of modulator in which these imperfections are reduced to $\sim 10^{-4}$, which corresponds to a systematic effect of $\sim 1\%$ of the parity signal. With this device, the circular polarization is not reversed, but continuously modulated with a characteristic frequency and phase. In addition, the phase can be reversed. The linear polarization is modulated at a different frequency (in fact, at 2 frequencies), and the intensity remains unmodulated. Spurious cross-modulations are $\lesssim 10^{-4}$. More details can be found in ref. ([45]).

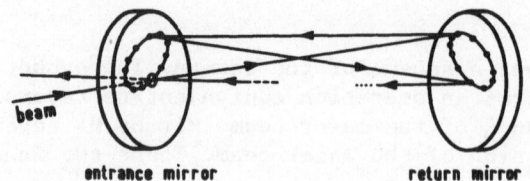

entrance mirror   return mirror

Fig. 6. Multipass of a laser beam between two spherical mirrors

Reversal of the laser beam is replaced by a <u>multipass</u> (fig. 6) : the beam follows a zigzag path between two spherical mirrors ([46]). In our case, the (even) number of passes is $\cong$ 120; the angles between passes are $\lesssim$ 1°. In the multipass, physical quantities odd under reversal of the laser beam direction (e.g. $\vec{P}^{(1)}$) approximately cancel, while even quantities (e.g. the PV polarization $\vec{P}^{PV}$ and the standard polarization $\vec{P}^{(2)}$) are multiplied by a factor approximately equal to the number of passes. This provides a vital reduction of the integration times by two orders of magnitude ! The approximate character of the beam reversal (due to the angles and to the reflection losses) must of course be taken into account.

Losses on windows placed between the mirrors would cause the light intensity in the multipass to decrease geometrically, killing both functions of the multipass. Therefore the mirrors are placed inside the sample cell (fig. 7), but far away from the reactive heated cesium vapor, and protected by a cooled buffer gas. More details can be found in ref. ([47]).

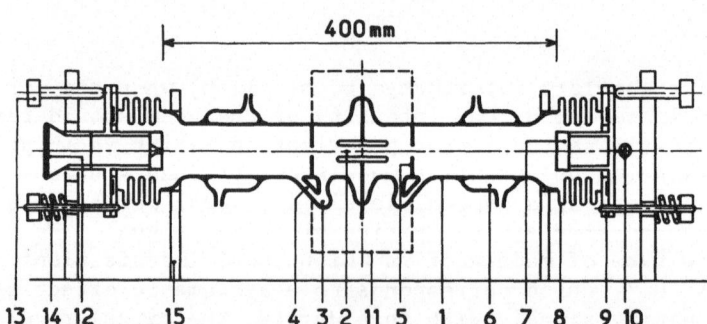

400 mm

13 14 12   15   4 3 2 11 5 1 6 7 8 9 10

Fig. 7. Sample cell in the cesium experiment. 1 : glass tube; 2 : observation region; 3 : cesium reservoirs; 4 : liquid cesium return tubes; 5 : diaphragms; 6 : water coolings; 7 : multipass mirrors; 8 : stainless steel bellows; 9 : flange; 10 : repumping valve: 11 : oven; 12 : beam entrance window; 13 : adjustment screws; 14 : springs; 15 : mount.

XIII. THE CRUCIAL PROBLEM : SYSTEMATIC EFFECTS

Systematic effects may arise from parity-conserving origins if
the right-left interchange is affected by geometrical imperfections
(imperfect orthogonality between the observation direction x, the
field direction y and the laser direction z; cf. fig. 3) or by imperfect parameter reversals. Such imperfections can be inserted into
the theory. A model of these systematics is then obtained. To compute
quantitative estimates, one must still measure which values the relevant parameters actually take on in the real set-up. This is in
most cases measured from atomic signals, simulstaneously with the
parity measurement.

Atomic signals providing for control of the relevant imperfections are in practice available for most systematic effects. This is
essentially because of the profusion of available atomic signals :
from the fluorescence intensity, one can extract the contribution
which is either unpolarized or polarized, which is modulated like
the linear or circular polarization or the intensity of the laser
beam, and which is even or odd under reversal of the field $\vec{E}$, or of
an auxiliary electric or magnetic field, etc... This gives a wide
number of possible combinations, most of which provide useful informations. (For more details, see ([28])). Finally, quantitative estimates are obtained for all systematics ... of the type described here,
which can be called expected systematics. The case of possible unexpected systematics will be considered in § XV.

For fear of boring the reader, we do not say much here concerning systematics. Yet they are both the central problem and the biggest difficulty. To illustrate the latter point, let us simply give
the example of the systematic effect associated with imperfections
in the reversal of the circular polarization of the laser beam. The
need of developing a new type of polarization modulator has already
been mentioned (§ XII.2). The next step consisted in measuring birefringence and circular dichroism in the multipass mirrors, at the
level of $\sim 10^{-6}$ rad per reflection. This was achieved using the enhancement of these imperfections by the multipass, and with our modulator as a source of clean polarization. Mastering the purity of
light polarization took 1.5 year. (The whole experiment took 9 years).

XIV. SIGNAL / NOISE RATIO

All sources of noise -laser frequency and intensity fluctuations;
detector (Ge photodiode) noise; detector's sensitivity to X and γ
rays and cosmic muons (!); thermal infrared radiation from the oven-
were successively reduced, so that in spite of the infrared detection
wavelength (1.36 μ) the ultimate noise was reached. This is the noise
associated with the quantized structure of light. Further improvement
of the S/N ratio then requires more photons (powerful laser, multipass,
detection optics of large aperture).

The integration time for unit S/N ratio was typically 5 hours, corresponding to about 8 hours of operation. (Some time is spent waiting for the experiment to stabilize again after each reversal of a parameter, such as electric field, etc...; some time is devoted to systematic effect controls that imply a change in the experimental configuration, e.g. the application of an auxiliary electric or magnetic field; finally, some time is lost in various readjustments, in laser dye changing, etc...). Each of the two parity results corresponds to several hundred hours of integration.

XV. UNEXPECTED SYSTEMATICS ?

As a protection against unexpected systematics, several quantities expected to vanish are averaged, and actually found to be consistent with zero. Among these are : i) the difference between the two PV channels (corresponding to modulations at two different frequencies present in the fluorescence intensity); ii) the result obtained when detecting (on each channel) $\pi/2$ out-of-phase with the PV signal; iii) the difference between the results obtained with the two $\lambda/2$ plates of the modulator rotating both clockwise or both counterclockwise; iv) the result obtained when treating the order (+- or -+) of the reversals as fixed while it is actually random. Cross-correlation between the two PV channels, which would imply a common fluctuating systematic effect on both channels, is not found. Serial correlation on either channel, which would imply a slowly drifting systematic effect on the corresponding channel, is absent. Finally, there is no clue for unexpected systematics.

One more protection against possible unexpected systematics is found in the cross-check between the two results, obtained in a $\Delta F = 0$ and a $\Delta F = 1$ hfs component respectively. In these two cases, the physical quantities involved in the PV measurement are different (§§ X and XI), as well as those involved in controlling most systematics. In addition, several experimental parameters and pieces of apparatus have undergone significant modifications. Consequently, the two experiments are sufficiently different to provide a genuine cross-check.

XVI. RESULTS ([28])

The results of the two measurements (in the $\Delta F = 0$ and $\Delta F = 1$ case respectively) are :

$$\text{Im } E_1^{pv}/\beta \text{ (mV/cm)} = -1.34 \pm 0.22 \pm 0.11 \text{ (after corrections}$$
$$\sim 0.5\%)$$
$$(13)$$
$$\text{and } -1.78 \pm 0.26 \pm 0.12 \text{ (without any correction)}$$
$$(14)$$

Every systematic has been reduced to the level of $\lesssim$ 1% of the observed value, so that practically no correction has to be performed; the overall systematic uncertainty is 7-8%. The statistics is in both cases better than 6 standard deviations. The two results are very consistent : even if the systematic uncertainty is ignored, they still agree at 1.3 statistical standard deviation. Thus they may be combined, which leads to

$$-1.56 \pm .17 \pm .12, \qquad\qquad\qquad (15)$$

i.e. 11% statistical uncertainty (and the same 8% systematic uncertainty). These results clearly show a significant parity violation.

The corresponding theoretical prediction involves the standard electroweak model (with radiative corrections estimated in ref. ([48])) plus an atomic calculation ([42]). The result is

$$-1.61 \pm 0.07 \pm 0.20$$

(The first uncertainty arises from the value of $\sin^2\theta$, the second one from the atomic model). This prediction is quite consistent with both results (13) and (14), and nearly coincides with the combined value (15). Thus the standard model is now verified for momentum transfer values in the MeV/c range (typical momentum transfer in the cesium experiment : 2 MeV/c ([49])).

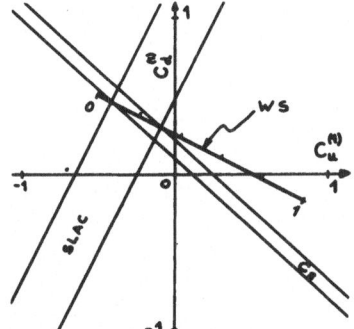

Fig. 8. Model-independent interpretation of the cesium result. The coordinates are the weak vector charges $C_u^{(1)}$ and $C_d^{(1)}$ of the up and down quarks. The region permitted by the cesium result is the strip labeled Cs, almost orthogonal to the strip (labeled SLAC) permitted by the high-energy result of ref. ([20]). The black segment labeled WS is the locus of the predictions of the standard model when $\sin^2\theta$ varies from 0 to 1. The intersection of the two strips turns out to be consistent with the prediction of the standard model for $\sin^2\theta \sim .2$.

Figure 8 compares, in the model-independent approach of § IV.1, the information concerning $C_u^{(1)}$ and $C_d^{(1)}$ from the Cs result, with the corresponding information from the SLAC experiment ([20]). Although the two experimental strips are obtained in a model-independent way, they turn out to intersect precisely in the region that corresponds to the prediction of the standard model for values of $\sin^2\theta$ around .2.

ACKNOWLEDGEMENTS

The present text owes much to M.A. BOUCHIAT's lectures at Les Houches in 1982 ([50]). Clarifying discussions with C. BOUCHIAT are gratefully acknowledged.

REFERENCES

[1] T.D. Lee, C.N. Yang, Phys. Rev. 104, 254 (1956)
[2] C.S. Wu, R.W. Hayward, D.D. Hoppes, R.P. Hudson, Phys. Rev. 105, 1413 (1957)
[3] R.L. Garwin, L.M. Lederman, M. Weinrich, Phys. Rev. 105, 1415 (1957)
[4] J.I. Friedman, V.L. Telegdi, Phys. Rev. 105, 1681 (1957)
[5] Ya.B. Zeldovich, Zh. Eksp. Teor. Fiz. 36, 964 (1959)
[6] F. Curtis-Michel, Phys. Rev. 138, B408 (1965)
[7] L.C. Bradley III, N.S. Wall, Nuovo Cim. 25, 48 (1962)
[8] R. Poppe, Physics (Utrecht) 50, 48 (1970)
[9] S.L. Glashow, Nucl. Phys. 22, 579 (1961)
[10] S. Weinberg, Phys. Rev. Lett. 19, 1264 (1967)
[11] A. Salam, in "Elementary particle theory", Proc. 8th Nobel Symposium, 367 (1968)
[12] G.'t Hooft, Nucl. Phys. B33, 173 (1971); and B35, 167 (1971)
[13] M.A. Bouchiat, C. Bouchiat, Phys. Lett. 48B, 111 (1974)
[14] M.A. Bouchiat, C. Bouchiat, J. Physique 35, 899 (1974); and 36, 493 (1975)
[15] L.M. Barkov, M.S. Zolotorev, Pis'ma Zh. Eksp. Teor. Fiz. 26, 379 (1978); and Phys. Lett. 85B, 308 (1979)
[16] P. Fayet, Phys. Lett. 96B, 83 (1980)
[17] E. Fermi, E. Segré, Rend. Accad. Sc. Fis. Mat. Nat. Soc. Reale Napoli 4, 131 (1933); and Z. Physik 82, 729 (1933)
[18] L.L. Foldy, Phys. Rev. 111, 1093 (1958)
[19] N. Fröman, P.O. Fröman, Phys. Rev. 6A, 2064 (1972)
[20] C.Y. Prescott et al., Phys. Lett. B77, 347 (1978); and Phys. Lett. 84B, 524 (1979)
[21] P.E.G. Baird, M.W.S.M. Brimicombe, R.G. Hunt, G.J. Roberts, P.G.H. Sandars, D.N. Stacey, Phys. Rev. Lett. 39, 798 (1977)
[22] R.C. Thompson, lecture at Les Houches (1982), to be published in "New Trends in Atomic Physics", Elsevier Science Publ. (1984)

(^{23}) L.L. Lewis, J.H. Hollister, D.C. Soreide, E.G. Lindahl, E.N. Fortson, Phys. Rev. Lett. 39, 795 (1977)

(^{24}) J.H. Hollister, G.R. Apperson, L.L. Lewis, T.P. Emmons, T.G. Vold, E.N. Fortson, Phys. Rev. Lett. 46, 643 (1981)

(^{25}) Yu.V. Bogdanov, I.I. Sobelman, V.N. Sorokin, I.I. Struk, Sov. Phys. JETP Lett. 31, 214 (1980); and 31, 522 (1980)

(^{26}) T.P. Emmons, J.M. Reeves, E.N. Fortson, Phys. Rev. Lett. 51, 2089 (1983)

(^{27}) R. Conti, P. Bucksbaum, S. Chu, E. Commins, L. Hunter, Phys. Rev. Lett. 42, 343 (1979); and P. Bucksbaum, E. Commins, L. Hunter, Phys. Rev. Lett. 46, 640 (1981)

(^{28}) M.A. Bouchiat, J. Guéna, L. Hunter, L. Pottier, Phys. Lett. 117B, 358 (1982); and 134B, 463 (1984)

(^{29}) R.R. Lewis, W.L. Williams, Phys. Lett. 59B, 70 (1975); and R.W. Dunford, R.R. Lewis, W.L. Williams, Phys. Rev. A18, 2421 (1978)

(^{30}) E.A. Hinds, V.W. Hughes, Phys. Lett. 67B, 487 (1977)

(^{31}) E.G. Adelberger, T.A. Trainor, E.N. Fortson, Bull. Am. Phys. Soc. 23, 546 (1978)

(^{32}) E.A. Hinds, Phys. Rev. Lett. 44, 374 (1980)

(^{33}) E.G. Adelberger, T.A. Trainor, E.N. Fortson, T.E. Chupp, D. Holmgren, M.Z. Iqbal, H.E. Swanson, Nucl. Instr. Meth. 179, 181 (1981)

(^{34}) L.P. Lévy, W.L. Williams, Phys. Rev. Lett. 48, 607 (1982)

(^{35}) J. Hoffnagle, V.L. Telegdi, A. Weis, Phys. Lett. 86A, 457 (1981)

(^{36}) S.L. Gilbert, R.N. Watts, C.E. Wieman, Phys. Rev. A29, 137 (1984)

(^{37}) L. Novikov, O. Sushkov, I. Khriplovich, Sov. Phys. JETP, 44, 872 (1976)

(^{38}) "A Novosibirsk group of theorists", quoted in the second ref. (^{15})

(^{39}) A. Martensson, E. Henley, L. Wilets, Phys. Rev. A24, 308 (1981)

(^{40}) P.G.H. Sandars, Phys. Scr. 284, 21 (1980)

(^{41}) D.V. Neuffer, E.D. Commins, Phys. Rev. A16, 844 (1977)

(^{42}) C. Bouchiat, C.A. Piketty, D. Pignon, Nucl. Phys. B221, 68 (1983)

(^{43}) B.P. Das, J. Andriessen, Mina Vajed-Samii, S.N. Ray, T.P. Das, Phys. Rev. Lett. 49, 32 (1982)

(^{44}) M.A. Bouchiat, J. Guéna, L. Hunter, L. Pottier, Optics Comm. 45, 35 (1983)

(^{45}) M.A. Bouchiat, L. Pottier, Optics Comm. 37, 229 (1981)

(^{46}) D. Herriott, H. Kogelnik, R. Kompfner, Appl. Opt. 3, 523 (1964)

(^{47}) M.A. Bouchiat, L. Pottier, G. Trénec, Revue Phys. Appl. 15, 785 (1980)

(^{48}) W.J. Marciano, A. Sirlin, Phys. Rev. D27, 52 (1983)

(^{49}) C. Bouchiat, C.A. Piketty, Phys. Lett. 128B, 73 (1983)

More detailed recent reviews of parity violation in atoms :

(^{50}) M.A. Bouchiat, lecture at Les Houches (1982, to be published in "New Trends in Atomic Physics", Elsevier Science Publ. 1984

(^{51}) E.N. Fortson, L.L. Lewis, to be published in Physics Reports (1984 ?)

NEUTRAL CURRENTS IN MUONIC ATOMS

J. Missimer

Schweizerisches Institut fuer Nuklearforschung
CH-5234 Villigen, Switzerland

Prof. Pottier has already introduced the principles
of parity violation in atomic systems, and described the
elegant experiment which measured a parity-odd effect in
Ce[1]. Proposals to measure similar effects in muonic atoms[2]
are the subject of this contribution.

Muonic atoms were proposed as promising systems to
detect neutral currents[3] at about the same time the
Bouchiats suggested heavy atoms. Since the muonic atom is
essentially an hydrogen-like system, the motivations for
experiments are similar to those for measuring parity-odd
effects in hydrogen atoms. First, the effects of neutral
currents in the simplest atomic system can be interpreted
with much more certainty than effects in heavy atoms. The
simplicity facilitates the precise determination of
neutral current couplings and observation of higher order
weak and electromagnetic corrections[4]. Second, the effects
of neutral currents in atomic systems are determined by a
different combination of couplings than effects measured
at high energy or in neutrino scattering at low energy.
Effects might be detected in atomic systems which are not
measurable at high energies[5]. Finally, comparison of
effects in muonic atoms and conventional atoms could
confirm the universality of neutral current couplings.

The original proposal to detect neutral current in
muonic atoms considered the magnetic dipole transition
between the 2S- and 1S-states[3]. Proposed was the measurement of an angular asymmetry of the emitted magnetic dipole radiation with respect to a muon polarized in the

initial (2S) or final (1S) state, or the circular polarization of the emitted radiation.

The motivation for this proposal is shown in Fig. 1. Due to the parity-odd piece of the weak current interaction, the 2S- and 2P-states are admixed. The one-photon transition amplitude between the 2S- and 1S-states is thus an admixture of magnetic dipole and electric dipole amplitude:

$$T_\gamma = M1 + \widetilde{E}1$$

The dominant part of the electric dipole amplitude is given by the product

$$\widetilde{E}1 = \eta \, E1$$

of the electric dipole amplitude, E1, describing the radiative transition between 2P- and 1S-states, and an admixture coefficient

$$\eta = \frac{\langle 2P | V_{\mu N} | 2S \rangle}{\Delta E}$$

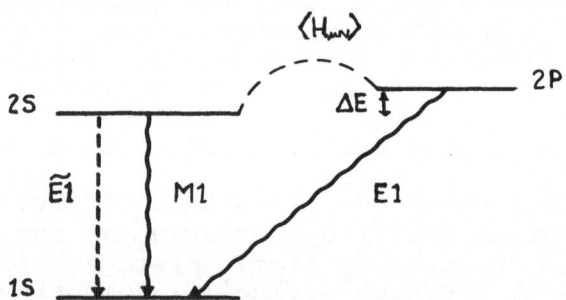

Fig. 1. Simplified level scheme showing admixture of 2S- and 2P-states produced by the weak neutral current interaction $H_{\mu N}$, and the consequent interference of magnetic (M1) and electric (E1) dipole transition amplitudes.

This coefficient contains the energy difference, ΔE, between the 2S- and 2P-states, and the matrix element of the weak potential:

$$V_{\mu N} = \frac{G}{\sqrt{2}}(ZC_{1p} + (A-Z)C_{1n})\rho(\vec{r})\gamma_5 - \frac{G}{\sqrt{2}}(C_{2p}\sum_i^Z \vec{\sigma}_i + C_{2n}\sum_j^{A-Z}\vec{\sigma}_j)\cdot\vec{\alpha}$$

derived from the parity-odd part of the neutral current interaction between muon and nucleons:

$$H_{\mu N} = \frac{G}{\sqrt{2}}\sum_{N=n}^{P}\{C_{1N}\,\overline{N}\gamma_\nu N\overline{\mu}\gamma^\nu\gamma_5\mu + C_{2N}\,\overline{N}\gamma_\nu\gamma_5 N\overline{\mu}\gamma_\mu^\nu\}$$

The nucleon numbers, Z and $A-Z$, and the normalized nucleon density $\rho(\vec{r})$ in the first term of the potential result from the hadronic vector current contribution to the interaction; the nucleon spin densities $\sum_i\vec{\sigma}_i$ occuring in the second term are due to the hadronic axial vector current. To the extent that the interaction is a simple product of current, the coupling constants $\{C_{iN}\}$ can be factored

$$C_{1N} = 2\,v_N a_\mu \qquad\qquad C_{2N} = 2\,a_N v_\mu$$

into nucleon and muon current coupling constants. In the standard model[6], these factors are expressed in terms of the Weinberg angle:

$$v_\mu = -\frac{1}{2} + 2\sin^2\theta_w \qquad a_\mu = \frac{1}{2}$$
$$v_p = \frac{1}{2} + 2\sin^2\theta_w \qquad a_p = -\frac{1}{2}g_A$$
$$v_n = -\frac{1}{2} \qquad\qquad a_n = \frac{1}{2}g_A$$

The interaction is written in terms of nucleon coordinates since the nucleus remains intact during the atomic transitions. The description in terms of valence quarks is completely equivalent.

The interference of magnetic and electric dipole amplitudes produces the angular asymmetry and circular polarization mentioned above. The magnitudes of these parity-odd observables is determined by the ratio:

$$A = \frac{2\text{Re}(M1\cdot\widetilde{E1}^*)}{|M1|^2 + |E1|^2} \simeq \eta\,\frac{E1}{M1}$$

The mixing coefficient η is typically of order $Gm_\mu^2 \sim 10^{-7}$, and the ratio of dipole transition amplitudes E1/M1 approaches 10^5 in light muonic atoms. In boron (Z=5), an example to be discussed in more detail below, the ratio, A, is predicted to attain a magnitude of 1 percent. This represents a considerable enhancement of the expected magnitude:

$$A = \frac{Gm_\mu^2}{e^2} \approx 10^{-5}$$

The enhancement results from the ratio of dipole transition amplitudes. However, this ratio is large because the magnetic dipole transition is hindered; it has yet to be observed in muonic atoms. The dominant competing transitions are shown in Fig. 2, and the corresponding transition rates in Table I. In an isolated light muonic atom (Z<12), the electron shell is stripped by muonic conversion (μ-Auger), conventional Auger and Coster-Krönig processes during the first few muonic transitions. The 2S-state is then depopulated by the two-photon transition to the ground state and by the electric dipole transition to the 2P-state. In conventional targets of high density, other electrons immediately refill the shell and can be ejected or rearranged during the muonic transition from the 2S- or the 2P- to 1S-states. If the electron L-shell refills, an electron will be ejected in a radiationless electric dipole transition of the muon to the 2P-state. If only the electron K-shell is populated, a radiationless electric monopole transition to the 1S-state is very probable. The dominance of these decay modes suppresses the yield of the magnetic dipole transition to 10^{-9}, which is far too small to be observed.

The problem posed by the small yield can be circumvented in two ways. The first is the choice of another system in which to measure parity-odd observables. The second is the development of experimental techniques to increase the yield.

One alternative system which has been studied extensively is the transition between the 3D- and 1S-states in heavy muonic atoms[7]. The mixing of 3P- and 3D-states induced by the weak interaction, $H_{\mu N}$, produces an admixture of electric dipole (E1) and quadrupole (E2) amplitudes in the one-photon transition amplitude:

$$A_\gamma = E2 + \tilde{E1} = E2 + \eta'E1$$

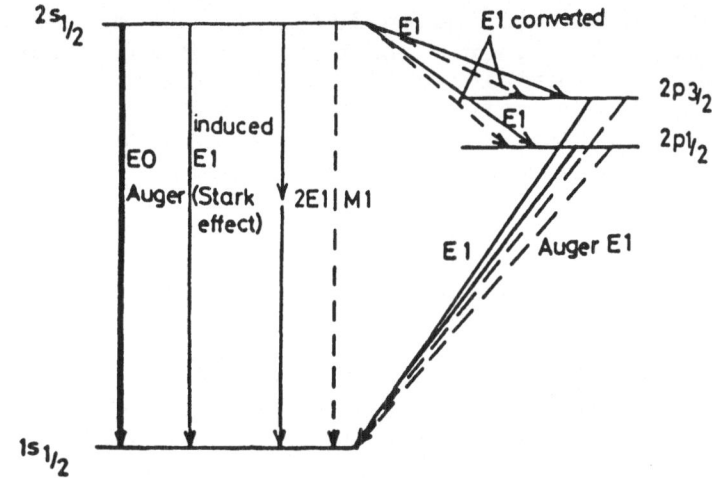

Fig. 2. Decay modes of the 2S- and 2P-states in light
muonic atoms. For Z>4, the Stark effect can be
neglected at pressures less than 760 Torr. As
shown in Table I, the magnetic dipole mode contributes the least to the 2S-transition rate.

Table I. Rates for transitions of the 2S- and 2P-states
in muonic boron (Z=5) and neon (Z=10).

Radiative transitions	B	Ne
$2S \rightarrow 1S + \gamma_{M1}$	$5 \cdot 10^3 \ s^{-1}$	$5 \cdot 10^6$
$2S \rightarrow 1S + 2\gamma_{E1}$	$3 \cdot 10^7$	$2 \cdot 10^9$
$2S \rightarrow 2P + \gamma_{E1}$	$1 \cdot 10^4$	$1 \cdot 10^9$
$2P \rightarrow 1S + \gamma_{E1}$	$6 \cdot 10^{13}$	$1 \cdot 10^{15}$

Auger and rearrangement processes

	B	Ne
$2S \rightarrow 1S + e_{E0}$	$2 \cdot 10^9$	$3 \cdot 10^9$
$2S \rightarrow 2P + e_{E1}$	$1 \cdot 10^{11}$	$1 \cdot 10^{12}$
$2S \rightarrow 2P + e_{rearr.}$	$1 \cdot 10^9$	–

119

η' is a mixing coefficient analogous to the one describing the admixture of magnetic and electric dipole transition amplitudes discussed above; El is the electric dipole amplitude for the transition between 3P- and 1S-states. A systematic study indicated that an angular asymmetry in thullium should be measurable. The asymmetry is predicted to be of order 10^{-5}. However, the reduction in the size of the effect is compensated by the increased yield of the 3D-transition, which approaches $5 \cdot 10^{-2}$. No technical reason forbids the measurement of the asymmetry in thullium.

Nevertheless, since the measurement in thullium was proposed, two experimental techniques have been developed which can facilitate a measurement of the asymmetry in the magnetic dipole transition.

The first is the use of X-ray absorption edges to discriminate transitions which cannot be resolved by detector resolution. J. Bailey proposed X-ray edges to distinguish between the one-photon and two-photon transitions[8]. The K-edge in terbium can be used for this purpose in muonic boron, and the K-edge in zinc can resolve transitions to different hyperfine components of the 2S-state. In muonic neon, the K-edge in lanthanum discriminates the transition between 3P- and 2S-states from that between 3D- and 2P-states; this discrimination permits transitions from the 2S-state to be distinguished from those of the 2P-state via coincidence. The existence of these X-ray absorption edges is the decisive factor in proposing boron and neon for the experiments to be discussed in greater detail below.

The second experimental development is the cyclotron trap[9], a device which produces high densities of stopped muons at low pressures. The low pressures are required to prevent the refilling of the electron shell of the muonic atom after it is stripped during the muonic cascade. Fig. 3 shows a computer simulation of an antiproton stopping in the cyclotron trap; the orbit of a stopping muon is very similar. First, the antiproton is slowed down by a moderator from a beam momentum of 300 MeV/c to the acceptance momentum of 110 MeV/c. It is now trapped in the target chamber by magnetic weak focussing. In the simulation, the chamber is filled with hydrogen gas at a pressure of 10 Torr. If it were not trapped by the magnetic field, the antiproton would require 180 m to stop in hydrogen gas at this pressure, and experience transversal and vertical dispersion of 1.8 and 1.0 meters respectively. In the cyclotron trap, the antiproton spirals inward and

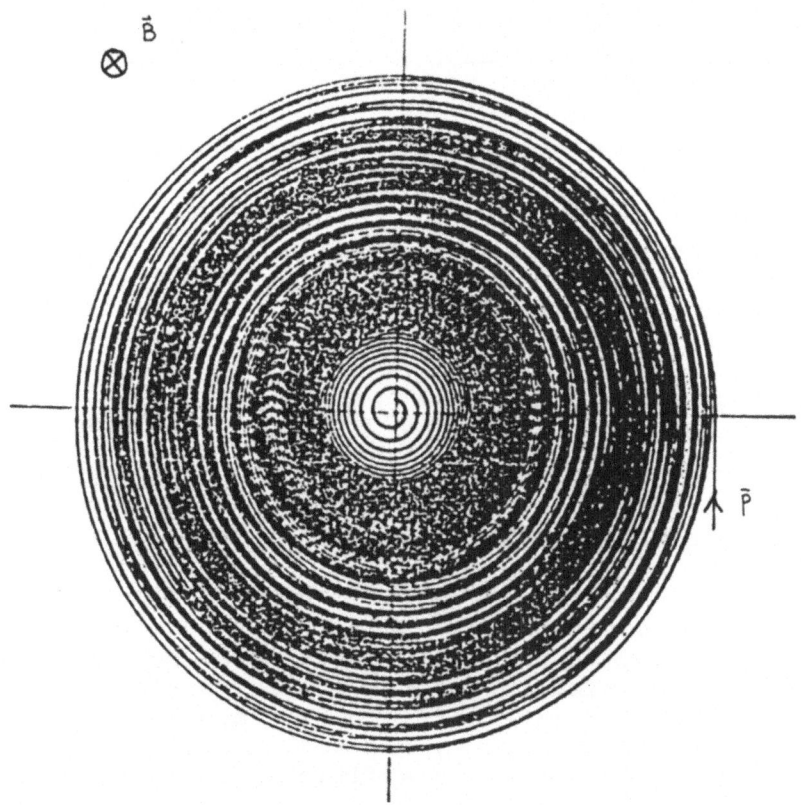

Fig. 3. Computer simulation of an antiproton with momentum 110 MeV/c stopping in the cyclotron trap. The
cyclotron field has a diameter of 280 mm; the weak
focussing magnetic field varies from 3.1 T at the
injection point to 3.9 T at the center.

stops in a volume of about 10^{-4} m^3 at the center of the
chamber; the stopping volume is therefore reduced by a
factor 10^4. Further characteristics of the cyclotron trap
are summarized in Table II.

The cyclotron trap is designed to produce high densities of stopped muons at the low pressures required to
prevent the refilling of the electron shell of the muonic
atom during the muonic cascade. Without an electron shell,
the conversion and rearrangement processes which dominate
the transition rate of the 2S-state cannot occur, and the
yield of the magnetic dipole transition increases to a
magnitude 10^{-5}. At low pressures, the application of X-ray
absorption edges and coincidence techniques should enable
the detection of the transition. The measurement of a

Table II. Characteristics of the cyclotron trap

	μ	$\bar{p}$
Beam momentum	170 MeV/c	300 MeV/c 12.83
		200 MeV/c 8.84
Acceptance momentum	110 MeV/c	
Beam acceptance	30 %	15 ± 5 %
Time of flight to stop	~200 ns	~200 ns
in target gas at pressure	Air/760 Torr	H /320 Torr
Operating gas pressure	100 Torr	30 Torr
Stopping density	$5 \cdot 10^4 \mu$/gm-sec	$2 \cdot 10^5 \bar{p}$/gm-sec

parity-odd observable is then feasible.

The angular asymmetry with respect to a muon polarization cannot be measured in the cyclotron trap, because the stopping process depolarizes the muon. Measurement of the circular polarization of the photon is also impracticable because polarimeter efficiencies are too small. However, in light muonic atoms, the weak decay of the muon in the 1S-state furnishes an additional asymmetry. The decay produces electrons whose momenta are correlated with the magnetic substate occupied by the muon; decay electrons with the maximum energy (~52.8 MeV) are emitted only opposite to the muon spin direction. Thus, the correlation, $\hat{k}_\gamma \cdot \hat{k}_e$ between the momenta of the photon emitted in the transition from the 2S-state and of the decay electron is equivalent to the correlation, $\vec{\sigma}_\mu \cdot \hat{k}_\gamma$. The equivalence is evident in the expression for the forward-background asymmetry:

$$\alpha_{M1}(\varepsilon) = \frac{d\Gamma(\varepsilon,1) - d\Gamma(\varepsilon,-1)}{d\Gamma(\varepsilon,1) + d\Gamma(\varepsilon,-1)} = \frac{1-2\varepsilon}{3-2\varepsilon} \frac{2\mathrm{Re}(M1 \cdot \tilde{E1}^*)}{|M1|^2 + |\tilde{E1}|^2}$$

where $d\Gamma(\varepsilon,1)$ is the differential decay rate for an electron with energy $E = 52.8 \, \varepsilon$ MeV to be emitted parallel to the photon, and $d\Gamma(\varepsilon,-1)$ the corresponding rate for an electron with the same energy to be emitted antiparallel. The factor involving electric and magnetic dipole amplitudes is just the A discussed above.

The existence of suitable X-ray absorption edges limits the proposed experiments to muonic boron, neon and sodium. Muonic neon appears to be most feasible proposal, because the magnetic dipole transition attains a yield of

$8 \cdot 10^{-5}$ in the absence of an electron shell, and because it is a monoatomic gas. The K-edge in lanthanum at 38.925 keV can be used in conjunction with the coincidence $(3 \rightarrow 2) \cap (2 \rightarrow 1)$ to discriminate the transition of the 2S-state to the ground state from that of the 2P-state. The contribution of the two-photon transition to the background intensity at the magnetic dipole transition energy of 207.8 keV can be suppressed by good energy resolution; a typical energy resolution of 1 keV yields a reduction of the two-photon background intensity to about 20 % of the magnetic dipole intensity. In neon, 60 % of the muons in the ground state decay into an electron and two neutrinos; the standard model then predicts an electron-photon correlation:

$$\alpha_{M1}(\varepsilon) = -\frac{1-2\varepsilon}{3-2\varepsilon} \, 1.6 \cdot 10^{-3}$$

Measurements in muonic boron appear more difficult than those in neon, but the effort would be rewarded. The principal difficulty is that the magnetic dipole transition attains a yield of only 10^{-5} in the absence of its electron shell. Also, boron in gaseous form (B_2H_6) is very poisoness. The most important advantage of muonic boron resides in the possibility of extracting the four coupling constants $\{C_{iN}\}$ which characterize the weak neutral current interaction between muon and nucleons.

This possibility was pointed out by Moskalev and Ryndin[10], and are evident in their calculation of the matrix elements. For $\mu^{10}B$, they obtain:

$$\langle 2P,3;\tfrac{7}{2}|V_{\mu N}|2S,3;\tfrac{7}{2}\rangle = i[2.7(C_{1p}+C_{1n})-0.2(C_{2p}+C_{2n})]\cdot 10^{-6}\text{eV}$$
$$\langle 2P,3;\tfrac{5}{2}|V_{\mu N}|2S,3;\tfrac{5}{2}\rangle = i[2.7(C_{1p}+C_{1n})+0.2(C_{2p}+C_{2n})]\cdot 10^{-6}\text{eV}$$

and for ^{11}B:

$$\langle 2P,\tfrac{3}{2};2|V_{\mu N}|2S;\tfrac{3}{2};2\rangle = i[2.7(C_{1p}+1.2C_{1n})-0.2C_{2p}]\cdot 10^{-6}\text{ eV}$$
$$\langle 2P,\tfrac{3}{2};1|V_{\mu N}|2S;\tfrac{3}{2};1\rangle = i[2.7(C_{1p}+1.2C_{1n})+0.2C_{2p}]\cdot 10^{-6}\text{ eV}$$

The hyperfine states $|nlj,I;F\rangle$ in the matrix elements are determined by the quantum numbers of the muon, nlj, the spin of the nucleus, I, and the total angular momentum, F. Thus, measuring asymmetries in transitions of the four hyperfine components of the two boron isotopes determines unambiguously the four coupling constants.

The resolution of the hyperfine components appears possible using the K-edge in zinc at 9.661 keV.

This edge could be used to determine a coincidence condition involving the $3P \to |2S_{1/2}, I; F>$ transition followed by the transition of the 2S-state to the ground state. In muonic boron, the rate of the 2P-transition is so much larger than that of the magnetic dipole transition, that they can be distinguished by timing; the K-edge in zinc is not necessary for their resolution. A K-edge in terbium at 51.996 keV is necessary to suppress the background intensity due to two-photon transition in which one photon carries almost all the energy; the edge in terbium permits the suppression of this background to 8 % of the magnetic dipole intensity.

Measuring the electron-photon correlation in muonic boron has two additional experimental advantages. First, the branching ratio for decay of the muon in orbit is 90 % Second, in the standard model, the average asymmetry is predicted to be ten times larger than in neon:

$$\alpha_{M1}(\varepsilon) = -\frac{1-2\varepsilon}{1-3\varepsilon} \, 1.6 \cdot 10^{-2}$$

The asymmetry has been averaged over hyperfine components.

Conclusion

The measurements of the photon-electron correlation in the magnetic dipole transitions of neon and boron are difficult. However, the application of X-ray absorption edges and the development of the cyclotron trap promise to enable the observation of the transitions despite their small yields, and thus to facilitate these measurements.

The demand that the measurements be made precisely is perhaps premature. However, muonic atoms are hydrogen-like systems whose simplicity can be exploited in precise measurements to extract the four coupling constants which characterize the weak neutral current interaction between muon and hadrons. The analysis of coupling constants requires, too, exact evaluation of matrix elements of the spin operators which occur in the admixture coefficient, and of the mean square radius which determines the hyperfine splittings of the 2S-state. Reliable evaluations of these observables will permit an analysis of higher order weak and electromagnetic corrections which constitute a decisive test of gauge theories of the electroweak interactions. Thus, the muonic atom could assume the same role in precise tests of unified gauge theories as it has in the verification of quantum electrodynamics.

References

1. L. Pottier, Parity violation in atomic systems, contribution to this meeting; M.A. Bouchiat et al., PL __117B__:358 (1982)
2. J. Missimer and L. Simons, Phys. Reports, to be published
3. J. Bernabeu, T.E.O. Ericson and C. Jarlskog, PL __50B__:467 (1974); G. Feinberg and M.Y. Chen, PRD__10__:190 (1974) A.N. Moskalev, JETP Lett. __19__:216 (1974)
4. W.J. Marciano and A. Sirlin, PRD__27__:552 (1983)
5. P. Fayet, PL __95B__:285 (1980)
6. S. Weinberg, PRL __19__:1264 (1967); __27__:1688 (1971); PR __D5__:1412 (1972); A. Salam, in Elementary Particle Theory: Relativistic groups and analyticity (Nobel Symposium No. 8), ed. N. Svartholm (Almquist and Wiksell, Stockholm, 1968), p. 367; S.L. Glashow, N.P. __22__:519 (1961)
7. L. Simons, Helv. Phys. Acta __48__:141 (1975); J. Missimer and L. Simons, N.P. __A316__:413 (1979); __A356__:317 (1981)
8. J. Bailey, private communication
9. L.M. Simons, The cyclotron trap: status of preparation and planned experiments, LEAR Workshop, Erice (1982)
10. A.N. Moskalev and R.M. Ryndin, Sov. J. Nucl. Phys. __22__:71 (1976)

RARE MUON DECAYS AND LEPTON-FAMILY NUMBER CONSERVATION

C. M. Hoffman

Los Alamos National Laboratory

Los Alamos, New Mexico

I. HISTORICAL SURVEY

A. Discovery of the Muon

The muon was discovered in cosmic radiation in 1937.[1] For several years it was believed to be the meson of Yukawa's theory that was the carrier of the strong nuclear force: its mass (105.5 MeV/c^2) is deceptively close to Yukawa's predicted meson mass. It was only after it was found that the muon did not interact strongly and another particle (the pi meson) did that the real puzzle presented itself: what role does the muon play? This mystery was succinctly expressed by Rabi: "The Muon, Who Ordered That?"

In fact, this remains one of the central questions in physics to this day, even though the language has changed. Today, one speaks of the family or generation problem, seeking to understand the apparent replication of quark and lepton generations, rather than only why the muon exists.

B. Interest in Neutrinoless Processes

Historically, neutrinoless processes such as $\mu^+ \rightarrow e^+\gamma$, $\mu^+ \rightarrow e^+e^+e^-$, and $\mu^-Z \rightarrow e^-Z$ were of great interest. It was believed that the muon and the electron had identical quantum numbers and so these processes should occur. One can show that a minimal electromagnetic $\mu^+ \rightarrow e^+\gamma$ transition [Fig. 1(a)] violates current conservation.

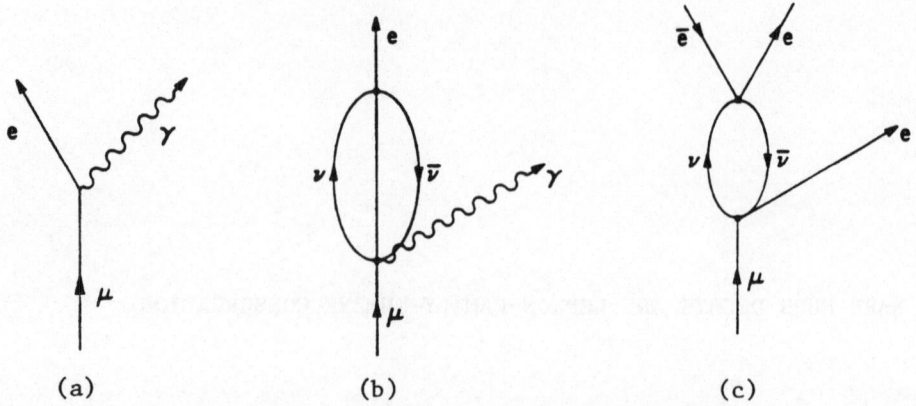

Fig. 1. (a) Diagram for a minimal $\mu \rightarrow e\gamma$ transition.
(b) Diagram for $\mu \rightarrow e\gamma$ mediated by a second-order weak interaction.
(c) Second-order weak-interaction diagram for $\mu \rightarrow ee\bar{e}$.

The interaction Lagrangian for Fig. 1(a) is

$$L_{int} = -ej^{\nu}A_{\nu} + \text{Hermitian Conjugate} \qquad (1)$$

where $j^{\nu} = \bar{\psi}_e\gamma^{\nu}\psi_{\mu}$.

Current conservation implies

$$\nabla_{\nu}j^{\nu} = 0 = \nabla_{\nu}(\bar{\psi}_e\gamma^{\nu}\psi_{\mu}) = \nabla_{\nu}(\bar{\psi}_e\gamma^{\nu})\psi_{\mu} + \bar{\psi}_e(\nabla_{\nu}\gamma^{\nu}\psi_{\mu}) \quad . \qquad (2)$$

The Dirac equation can be written

$$i\nabla_{\nu}\bar{\psi}\gamma^{\nu} + m\bar{\psi} = 0 \qquad (3a)$$

and also as

$$i\nabla_{\nu}\gamma^{\nu}\psi - m\psi = 0 \quad , \qquad (3b)$$

where m is the mass of the spin $-\frac{1}{2}$ particle involved. Putting (3a) and (3b) into (2), we find

128

$$0 = \nabla_\nu(\overline{\Psi}_e \gamma^\nu)\psi_\mu + \overline{\Psi}_e(\nabla_\nu \gamma^\nu \psi_\mu)$$

$$0 = -im_e\overline{\Psi}_e\psi_\mu + im_\mu\overline{\Psi}_e\psi_\mu = i(m_\mu - m_e)\overline{\Psi}_e\psi_\mu \quad , \tag{4}$$

which is inconsistent. Thus, the interaction of Fig. 1(a) violates current conservation and so cannot occur. However, the interaction shown in Fig. 1(b) is allowed. Note that this process is second order in the weak interaction. The calculation of this process yields,[2] for the branching ratio,

$$B_{\mu e\gamma} = \frac{\Gamma(\mu^+ \to e^+\gamma)}{\Gamma(\mu^+ \to e^+\nu\overline{\nu})} = \frac{2\alpha}{3\pi^5} G^2 m_p^4 \left(\frac{\Lambda}{m_p}\right)^4 \left(\ln\frac{2\Lambda}{m_p}\right)^2$$

$$\simeq 10^{-15}\left(\frac{\Lambda}{m_p}\right)^4 \left(\ln\frac{2\Lambda}{m_p}\right)^2 \quad , \tag{5}$$

where Λ is the momentum at which the divergent integral over the neutrino momentum is cut off. It is interesting to note that in the current × current model, the process $\mu^+ \to e^+e^+e^-$ [Fig. 1(c)] is second-order weak, whereas $\mu^+ \to e^+\gamma$ [Fig. 1(b)] is further suppressed by order α. The small rate implied by (5) did not confront the experimental upper limit, $B_{\mu e\gamma} < 2 \times 10^{-5}$, in 1957.[3]

A fundamental problem with the current × current model of the weak interaction is that the interaction grows with energy and ultimately violates unitarity. A solution to this problem was the introduction of the intermediate vector boson (now called $W^\pm$) proposed by Schwinger in 1957.[4] In this theory, the decay $\mu^+ \to e^+\gamma$ can proceed via the diagram shown in Fig. 2. The rate for this

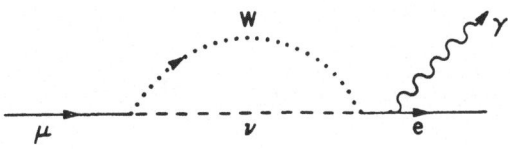

Fig. 2. Diagram for $\mu \to e\gamma$ in intermediate vector boson model.

diagram is second order in the semiweak coupling constant of the W to the leptons and so is first-order weak. $B_{\mu e \gamma}$ is given by[5]

$$B_{\mu e \gamma} = \frac{3}{8\pi} \alpha N^2(m_W, \Lambda) \tag{6}$$

$N^2 = 1$ for $\Lambda = m_W$ provided that the W has no anomalous magnetic moment. In general, N varies rapidly with Λ/m_W.

The large $B_{\mu e \gamma}$ implied by (6) was in conflict with the measured upper limit. In order to explain this, Ebel and Ernst[5] noted that for an anomalous moment near 0.75, $B_{\mu e \gamma}$ is nearly zero for $0.1 \lesssim \Lambda/m_W \lesssim 100$. However, the calculated rates for $\mu^+ \rightarrow e^+ e^+ e^-$ and $\mu^- Z \rightarrow e^- Z$ are not similarly suppressed. Thus, a dilemma existed. We could conclude that neutrinoless μ-e transitions were either suppressed by some dynamical mechanism or forbidden by some conservation law. Incidentally, this historical example of the need to measure all possible neutrinoless μ-e transitions will be repeated.

C.   Lepton-Number-Conservation Laws

The first lepton-number conservation law was proposed by Konopinski and Mahmoud in 1953.[6] In this scheme, e^- and μ^+ are each assigned $L = +1$, whereas e^+ and μ^- have $L = -1$. The assignments are summarized in Table I. There is a single conserved quantity, ΣL. This is an extremely economical scheme that forbids $\mu^+ \rightarrow e^+ \gamma$, $\mu^+ \rightarrow e^+ e^+ e^-$, $\mu^- Z \rightarrow e^- Z$, $K^0 \rightarrow \mu^\pm e^\mp$ but allows $\mu^- Z \rightarrow e^+(Z-2)$. An early problem with this arrangement was that it predicted that identical neutrinos emerge from muon decay ($\mu^+ \rightarrow e^+ \nu \nu$), giving a value of zero to the Michel parameter ρ, in contradiction to the measured $\rho \simeq 3/4$. The scheme can be patched up by allowing distinct ν_e and ν_μ, but it cannot be extended to a third lepton generation.

The additive lepton number scheme was introduced in 1957 by Schwinger, by Nishijima, and by Bludman.[7] The assignments are shown in Table II. The conserved quantities are ΣL_μ and ΣL_e. This forbids all neutrinoless μ-e transitions, as well as $K_L \rightarrow \mu^\pm e^\mp$ and $K^+ \rightarrow \pi^+ \mu^\pm e^\mp$. The extension of this scheme to include the τ lepton generation is straightforward.

The multiplicative scheme was put forth by Feinberg and Weinberg[8] in 1961. The assignments are given in Table III. The conserved quantities here are ΣL and ΠL_p. This choice forbids all neutrinoless processes but allows $\mu^+ \rightarrow e^+ \bar{\nu}_e \nu_\mu$, as well as $\mu^+ \rightarrow e^+ \nu_e \bar{\nu}_\mu$.

These schemes require distinct muon- and electron-type neutrinos. In 1959-60, Pontecorvo and, independently, Schwartz[9] proposed

TABLE I. Konopinski–Mahmoud Lepton–Number Assignments

_____

$\underline{L}$

+1	e^-, μ^+, ν	
-1	e^+, μ^-, $\bar{\nu}$	Conservation Law: ΣL = Constant
0	Everything else	

Forbids: $\mu^+ \rightarrow e^+\gamma$ Allows: $\mu^-Z \rightarrow e^+(Z-2)$

$\mu^+ \rightarrow e^+e^+e^-$ $(\mu^+e^-) \rightarrow (\mu^-e^+)$

$\mu^-Z \rightarrow e^-Z$ $K^+ \rightarrow \mu^+e^+\pi^-$

$K^0 \rightarrow \mu^\pm e^\mp$ $\mu^+ \rightarrow e^+\nu\nu$

TABLE II. Additive Lepton–Number Assignments

_____

$\underline{L_e}$		$\underline{L_\mu}$		$\underline{L_\tau}$	
+1	e^-, ν_e	+1	μ^-, ν_μ	+1	τ^-, ν_τ
-1	e^+, $\bar{\nu}_e$	-1	μ^+, $\bar{\nu}_\mu$	-1	τ^+, $\bar{\nu}_\tau$
0	Everything else	0	Everything else	0	Everything else

Conservations Laws: ΣL_e = Constant

ΣL_μ = Constant

ΣL_τ = Constant

Forbids: $\mu^+ \rightarrow e^+\gamma$ Allows: $\mu^+ \rightarrow e^+\nu_e\bar{\nu}_\mu$

$\mu^+ \rightarrow e^+e^+e^-$

$\mu^-Z \rightarrow e^-Z$

$\mu^-Z \rightarrow e^+(Z-2)$

$K^0 \rightarrow \mu e$

$\mu^+ \rightarrow e^+\bar{\nu}_e\nu_\mu$

accelerator experiments to determine if these neutrinos are indeed
distinct. Antineutrinos produced in reactors result from β decay
and so are electron–antineutrinos. When they interact with matter
they produce positrons. Neutrinos produced at accelerators come
predominantly from $\pi^+ \rightarrow \mu^+\nu_\mu$ and so should produce muons when they

TABLE III. Multiplicative Lepton-Number Scheme

L		L_p	
+1	$\mu^-,\ e^-,\ \nu_e,\ \nu_\mu$	+1	$e^+,\ \overset{(-)}{\nu_e}$
-1	$\mu^+,\ e^+,\ \bar{\nu}_e,\ \bar{\nu}_\mu$	-1	$\mu^+,\ \overset{(-)}{\nu_\mu}$
0	Everything else	+1	Everything else

Conservation Laws: ΣL = Constant

ΠL_p = Constant

Forbids: $\mu^+ \rightarrow e^+\gamma$ Allows: $\mu^+ \rightarrow e^+\nu_e\bar{\nu}_\mu$

$\mu^+ \rightarrow e^+e^+e^-$ $\mu^+ \rightarrow e^+\bar{\nu}_e\nu_\mu$

$\mu^-Z \rightarrow e^-Z$ $(\mu^+e^-) \rightarrow (\mu^-e^+)$

$\mu^-Z \rightarrow e^+(Z-2)$

$K^0 \rightarrow \mu e$

interact. This was indeed found to be the case in the famous "two-neutrino experiment" in 1962.[10] This stunning result validated the concept of lepton-number conservation and virtually closed the book on further searches for neutrinoless μ-e transitions for some time.

The situation was summarized in a 1963 paper by S. Frankel et al.[11]: "The results of the neutrino experiments...indicate that the normal weak interaction channels are closed to this decay mode ($\mu \rightarrow e\gamma$). Since it now appears that this decay is not lurking just beyond present experimental resolution, any further search...seems futile." The experimental status as of 1964 and 1975 is given in Table IV.

II. PRESENT VIEW OF LEPTON-NUMBER CONSERVATION

A. Nature of Conservation Laws

There are two kinds of conservation laws: those which are related to space-time translations or rotations and those which are not. Examples of the first kind are conservation of energy, momentum, and angular momentum. The second type includes conservation of electric charge, baryon number, and lepton number. We believe the first type of conservation law is fundamental. In 1933 Pauli[12] postulated the existence of what appeared to be an unobservable

TABLE IV. Experimental Status of Various Lepton-Family-Violating
Processes in 1964 and 1975

Process	Upper Limit in 1964 (90% CL)	Upper Limit in 1975 (90% CL)
$BR(\mu^+ \rightarrow e^+\gamma)$	2.2×10^{-8}	2.2×10^{-8}
$BR(\mu^+ \rightarrow e^+e^+e^-)$	1.2×10^{-7}	6.2×10^{-9}
$BR(\mu^+ \rightarrow e^+\gamma\gamma)$	1.6×10^{-5}	4×10^{-6}
$\dfrac{\Gamma(\mu^-Z \rightarrow e^-Z)}{\Gamma(\mu^-Z \rightarrow \nu[Z-1])}$	2.4×10^{-7} (Cu)	2×10^{-8} (Cu)

particle (the neutrino), rather than give up conservation of energy
and angular momentum in beta decay.

In general, the second kind of conservation law is regarded as
less fundamental. However, the conservation of electric charge is
related to gauge invariance of the electromagnetic field and its
associated massless gauge boson (the photon). No such massless
gauge boson exists for baryon or lepton number, and so it has been
argued that exact conservation of these quantities is absurd. A
heuristic argument presented by de Rujula, Georgi, and Glashow[13]
demonstrates this.

Consider a region of space from which you are excluded. If
someone throws an electrically charged object into the region, can
you detect this event by observations made outside the region?
Yes, you can. The memory of the electric charge is preserved by
the electric field outside the region: the fact that the photon is
massless implies that the field extends over all space. Now, if
someone throws a lepton (or a baryon) into the region, can you
tell? No! The lepton leaves no trace at all. Thus, we do not
expect lepton number to be an exact global symmetry.

How do we know that there is no massless gauge boson coupled
to lepton number? Of course, we have not seen one but that does not
constitute proof. The best theoretical argument was given by Lee
and Yang in 1955.[14] They showed that such a massless gauge boson
would violate the gravitational equivalence principle (Eötvös
experiment). Consider a massless vector field coupled to lepton
number with coupling constant η. The force between two massive
objects then includes a contribution from the Coulomb-like force
between the leptons:

$$F_G = - \frac{GM_1M_2}{R^2} + \frac{\eta^2 Z_1 Z_2}{R^2} = - \frac{G}{R^2} M_{1G} M_{2G} \quad . \tag{7}$$

The gravitational mass, M_{iG}, is supposed to be equivalent to the inertial mass, M_i. But M/Z varies greatly from substance to substance. Thus, the equivalence principle cannot hold for all substances. The experimental precision on the equivalence principle corresponds to an upper limit on $\eta/\sqrt{G}$.

Thus, we do not expect lepton or baryon numbers to be exactly conserved. Nevertheless, the experimental evidence shows that any violation of conservation of these quantities is extremely weak. In fact, it is a theoretical problem to explain why the violation is so weak if it is not zero.

B.   The "Standard Model"

The standard model is a spontaneously broken gauge theory of electroweak and strong interactions with a group structure $SU(3)_c \times SU(2)_L \times U(1)$.[15]   The fermions are placed in left-handed doublets and right-handed singlets. In this model, the neutrinos are massless, and there is no lepton number (either total lepton number or separate electron, muon and tau number) violation. The standard model is a minimal model in the sense that it includes only those elements which are required by present data. Many elements (such as exact V−A weak interactions) are put in explicitly, and there is no explanation for the replication of the family structure, nor is there any way to calculate the many parameters (coupling constants, masses, mixing angles) of the theory. Nevertheless, this model is in agreement with all observations.

C.   Heavy Neutrinos

The simplest extension of the minimal Weinberg-Salam-Glashow model would be the inclusion of neutrino masses. In such a model, the neutrinos could mix and give rise to (for example) $\mu \to e\gamma$.[16] The branching ratio, $B_{\mu e \gamma}$, is given by

$$B_{\mu e \gamma} = \frac{3\alpha}{32\pi} \, |U_{23}^* U_{13}|^2 \frac{(m_3 - m_2)^2 (m_3 - m_1)^2}{m_W^4} \quad , \tag{8}$$

where m_i is the mass of the neutrino in the ith generation and U_{ij} is an element of the unitary neutrino mass mixing matrix. Using

134

the present upper limit for the tau-neutrino mass $(m_{\nu_\tau} < 250$ MeV$)$ and $|U^*_{23}U_{13}|^2 < 3 \times 10^{-3}$, we find $B_{\mu e\gamma} < 1 \times 10^{-16}$. For a further lepton generation with maximal mixing, a neutrino mass of 1.8 GeV would result in $B_{\mu e\gamma} = 10^{-10}$. Note that this model has succeeded in suppressing muon number violating processes without needing to impose a conservation law.

In this model, $\mu^+ \rightarrow e^+e^+e^-$ is dominated by Z^0 exchange. Here,

$$\Gamma(\mu \rightarrow ee\bar{e})/\Gamma(\mu \rightarrow e\gamma) = \frac{2\alpha}{\pi} \ln^2 \left[\frac{(m_3 - m_2)(m_3 - m_1)}{m_W^2} \right] . \tag{9}$$

Models with doubly charged leptons can also have muon-number violation but with $B_{\mu 3e}$ larger than $B_{\mu e\gamma}$.

D.   Expanded Higgs Sector

Another straightforward extension of the standard model is the inclusion of more than one doublet of Higgs bosons.[17] The standard model requires at least one Higgs doublet to generate masses for the vector bosons. However, there is considerable freedom in the nature of the Higgs sector. Muon-number violations can be mediated by the multiple Higgs doublets.

The Higgs can contribute through one-loop diagrams and two-loop diagrams. The two-loop is larger than the one-loop contribution if the Higgs mass, m_H, is greater than 3 GeV. If $m_H \ll m_W$, then

$$B_{\mu e\gamma} \simeq 3\left(\frac{\alpha}{\pi}\right)^3 \simeq 4 \times 10^{-8} , \tag{10}$$

which exceeds present experimental bounds. If $m_H > m_W$,

$$B_{\mu e\gamma} \sim \left(\frac{\alpha}{\pi}\right)^3 \left(\frac{m_W}{m_H}\right)^4 . \tag{11}$$

For $m_H < 1$ TeV (this is current theoretical prejudice), $B_{\mu e\gamma} > 4 \times 10^{-13}$. Note that by studying very rare processes, one is

135

exploring extremely large mass scales. Finally, the rate for $\mu \rightarrow e\bar{e}e$ is smaller than the rate for $\mu \rightarrow e\gamma$ but by less than a factor of α/π as would be expected from Dalitz pair production.

E.   Other Theoretical Models

Space does not permit a discussion of the many other extensions to the standard model which result in lepton-family-number nonconservation. Examples include existence of flavor-changing neutral gauge bosons (for example, the gauge bosons associated with horizontal gauge interactions,[18] or the gauge bosons present in extended technicolor theories[19]); composite models[20]; muon-number violation mediated by light lepto-quarks (present in some grand unified theories[21] and in extended technicolor theories[19]); muon-number violation mediated by supersymmetric partners of the usual $SU(2)_L \times U(1)$ gauge bosons[22]; existence of new electroweak interactions.[23] In general, these different sources of lepton-number nonconservation predict different relative strengths for the various neutrinoless transitions. This result underscores the importance of searching for all of these processes.

Another process which violates lepton-family-number conservation is neutrino oscillations.[24] Oscillations explicitly require massive neutrinos, whereas this is not the case for the processes discussed above. However, oscillation experiments can be sensitive to very small neutrino masses (<1 eV), whereas effects in the neutrinoless transitions caused by those masses alone would be negligibly small.

III. THE SEARCH FOR $\mu \rightarrow e\gamma$

A.   Rare Decays

In order to search for any rare decay, one needs a copious source of the decaying particle and an apparatus that can detect the decay products. Furthermore, the experiment must be capable of eliminating possible background processes and of identifying the desired reaction.

In the absence of background or signal, the 90% confidence level upper limit is given by

$$B(90\% \text{ CL}) = \frac{2.3}{N_D \times \Omega/4\pi \times \varepsilon} , \qquad (12)$$

where N_D = number of parent particles decaying, $\Omega/4\pi$ = fractional solid-angle acceptance of the detector, and ε = overall detector efficiency.

N_D increases linearly with running time, so the limit improves linearly with running time until some background is encountered. In the presence of background, the limit improves with the square root of the running time because one must subtract the number of background events from the number of observed events, and the statistical uncertainties in these numbers determine the limit.

B. History

The first searches for the $\mu \rightarrow e\gamma$ decay were performed by Hincks and Pontecorvo, and Sard and Althaus in 1948[25] before the true nature of muon decay was understood. The energy of the particles was measured using Geiger-Mueller tubes and absorbers. An upper limit of $B_{\mu e\gamma} < 5 \times 10^{-2}$ was obtained. Subsequent measurements[3,26] used a variety of techniques (range measurements with scintillation counters, water Čerenkov counters, spark chambers, a Freon bubble chamber, and energy measurements with NaI crystals). The upper limit, as a function of time, is shown in Fig. 3.

Due to the long lifetime of the muon (2.2 μs), all studies of muon decay have been done with stopped muons. The decay $\mu^+ \rightarrow e^+\gamma$ (μ^+ are studied because stopped μ^- can be captured in the stopping target) is characterized by a monochromatic positron and photon (Energy = 52.83 MeV), which are emitted simultaneously at 180° with respect to each other. It is these distinguishing attributes which must be used to search for $\mu \rightarrow e\gamma$.

Early measurements used range to determine the energy of the positron and the converted photon. Later experiments used NaI crystals or a magnetic spectrometer. The precision of the energy measurements was poor. As a result, these experiments were limited by the background from the decay $\mu^+ \rightarrow e^+\nu_e\bar{\nu}_\mu\gamma$ where the two neutrinos have little energy and at $e^+-\gamma$ are nearly acollinear. The branching ratio in this configuration[27] is

$$B_{rad} = \frac{\Gamma(\mu^+ \rightarrow e^+\nu\bar{\nu}\gamma, \theta_{e\gamma} \approx 180^\circ)}{\Gamma(\mu^+ \rightarrow e^+\nu\bar{\nu})} =$$

$$\frac{\alpha}{2\pi}[(1 - x)^2 + 4(1 - x)(1 - y)]y \, dy \, dx \, d(\cos \theta_{e\gamma}) \quad , \tag{13}$$

where $\theta_{e\gamma}$ is the angle between the positron and the photon and x and y are the positron and photon energies, respectively, in units of $m_\mu/2$. Thus, for $\Delta\theta_{e\gamma} = 0.1$, $\Delta x = 0.1$, and $\Delta y = 0.5$ (typical of experiments performed around 1963), $B_{rad} = 1.4 \times 10^{-8}$. This is about as well as these experiments did. However, the modest μ^+ intensities (derived from stopped π^{+}'s) did not permit more sensitive searches to be made. The advent of high intensity pion beams and muon beams at meson factories, coupled with substantial detector improvements, allowed more sensitive experiments to be performed.

C.   The SIN Experiment of 1977

This experiment[28] is largely responsible for the renaissance in interest in processes violating muon number conservation. An ill-founded rumor of a $\mu \to e\gamma$ signal, which proved to be incorrect,

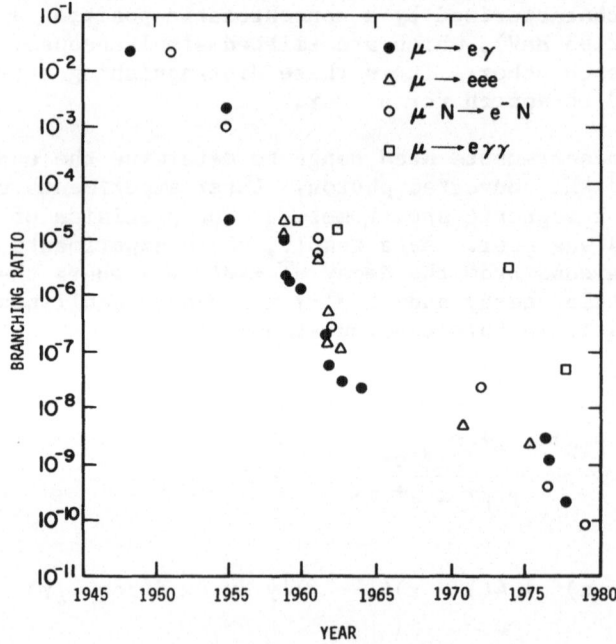

Fig. 3.   Upper limit for $B_{\mu e\gamma}$ and several other muon-number violating processes as a function of time.

was the cause. This experiment was an order of magnitude more sensitive than any previous experiment. The apparatus is shown in Fig. 4. The major detecting device is two large NaI crystals located on opposite sides of the stopping target. This design is essentially identical to that used by Frankel et al.[26] in 1963. This arrangement had a relatively small acceptance ($\Omega/4\pi = 1.2\%$ for $\mu \rightarrow e\gamma$ events). The experiment utilized a 90 MeV/c μ^+ beam with a moderator before the scintillation target. The stopping rate was ~5×10^5 μ/s. The NaI energy resolution was (4.6 ± 0.4) MeV (FWHM). Good shower containment was achieved by using large NaI detectors (27.7 cm diameter, 33 cm long) and collimating the entrance aperture. The improved energy resolution implies that the internal bremsstrahlung background is comfortably below 10^{-9}. An upper limit $B_{\mu e\gamma} < 1.0 \times 10^{-9}$ (90% CL) was achieved.

A nearly identical experiment was performed at the same time at TRIUMF.[29] The beam intensity was 2×10^5 π^+/s. An upper limit of $B_{\mu e\gamma} < 3.6 \times 10^{-9}$ (90% CL) was obtained.

These two experiments improved the upper limit for $B_{\mu e\gamma}$ by an order of magnitude using an experimental design that is essentially identical to previous experiments. The improvement was possible because of improvements in the incident beam, in the NaI detectors, and in the electronics. Nevertheless, this limit is about as low as one can set with this technique. Further improvement requires different techniques.

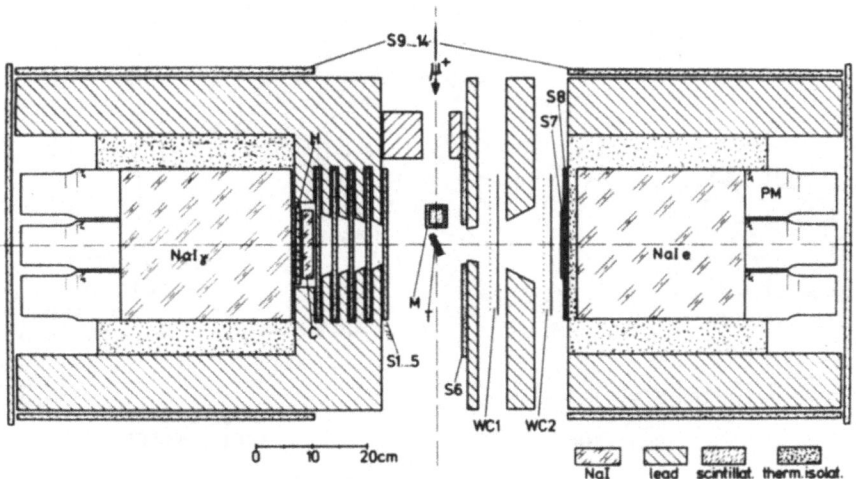

Fig. 4. Schematic diagram of the SIN $\mu \rightarrow e\gamma$ experimental apparatus.

D.   The LAMPF Experiment of 1978-79

A plan view of the apparatus for this experiment[30] is shown in
Fig. 5.   There were several new techniques used in this experiment,
which will be described below.   These are a magnetic spectrometer
to measure the e^+ vector momentum, a segmented NaI detector to measure the photon energy and conversion point, and a muon beam of
very low energy ("surface" beam).

The goal of this experiment was to be sensitive to $B_{\mu e \gamma}$ at the
10^{-10} level.   Since $\Omega/4\pi$ for this detector is ~2%, more than 10^{12}
muon decays were required.   Furthermore, a thin stopping target was
required to minimize multiple scattering, bremsstrahlung, and annihilation of the outgoing positron.

The experiment used a surface muon beam.   The muons in such a
beam originate from pions that come to rest near the surface of the
pion production target.   These muons come from a small, well-
defined source.   They may be imaged by simple beam optics.   The
muon acceptance is high, the momentum is low (28 MeV/c), the momentum dispersion ($\Delta p/p$) and spot size are both small, and the intensity is high.   The range of the surface muons in polyethylene is

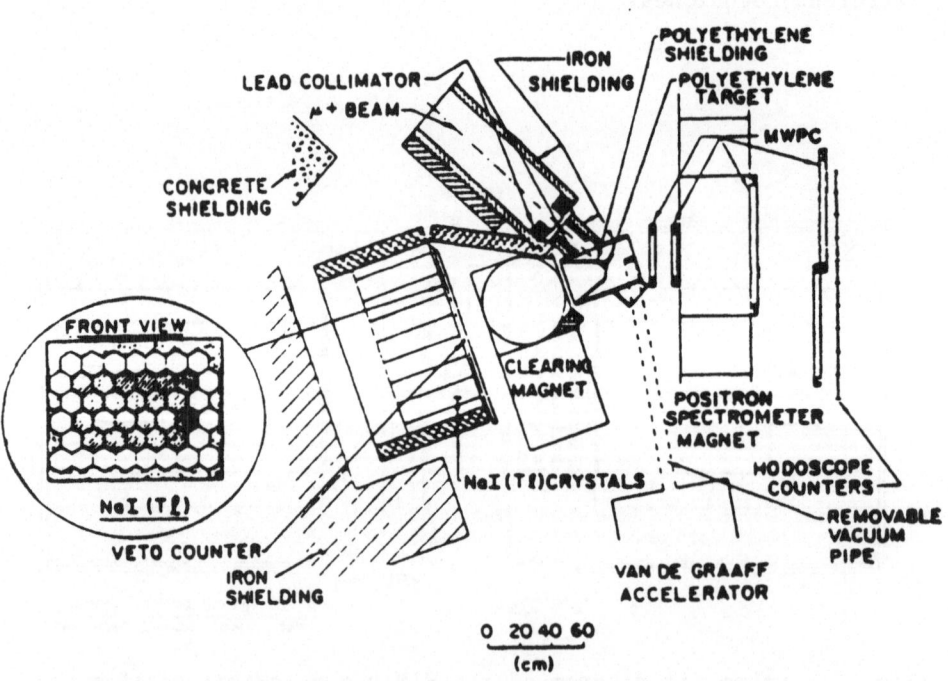

Fig. 5.   The apparatus for the LAMPF $\mu \to e\gamma$ experiment.

69 mg/cm^2 with a full width at half maximum (FWHM) of only 20 mg/cm^2. The stopping intensity was 2.4×10^6 muons/s. Because of a long beam line and the low beam momentum, almost all of the pions accepted by the beam decay before reaching the experimental target. On the other hand, the channel also transports a large contamination of positrons created by pair production from γ rays in the production target. These positrons were separated from the muons by the use of a degrader in the beam, upstream of a bending magnet.

The magnetic spectrometer used MWPC's to determine the positron trajectory and a scintillation counter hodoscope to measure its time of arrival. This was not the first use of a magnetic spectrometer in a $\mu \to e\gamma$ experiment.[31]

The most novel technique used in this experiment was the photon detector. The photon telescope was a total absorption NaI detector, but it used a segmented array of NaI crystals. This permitted the measurement of the photon conversion point. In addition, a sweeping magnet was placed between the stopping target and the NaI. No positron from muon decay at the target could reach the NaI. This arrangement reduced the singles rates in the photon detector by about two orders of magnitude and allowed the experiment to use a high muon rate.

Table V shows some of the parameters of the experiment. An upper limit, $B_{\mu e\gamma} < 1.7 \times 10^{-10}$ (90% CL) was set by this experiment. The major background in this experiment was the random

TABLE V. Characteristics of Several $\mu \to e\gamma$ Experiments

	Ref. 30	Ref. 32	Ref. 33
$\left(\frac{\Omega}{4\pi}\right)$	1.8%	50%	16%
$\dfrac{\Delta E_\gamma}{E_\gamma}$	8% (FWHM)	6% (FWHM)	6% (FWHM)
$\dfrac{\Delta E_e}{E_e}$	9% (FWHM)	6% (FWHM)	0.6% (FWHM)
$\Delta t_{e\gamma}$	2ns (FWHM)	0.7 NS (FWHM)	0.6ns (FWHM)
$\Delta\theta_{e\gamma}$	5° (FWHM)	~6° (FWHM)	1.5° (FWHM)
μ^+ rate	2.5×10^6/s (average)	$5 \ 10^5$/s (average)	$>10^7$/s (average)
Sensitivity	1.7×10^{-10}	10^{-11}	$<10^{-12}$

coincidence between uncorrelated positrons and photons. The internal bremsstrahlung background was at the few × 10⁻¹² level.

E.   Future Measurements

At LAMPF one μ → eγ measurement is in progress and another is planned. The measurement in progress uses a large, multipurpose detector called the Crystal Box.[32] A schematic diagram of the apparatus is shown in Fig. 6. Surface muons stop in a thin polystyrene target in the middle of a cylindrical drift chamber. The drift chamber is surrounded by a hodoscope of plastic scintillation counters and by a large, modular, sodium iodide array. The parameters of the detector are given in Table V. The apparatus is being used to search for μ⁺ → e⁺γ, μ⁺ → e⁺e⁺e⁻, and μ⁺ → e⁺γγ simultaneously. The single particle acceptance is Ω/4π = 0.5, which is very large. The e⁺ and γ are back-to-back for μ⁺ → e⁺γ, so this is also the acceptance for this two-body decay. Since there is no magnetic field in the apparatus, the sodium iodide is exposed to the full flux of positrons from the muons decaying in the middle of the detector. In order to minimize the effects of pile-up (two particles depositing energy in an NaI crystal within the sensitive time of

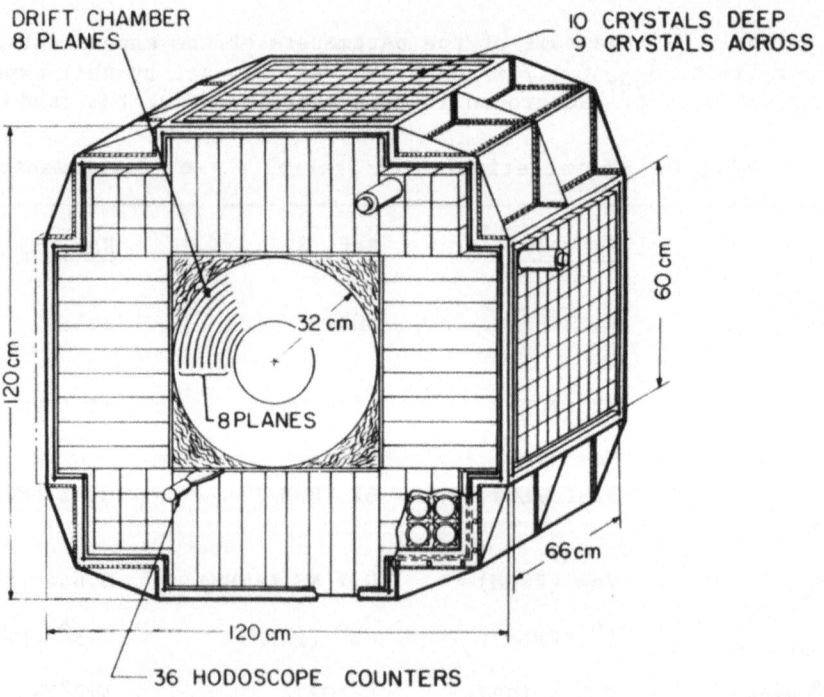

Fig. 6. Diagram of the apparatus for the Crystal Box experiment.

the energy measurement, ~200 ns), the μ^+ stopping rate is 5×10^5 μ/s, with a duty factor of ~7.5%.

The drift chamber determines the e^+ trajectory, which is traced back to the stopping target to find the muon decay point. The line from this point to the photon conversion point (determined from the energy distribution in the individual sodium iodide crystals) defines the photon trajectory. The largest background comes from the random coincidence between a positron and a photon. This should enter at a branching ratio level of a few $\times 10^{-11}$. An ultimate sensitivity of 10^{-11} for $\mu \rightarrow e\gamma$ is anticipated.

A third LAMPF $\mu \rightarrow e\gamma$ experiment is planned[33] after the Crystal Box experiment is completed, assuming a null result is obtained. The NaI will be reconfigured and combined with a magnetic spectrometer, as shown in Fig. 7. The parameters of this arrangement are given in Table V. In most respects, this is the logical extension of the first LAMPF $\mu \rightarrow e\gamma$ experiment. Better resolutions in all parameters, a higher muon flux, and a larger acceptance should result in a background-free measurement at the level of several parts in 10^{13}.

If this experiment achieves its goal but does not detect $\mu \rightarrow e\gamma$, a new design will be required to extend the sensitivity. A substantial increase in solid angle acceptance and/or muon flux

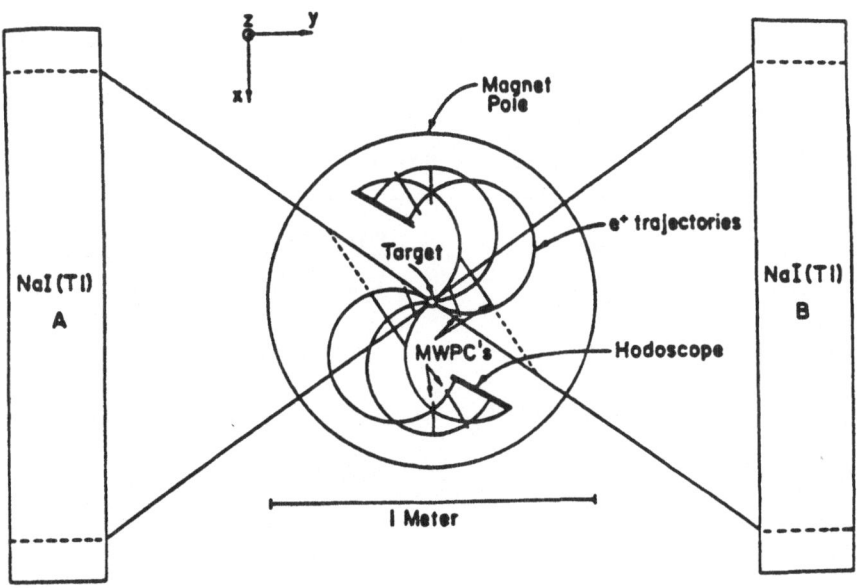

Fig. 7. Apparatus for the planned $\mu \rightarrow e\gamma$ experiment at LAMPF.

will certainly be required. As difficult as it may be to envision this at present, history indicates that it should occur. Figure 3 shows the experimental limits for various muon-number-nonconserving processes as a function of time. This figure shows the trend of an order of magnitude improvement approximately every five years.

IV. STATUS OF OTHER EXPERIMENTS

A. $\mu^+ \to e^+ e^+ e^-$

The present upper limit for this process comes from a 1976 experiment at Dubna.[34] This experiment utilized a multilayer cylindrical spark chamber with an axial magnetic field. The parameters of this experiment are given in Table VI. measurement are given in Table VI. The worst background came from an electron traversing the apparatus from outside (thus looking like an e^+ and an e^- emerging from the target at a relative angle of 180°) in random coincidence with an e^+ from a μ^+ decaying in the target. The constraints in this experiment are the vertex (the fact that all three charged particles must emerge from a common point in the target), timing, and conservation of energy and momentum. An intrinsic background comes from the decay $\mu^+ \to e^+ e^+ e^- \nu_e \bar\nu_\mu$. While this process satisfies the first two constraints, the expected branching ratio for events with the sum of the energies of the electron and positrons near M_μ is quite small. For example, the branching ratio for detected energy above 70 MeV is $\sim 10^{-9}$ and falls exponentially with increasing energy.[35]

Two new experiments are under way to search for $\mu \to e^+ e^+ e^-$. One is the aforementioned Crystal Box experiment at Los Alamos. The parameters of this measurement are given in Table VI. With no

TABLE VI. Characteristics of Several $\mu^+ \to e^+ e^+ e^-$ Experiments

	Ref. 34	Ref. 32	Ref. 36
$(\Delta\Omega/4\pi)$	5%	20%	16%
$\Delta E_e/E_e$	15% (FWHM)	6% (FWHM)	8% (FWHM)
$\Delta t_{ee'}$	6ns (FWHM)	0.5ns (FWHM)	0.5ns (FWHM)
Target Thickness	22 mm	1.5 mm	1.1 mm
μ^+ rate	2×10^4/s	5×10^5/s (average)	2.5×10^6/s
Sensitivity	1.9×10^{-9}	$\sim 2 \times 10^{-11}$	1.6×10^{-10} (achieved) 10^{-12} (planned)

magnetic field, there is no way to distinguish e^+ and e^-. The main background comes from three positrons from three separate muon decays in accidental coincidence. A sensitivity of $\sim 2 \times 10^{-11}$ should be reached in the next year.

The second new experiment is SINDRUM[36] at SIN. The detector, shown in Fig. 8, consists of cylindrical arrays of multiwire proportional chambers in an axial magnetic field. This design is an update of the apparatus of Korenchenko et al. (Ref. 34). The parameters of this experiment are shown in Table VI. This experiment benefits from the 100% duty factor at SIN. A preliminary run with only four of the five chambers yielded an upper limit $B_{\mu 3e} <$ 1.6×10^{-10} (90% CL). A longer run is now under way.

B. $\mu^- Z \rightarrow e^- Z$

When negative muons stop in matter, they can either decay or be captured by the nucleus, initiating the reaction $\mu^- Z \rightarrow \nu_\mu (Z-1)$. A muon-number nonconserving process which might occur is $\mu^- Z \rightarrow e^- Z$ (called μ-e conversion). This process can be coherent (in which the nucleus remains in the ground state and so the electrons are monoenergetic with $E_e \simeq m_\mu$) or incoherent (in which the nucleus is excited). Coherent μ-e conversion involves only scalar and vector quark densities.[37] Incoherent μ-e conversion involves pseudoscalar and axial vector quark densities.[38] The published limits on

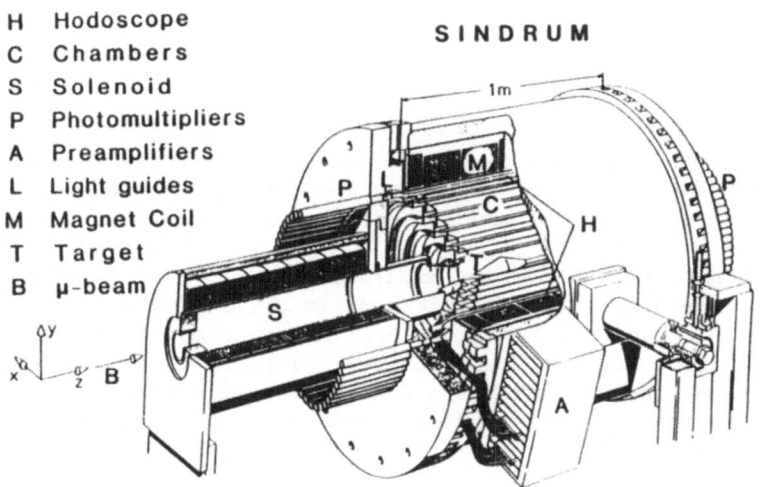

H Hodoscope
C Chambers
S Solenoid
P Photomultipliers
A Preamplifiers
L Light guides
M Magnet Coil
T Target
B μ-beam

SINDRUM

Fig. 8. Diagram for the SINDRUM experiment.

TABLE VII. Limits for Coherent μ⁻ – e⁻ Conversion

Nucleus	Limit (90% CL)	Reference
Copper	2×10^{-8}	Bryman et al., Ref. 41
Sulfur	7×10^{-11}	Badertscher et al., Ref. 39
Titanium	2×10^{-11}	Blecher et al., Ref. 40

coherent μ–e conversion for various nuclei are given in Table VII. Note that the coupling for ^{32}S must be isoscalar, whereas this is not necessarily the case for the other nuclei.

In the search for coherent μ–e conversion, one is performing a kinematically incomplete experiment to observe a rare process. The signature is a single electron with energy near 100 MeV. Backgrounds include muon decay in orbit (in which the whole atom can recoil), radiative muon capture with the production of an asymmetric $e^+ - e^-$ pair, and many possible processes induced by π^-, should any be present. The first two backgrounds fall approximately as $(E_{max} - E)^4$.

The SIN experiment[39] on ^{32}S used a streamer chamber inside an axial magnetic field to measure the electron momentum accurately. An upper limit,

$$R_{\mu e}^{coh}(^{32}S) = \frac{\Gamma(\mu^- Z \to e^- Z)_{coh}}{\Gamma(\mu^- Z \to \nu_\mu [Z - 1])} < 7 \times 10^{-11} \quad (90\% \text{ CL}) \quad , \quad (14)$$

was obtained. The beam in the SIN accelerator was pulsed at 400 kHz. Electrons emitted during the beam-off period were detected long after any pions had decayed or been captured.

The apparatus for a new experiment at TRIUMF[40] is shown in Fig. 9. The heart of the experiment is a time projection chamber located in a magnetic field. A preliminary limit $R_{\mu e}^{coh}(^{48}Ti) < 2 \times 10^{-11}$ (90% CL) has been obtained. The experiment is still running and hopes to achieve a sensitivity of several parts in 10^{12}. The μ–e conversion experiments at SIN, TRIUMF, and an earlier experiment at SREL[41] are compared in Table VIII.

The limits on incoherent μ–e conversion are much poorer due to the fact that the measurement is much more difficult and because no

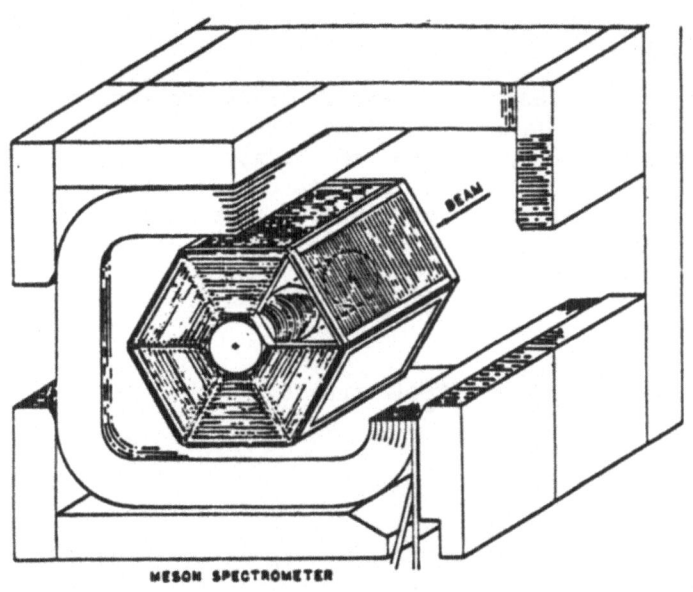

Fig. 9. Apparatus in use at TRIUMF to search for $\mu^- - e^-$ conversion.

TABLE VIII. Characteristics of Several μ^-e Conversion Experiments

	SREL (Ref. 41)	SIN (Ref. 39)	TRIUMF (Ref. 40)
$\left(\dfrac{\Omega}{4\pi}\right)$	0.6%	5%	40%
$\dfrac{\Delta E_e}{E_e}$(FWHM)	~15%	7%	4%
μ^- stopping rate	~10^5	3×10^5	10^6
Sensitivity (90% CL)	1.6×10^{-8}	7×10^{-11}	2×10^{-11} (achieved)
			few $\times 10^{-12}$ (planned)

experiment has specifically searched for this process. Recently the SIN experiment was reanalyzed and a limit[38] $R_{\mu e}^{incoh}(^{32}S) \lesssim 8 \times 10^{-9}$ (90% CL) was obtained.

C. $\mu \rightarrow e\gamma\gamma$

The $\mu \rightarrow e\gamma\gamma$ decay can occur as bremsstrahlung from external muon and electron lines for $\mu \rightarrow e\gamma$: this would lead to an additional suppression by a factor of $\sim(\alpha/\pi)$. However, there are gauge models in which the $\mu \rightarrow e\gamma\gamma$ rate can exceed the $\mu \rightarrow e\gamma$ rate. This is the case, for example, in some theories in which the mediating heavy leptons are charged.[42] One expects a differential decay distribution given by

$$\frac{d^2\Gamma}{dE_1 dE_2} = \frac{G(a,b)}{16\pi^3 m^6} E_e E_1^2 E_2^2 (1 - \cos\theta)^2 \quad , \tag{15}$$

where $E_{1,2}$ are the photon energies, E_e is the electron energy, θ is the angle between the photons, and G describes the couplings.

In the $\mu \rightarrow e\gamma$ experiments with two large NaI detectors[28,29] $\mu \rightarrow e\gamma\gamma$ could appear because an extra photon would simply increase the measured energy without being distinguished. Thus, upper limits of $B_{\mu e\gamma\gamma} < 5 \times 10^{-8}$ (90% CL) (Ref. 42) and $B_{\mu e\gamma\gamma} < 8.4 \times 10^{-9}$ (90% CL) (Ref. 43) have been deduced.

The Crystal Box experiment[32] expects to be sensitive to $\mu \rightarrow e\gamma\gamma$ at a level of a few parts in 10^{11}. This large solid-angle detector has a much larger acceptance for $\mu \rightarrow e\gamma\gamma$: the back-to-back configuration of the existing measurements is not favored by the decay distribution of Eq. (15).

D. Some Other Rare Muon Processes

There are several other rare processes which have been studied as tests of either the Konopinski-Mahmoud or the multiplicative lepton number schemes. These processes are

$$\mu^- Z \rightarrow e^+ (Z-2) \quad , \tag{16}$$

$$\mu^+ \rightarrow e^+ \bar{\nu}_e \nu_\mu \quad , \tag{17}$$

$$(\mu^+ e^-) \rightarrow (\mu^- e^+) \text{ and } e^- e^- \rightarrow \mu^- \mu^- \quad . \tag{18}$$

Present limits are (all 90% confidence levels)

$$\frac{\Gamma(\mu^- Z \rightarrow e^+[Z-2])}{\Gamma(\mu^- Z \rightarrow \nu_\mu[Z-1])} \ (Z = {}^{32}S) < 9 \times 10^{-10} \ \text{(Ref. 39)} \ , \tag{19}$$

and

$$\frac{\Gamma(\mu^+ \rightarrow e^+ \bar{\nu}_e \nu_\mu)}{\Gamma(\mu^+ \rightarrow e^+ \nu_e \bar{\nu}_\mu)} < 0.09 \ \text{(Ref. 44)} \ . \tag{20}$$

Limits on the processes in Eq. (18) may be found in Ref. 45.

One might think that the process $\pi^0 \rightarrow \mu e$ would also be a good place to look for muon-number nonconservation. Two recent papers[38,46] have reanalyzed old experiments and found

$$B_{\pi^0 \mu e} \equiv \frac{\Gamma(\pi^0 \rightarrow \mu^+ e^-) + \Gamma(\pi^0 \rightarrow \mu^- e^+)}{\Gamma(\pi^0 \rightarrow \text{all})} < 7 \times 10^{-8} \ \text{(90\% CL)} \ . \tag{21}$$

While this is a small number, it does not impose a meaningful con-straint. The $\pi^0 \rightarrow \mu e$ decay is mediated by pseudoscalar and axial vector quark densities, as is incoherent μ-e conversion. The con-straints from the incoherent conversion imply an upper limit of $\sim 10^{-15}$ for $\pi^0 \rightarrow \mu e$.[38] The reason $B_{\pi^0 \mu e}$ is so small is that the denominator in Eq. (21) is an electromagnetic rate, not a weak interaction rate.

Finally, I should mention the reaction $e^+ e^- \rightarrow \mu e$.

Because one would expect an effective four-fermion interac-tion, the cross section should be proportional to s. Assuming that the muon-number violation is characterized by a coupling constant, G_x, we find[47]

$$\frac{\sigma_{ee \rightarrow \mu e}}{B_{\mu 3e}} \simeq \frac{G_x^2 s}{G_x^2 / G_F^2} = s G_F^2 \ . \tag{22}$$

The present limit on $B_{\mu 3e}$ implies

$$\sigma_{ee} \lesssim sG_F^2[10^{-9}] = 4 \times 10^{-43} \text{ cm}^2 \text{ for } s = 10^4 \text{ GeV}^2 \quad , \qquad (23)$$

which implies an event rate of 4×10^{-11}/s at a luminosity of $10^{32}/\text{cm}^2\text{-s}$.

E.   Strangeness-Changing Muon-Number-Nonconserving Decays

For some theories of muon-number nonconservation, including horizontal gauge theories[48] (in which horizontal gauge interactions connect the different generations) strangeness-changing processes are the most sensitive. Present limits for some of these processes are

$$B_{K\mu e} \equiv \frac{\Gamma(K_L \to \mu e)}{\Gamma(K_L \to all)} < 2 \times 10^{-9} \text{ (90\% CL) (Ref. 49)} \qquad (24)$$

and

$$B_{K\pi\mu e} \equiv \frac{\Gamma(K^+ \to \pi^+\mu e)}{\Gamma(K^+ \to all)} < 5 \times 10^{-9} \text{ (90\% CL) (Ref. 50)} \quad . \qquad (25)$$

Experiments have been approved at the AGS[51] to improve these limits by several orders of magnitude. This is an area of great interest and activity. However, it lies somewhat outside the scope of this paper. The interested reader should consult Ref. 48 and references therein for more details.

F.   Tau Decays

A search for lepton-number-nonconserving τ decays[52] has set upper limits at $\simeq 5 \times 10^{-4}$ for processes such as $\tau \to e\gamma$, $\tau \to \mu\gamma$, $\tau \to e\mu\mu$, etc. The sensitivity of these measurements is determined by the number of available τ's. There is no obvious way to increase this sample at present. Thus, although these decays deserve as much attention as the rare muon processes, it appears that there is no way to achieve sensitivities that are as definitive as those that have been attained with muons.

V. CONCLUSIONS

The minimal standard model of the electroweak interactions has been enormously successful in accounting for a wide variety of phenomena. Nevertheless, theoretical prejudice indicates that this is not a complete theory. In the presence of any of the present possibilities beyond the minimal standard model, some of the important features of this standard model will be altered. One such feature is the conservation of lepton-family number. The cutting edge of elementary particle physics is the measurement of the parameters of the standard model and searches for extensions of the standard model. These studies can be pursued at the high-energy frontier and at the high-precision frontier. In the case of lepton-family-number conservation, the high-precision approach is the appropriate one.

Historically, the study of neutrinoless μ-e transitions has played a major role in our understanding of the fundamental interactions. The more sensitive experiments which are running or are planned will extend this understanding.

REFERENCES

1. C. D. Anderson and S. H. Neddermeyer, Phys. Rev. $\underline{51}$, 884 (1937); and C. Street and E. Stevenson, Phys. Rev. $\underline{51}$, 1005 (1937).

2. B. L. Ioffe, Sov. Phys. JETP $\underline{11}$, 1158 (1960); and H. Primakoff and S. P. Rosen, Phys. Rev. D $\underline{5}$, 1784 (1972).

3. S. Lokanathan and J. Steinberger, Phys. Rev. $\underline{98}$, 240(A) (1955).

4. J. Schwinger, Ann. Phys. $\underline{2}$, 407 (157).

5. G. Feinberg, Phys. Rev. $\underline{110}$, 1482 (1958); P. L. Meyer and G. Salzman, Nuovo Cimento $\underline{14}$, 4214 (1959); and M. E. Ebel and F. J. Ernst, Nuovo Cimento $\underline{15}$, 173 (1960).

6. E. J. Konopinski and H. M. Mahmoud, Phys. Rev. $\underline{92}$, 1045 (1953).

7. J. Schwinger, Ref. 4; K. Nishijima, Phys. Rev. $\underline{108}$, 907 (1957); and S. Bludman, Nuovo Cimento $\underline{9}$, 433 (1958).

8. G. Feinberg and S. Weinberg, Phys. Rev. Lett. $\underline{6}$, 381 (1961).

9. B. Pontecorvo, JETP $\underline{10}$, 1236 (1960); and M. Schwartz, Phys. Rev. Lett. $\underline{4}$, 306 (1960).

10. G. Danby, J.-M. Gaillard, K. Goulianos, L. M. Lederman, N. Mistry, M. Schwartz, and J. Steinberger, Phys. Rev. Lett. 9, 36 (1967).

11. S. Frankel, W. Frati, I. Halpern, L. Holloway, W. Wales, and O. Chamberlain, Nuovo Cimento 27, 894 (1963).

12. W. Pauli, 1933 (unpublished). Used by E. Fermi, Z. Physics 88, 161 (1934).

13. A. de Rujula, H. Georgi, and S. L. Glashow, Phys. Rev. D 12, 147 (1975).

14. T. D. Lee and C. N. Yang, Phys. Rev. 98, 101 (1955).

15. S. L. Glashow, Nucl. Phys. 22, 579 (1961); A. Salam, in Elementary Particle Theory: Relativistic Groups and Analycity, Nobel Symposium No. 8, N. Svartholm, Ed. (Almquist and Wiksell, Stockholm, 1968), p. 367; and S. Weinberg, Phys. Rev. Lett. 19, 1264 (1967).

16. T.-P. Cheng and L.-F. Li, Phys. Rev. D 16, 1565 (1977); B. W. Lee, S. Pakvasa, R. E. Shrock, and H. Sugawara, Phys. Rev. Lett. 38, 937 (1977); B. W. Lee and R. E. Shrock, Phys. Rev. D 16, 1444 (1977); G. Altarelli, L. Baulieu, N. Cabibbo, L. Maiani, and R. Petronzio, Nucl. Phys. B 125, 285 (1977); and W. J. Marciano and A. I. Sanda, Phys. Rev. Lett. 38, 1512 (1977).

17. J. D. Bjorken and S. Weinberg, Phys. Rev. lett. 38, 622 (1977); and G. C. Branco, Phys. Lett. 68B, 455 (1977).

18. Muon-number violation in models with horizontal gauge symmetries has been studied by T. Maehara and T. Yanagida, Lett. Nuovo Cimento 19, 424 (1977), Prog. Theor. Phys. 60, 822 (1978), and Prog. Theor. Phys. 61, 1434; M. A. B. Beg and A. Sirlin, Phys. Rev. Lett. 38, 1113 (1977); R. Cahn and H. Harari, Nucl. Phys. B 176, 135 (1980); I. Montvay, Z. Phys. C 7, 45 (1980); O. Shanker, Phys. Rev. D 23, 1555 (1981), Nucl. Phys. B 185, 382 (1981); P. Herczeg, in Proceedings of the Workshop on Nuclear and Particle Physics at Energies Up to 31 GeV, Los Alamos, New Mexico, 1981, J. D. Bowman, L. S. Kisslinger, and R. R. Silbar, Eds. (Los Alamos National Laboratory document LA-8755-C, 1981), p. 58; and D. R. T. Jones, G. L. Kane, and J. P. Leveille, Nucl. Phys. B 198, 45 (1982). See also O. Shanker, TRIUMF preprint TRI-PP-81-10 (1981). References to work dealing with other aspects of horizontal gauge symmetries can be found in the above papers.

19. Implications of extended technicolor theories on rare proc-
    esses and some associated problems of these schemes are dis-
    cussed in J. Ellis, M. K. Gaillard, D. V. Nanopoulos, and
    P. Sikivie, Nucl. Phys. B 182, 529 (1981); S. Dimopoulos and
    J. Ellis, Nucl. Phys. B 182, 505 (1981); J. Ellis, D. V. Nano-
    poulos, and P. Sikivie, Phys. Lett. 101B, 387 (1981); S. Dimo-
    poulos, S. Raby, and G. L. Kane, Nucl. Phys. B 182, 77 (1981);
    J. Ellis and P. Sikivie, Phys. Lett. 104B, 141 (1981); and
    A. Masiero, E. Papantonopoulos, and T. Yanagida, Phys. Lett.
    115B, 229 (1982).

20. Y. Tomozawa, Phys. Rev. D 25, 1448 (1982); and E. J. Eichten,
    K. D. Lane, and M. E. Peskin, Phys. Rev. Lett. 50, 811 (1983).

21. Such is a subclass of models due to Pati and Salam based on
    $[SU(2n)]^4$ (n $\geqslant$ 3) [cf. V. Elias and S. Rajpoot, Phys. Rev. D
    20, 2445 (1979)]; a recent view of the Pati-Salam models is
    given in J. C. Pati's invited talk at the Int. Conf. on Baryon
    Nonconservation, Tata Institute of Fundamental Research, Bom-
    bay, India, 1982 (University of Maryland report 82-151, 1982).
    The unification scales to two loops have been calculated in
    these models by T. Goldman in Particles and Fields - 1981:
    Testing the Standard Model, proceedings of the meeting of the
    Division of Particles and Fields of the APS, Santa Cruz,
    California, C. A. Heusch and W. T. Kirk, Eds. (AIP, New York,
    1982). He finds that for n = 4,5, some muon-number-violating
    K decays may have measurable rates.

22. J. Ellis and D. V. Nanopoulos, Phys. Lett. 110B, 44 (1982).

23. R. N. Mohapatra and G. Senjanović, Phys. Rev. D 23, 165
    (1981); and Riazuddin, R. E. Marshak, and R. N. Mohapatra,
    Phys. Rev. D 24, 1310 (1981).

24. See, for example, S. M. Bilenky and B. Pontecorvo, Phys. Rev.
    41, 225 (1978).

25. E. P. Hincks and B. Pontecorvo, Phys. Rev. 73, 257 (1948); and
    R. D. Sard and E. J. Althaus, Phys. Rev. 74, 1364 (1948).

26. H. F. Davis, A. Roberts, and T. F. Zipf, Phys. Rev. Lett. 2,
    211 (1959); D. Berley, J. Lee, and M. Bardon, Phys. Rev. Lett.
    2, 357 (1959); T. O'Keefe, M. Rigby, and J. Wormaid, Proc.
    Phys. Soc. (London) 73, 951, (1959); V. Krestnikov, IX Annual
    Int. Conf. on High Energy Physics, Kiev (1959), unpublished;
    J. Askin et al., Nuovo Cimento 14, 1266 (1959); S. Frankel,
    V. Hagopian, J. Halpern, and A. L. Whetstone, Phys. Rev. 118,
    589 (1960); R. R. Crittenden, W. D. Walker, and J. Ballam,
    Phys. Rev. 121, 1823 (1961); S. Frankel et al., Nuovo Cimento
    27, 894 (1963); S. Parker, H. L. Anderson, and C. Rey, Phys.

Rev. 133, B768 (1964); and S. M. Korenchenko et al., Yad. Fiz, 13, 341 (1971).

27. S. Frankel, in Muon Physics, Vol. II, Academic Press, New York (1973).

28. H. P. Povel et al., Phys. Lett. 72B, 183 (1971); and A. Schaaf et al., Nucl. Phys. A 340, 249 (1980).

29. P. Depommier et al., Phys. Rev. Lett. 39, 1113 (1977).

30. J. D. Bowman et al., Phys. Rev. Lett. 42, 556 (1979); and W. W. Kinnison et al., Phys. Rev. D 25, 2846 (1982).

31. S. M. Korenchenko et al., Ref. 26.

32. LAMPF Experiments 400/445, C. M. Hoffman, J. D. Bowman, and H. S. Matis, spokesmen. The collaborators are R. Bolton, J. D. Bowman, M. Cooper, J. Frank, D. Grosnick, A. Hallin, P. Heusi, V. Highland, C. Hoffman, G. Hogan, E. B. Hughes, F. Mariam, H. Matis, R. Mischke, D. Nagle, V. Sandberg, G. Sanders, V. Sennhauser, R. Werbeck, R. Williams, S. Wilson, and S. C. Wright.

33. See, for example, R. E. Mischke, "Future LAMPF Experiments on Lepton-Number Nonconservation," Procedures of Neutrino '81, Maui (1981).

34. S. M. Korenchenko et al., JETP 43, 1 (1976).

35. D. Yu Bardin, Ts. G. Istatkov, and G. B. Mitsel'Makher, Sov. J. Nucl. Phys. 15, 161 (1972); and P. Vogel, SINDRUM Note 5, SIN (1981) (unpublished).

36. W. Bertl et al., SIN preprint PR-84-01 (1984).

37. O. Shanker, Phys. Rev. D 20, 1608 (1979).

38. P. Herczeg and C. M. Hoffman, LAUR-83-3573 (1983) and Phys. Rev. D (to be published).

39. A. Badertscher et al., Phys. Rev. Lett. 39, 1385 (1977), Lett. Nuovo Cimento 28, 401 (1980), and Nucl. Phys. A 377, 406 (1982).

40. M. Blecher et al., 1983 Annual Meeting of Division of Particles and Fields, Blacksburg, Virginia.

41. D. A. Bryman et al., Phys. Rev. Lett. 208, 1409 (1972).

42. J. D. Bowman, T.-P. Chang, L.-F. Li, and H. S. Matis, Phys. Rev. Lett. $\underline{41}$, 442 (1978).

43. G. Azuelos et al., Phys. Rev. Lett. $\underline{51}$, 164 (1983).

44. S. Willis et al., Phys. Rev. Lett. $\underline{44}$, 522 (1980).

45. W. C. Barber, B. Gittelman, D. C. Cheng, and G. K. O'Neill, Phys. Rev. Lett. $\underline{22}$, 902 (1969); and J. J. Amato, P. Crane, V. W. Hughes, J. E. Rothberg, and P. A. Thompson, Phys. Rev. Lett. $\underline{21}$, 1709 (1968).

46. D. Bryman, Phys. Rev. D $\underline{26}$, 2538 (1983).

47. C. M. Hoffman, "Prospects in Lepton-Flavor Violation," Proc. of the Elementary Particle Physics and Future Facilities Summer Study, Snowmass, Colorado (1982).

48. See, for example, P. Herczeg, "Symmetry-Violating Kaon Decays," Proc. of the Kaon Factory Workshop, Vancouver (1979); and R. E. Shrock, "Rare K Decays as Probes of New Physics," Proc. of the LAMPF II Workshop, Los Alamos (1983).

49. A. R. Clark et al., Phys. Rev. Lett. $\underline{26}$, 1667 (1971).

50. A. M. Diamant-Berger et al., Phys. Lett. $\underline{62B}$, 485 (1976).

51. AGS Proposal 777, M. E. Zeller, spokesman ($K^+ \to \pi^+ \mu e$), and AGS Proposal 780, M. P. Schmidt, spokesman ($K^0_L \to \mu e$).

52. K. G. Hayes et al., Phys. Rev. D $\underline{25}$, 2869 (1982).

NEUTRINO MASSES AND MIXING FROM NEUTRINO OSCILLATIONS

Milla Baldo-Ceolin

Dipartimento di Fisica "G. Galilei", Università di Padova
Padova - Italy

Istituto Nazionale di Fisica Nucleare, Sezione di Padova
Padova - Italy

INTRODUCTION

The neutrino mass measurement is one of the most challenging problems for experimental physics: the knowledge of its value might be of fundamental importance in many open questions in the field of elementary particles as well as in cosmology and astrophysics.

In the standard electroweak theory neutrino are massless, however there is no fundamental reason requiring this. Grand unified theories admit neutrino masses in a broad range, $10^{-6} \mathrm{eV} \lesssim m_\nu \lesssim 10 \mathrm{eV}$, unfortunately they do not provide any predictions about the values.

For neutrinos with non-zero masses one would expect a mixing to exist between different flavour weak eigenstates, so that all the mass eigenstates are admixed in the flavour eigenstates.

There are several experimental effects related to a non vanishing neutrino mass and to flavour or lepton number non conservation, so that one can measure neutrino masses and mixing using different methods, as was presented in the Lusignoli's lectures. Most methods are based on kynematical effects as, for example, the direct determination of the electron momenta in two or three body beta decays or the search for secondary peaks in two-body decays and kinks in three-body decay spectra.

In this lecture I will discuss mainly the possibility of detecting and measuring neutrino masses by means of neutrino oscillations and review the experimental status and perspectives.

Up to now the only experimental evidence for neutrino mass remains the shape of the electron spectrum near the end point of tritium beta-decay, observed by the ITEP group[1], that gives $m_{\bar{\nu}_e} > 20eV$. No other experiments contradict or provide direct or indirect confirmation to this result.

Negative results from double beta decay experiments[2] may be interpreted as the indication that $m_{\bar{\nu}_e} <$ few eV[3]. This lower value for the neutrino mass may, in turn, be explained owing to the fact that the contributions of different mass-eigenstates to neutrinoless double beta decay amplitude could cancel one another[4].

Astrophysics suggests that there can be 3 or 4 distinct neutrino species and that the sum of their mass is less than 100eV[5].

As far as neutrino oscillations are concerned, in addition to the solar neutrino puzzle which may be explained by neutrino oscillations, some hints that oscillation effects may be nearby have been recently circulated[6]. In the following I will discuss the present experimental situation and the perspectives for neutrino oscillation detection.

PHENOMENOLOGY OF NEUTRINO OSCILLATIONS

Neutrino oscillations may arise if neutrinos have non vanishing masses and flavour leptonic number is not conserved[7]. In this hypothesis neutrinos, produced in weak interaction with a given flavour, may interact, after a time t, with a different flavour.

According to the experimental evidence for three types of charged leptons three neutrino flavour states ν_μ, ν_e and ν_τ are assumed in the following.

In the presence of neutrino mixing the weak interaction eigenstates $|\nu_\alpha>$, ($\alpha = e,\mu,\tau$), are a linear combination of the mass eigenstates $|\nu_i>$, (i = 1,2,3), such that

$$|\nu_\alpha> = \Sigma_i U_{\alpha i} |\nu_i>$$

where the mixing matrix U is unitary.

If the ν_i masses are different, an initial ν_α state becomes, after a time t, a superposition of ν_e, ν_μ, ν_τ states.

The most general expression for the probability for a ν_α to oscillate in a ν_β after a propagation time t is given by

$$P(\nu_\alpha \rightarrow \nu_\beta) = \Sigma_i \left(|U_{\alpha i} U_{i\beta}^+|^2 + 2\mathrm{Re} \Sigma_{j>i} U_{\alpha i} U_{i\beta}^+ U_{\beta j} U_{j\alpha}^+ \exp\left(i \frac{m_j^2 - m_i^2}{2} \cdot \frac{L}{E} \right) \right) \quad (1)$$

where:

a) $(m_j^2 - m_i^2) L/E \simeq (E_j - E_i)t$.

b) The first term $\Sigma_i |U_{\alpha i} U_{i\beta}^+|^2$ on the r.h.s. represents the average transition probability.

c) The second term on the r.h.s. represents the time dependent neutrino oscillations where the amplitude $A = \Sigma_{ji} U_{\alpha i} U_{i\beta}^+ U_{\beta j} U_{j\alpha}^+$ and the frequency $\omega = \left(\Delta m^2 \cdot \frac{L}{E} \cdot 1.27 \right)$.

Eq. (1) illustrates that

159

i) ν oscillations come from quantum-mechanical interference;

ii) ν oscillations are periodical in L/E and the frequency is proportional to $\Delta m^2_{ij} = m^2_i - m^2_j$;

iii) oscillation amplitudes depend on the mixing matrix;

iv) by measuring average (in time or energy) transition probabilities only the mixing matrix parameters are determined.

Moreover, eq. (1) shows how the mixing of three neutrino flavours is described by six parameters: the neutrino mass-squared differences, related to the oscillation frequencies, and three mixing angles, related to the oscillation amplitudes, plus a CP violating phase.

The experimental determination of these parameters is not trivial, and in general for the description of single experiments simplifying hypotheses have to be introduced, as for example CP conservation and the mixing of only two types of neutrino, for instance $\nu_\mu \rightleftarrows \nu_e$, so that

$$\nu_e = \cos\theta\ \nu_1 + \sin\theta\ \nu_2 \quad ; \quad \nu_\mu = -\sin\theta\ \nu_1 + \cos\theta\ \nu_2$$

with

$$m_{1,2} = \frac{1}{2}\left[m_{\nu_e} + m_{\nu_\mu}\right] \pm \sqrt{(m_{\nu_e} - m_{\nu_\mu})^2 + \delta\nu_\mu\nu_e}$$

and

$$\delta\nu_\mu\nu_e = \langle\nu_e|H'|\nu_\mu\rangle$$

That way

$$P(\nu_e \rightarrow \nu_\mu) = P(\nu_\mu \rightarrow \nu_e) = \sin^2(2\theta)\sin^2(1.27\ \Delta m^2 L/E) \tag{2}$$

and

$$P(\nu_e \rightarrow \nu_e) = P(\nu_\mu \rightarrow \nu_\mu) = 1 - P(\nu_\mu \rightarrow \nu_e)$$

where $\Delta m^2 = |m^2_1 - m^2_2|$ is expressed in eV^2, L is the neutrino propagation length in meters, E the neutrino energy in MeV, and

$$\sin^2(2\theta) = \frac{(\delta\nu_\mu\nu_e)^2}{(\delta\nu_\mu\nu_e)^2 + |m_{\nu_e} - m_{\nu_\mu}|^2} \ .$$

From eq. (2) it appears that even with this simplest hypothesis one has to determine, in an experimental measurement, two parameters: $\sin^2(2\theta)$ and Δm^2.

EXPERIMENTAL METHODS AND LIMITS IN NEUTRINO OSCILLATION DETECTION

There are essentially two methods for detecting and measuring neutrino oscillations:

a) the "appearance method" where one searches for flavour changes in an initially flavour defined ν-beam, for instance the presence ν_e in a ν_μ beam, $N(\nu_e)/N(\nu_\mu) \neq 0$;

b) the "disappearance method" where the neutrino beam intensity is measured at various distances from the source: deviations from the expected neutrino flux, $R = N(\nu_\mu)$ observed$/N(\nu_\mu)$ expected $\neq 1$, show the presence of neutrino oscillations.

In the simplest case of only two neutrino mixing, from eq. (2) one has

$$N(\nu_e, E, L, \Delta m^2, \theta) = N(\nu_\mu, E, 0) \ \sin^2(2\theta) \ \sin^2(1.27 \cdot \Delta m^2 \cdot L/E)$$

$$\tag{2'}$$

$$N(\nu_\mu, E, L, \Delta m^2, \theta) = N(\nu_\mu, E, 0) \left[1 - \sin^2(2\theta) \cdot \sin^2(1.27 \cdot \Delta m^2 \cdot L/E)\right]$$

From Fig. 1, where the transition probability $\nu_\alpha \to \nu_\beta$ is shown as a function of the oscillation frequency assuming an oscillation amplitude $A = \sin^2 2\theta = 0.2$, it appears that $\sin^2(2\theta)$ and Δm^2 may be separately determined, provided the experiments are able to measure the oscillatory behaviour of the ratios $\frac{N(\nu_e)}{N(\nu_\mu)}$ or $N(\nu_\mu)$ observed$/N(\nu_\mu)$ predicted as a function of $\frac{L}{E}$.

161

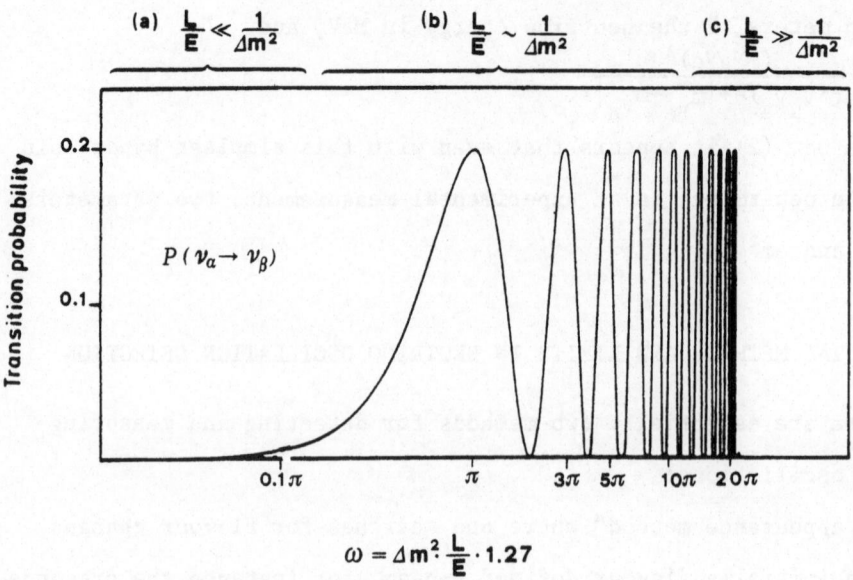

$$\omega = \Delta m^2 \, \frac{L}{E} \cdot 1.27$$

Fig. 1 - Transition probability $P(\nu_\alpha \to \nu_\beta)$, versus oscillation frequency.

The experimental sensitivity for detecting and measuring neutrino oscillations mainly depends on:

a) systematic uncertainties in beam impurities and background ($>10^{-3}$) in "appearance type experiments";

b) uncertainties in the predictions of event rates $\sim(5 \div 10)\%$ for "disappearance experiments";

c) the detector resolution and the neutrino source dimensions (the oscillation effects practically average when $\omega = \Delta m^2 \cdot 1.27 \cdot \frac{L}{E} > \pi$, namely for $\Delta m^2 > 2.5 \cdot \frac{E}{L}$ so that only values

$$\Delta m^2 < 2.5 \, \frac{E}{L} \left\{ \left(\frac{\Delta L}{L} \right)^2 + \left(\frac{\Delta E}{E} \right)^2 \right\}^{-\frac{1}{2}} \lesssim 2.5 \, \frac{E}{\Delta L}$$

are measurable).

Therefore in order to perform a sensitive neutrino-oscillation measurement:

i) the neutrino beam has to be as intense as possible, well known in composition, intensity and energy, furthermore the neutrino source dimensions should be negligible in comparison with the propagation length;

ii) the neutrino detector has to have a good event identification and to allow neutrino energy measurement with high resolution and background rejection.

EXPERIMENTAL RESULTS AND PERSPECTIVES

New results have been reached recently in experiments with reactor and accelerator neutrinos, moreover a few hundred atmospheric neutrions have been collected in the proton-decay large underground detectors.

The present situation then appears as follows:

a) there is still the problem of solar neutrinos where the number of electron neutrinos from the sun measured on the earth[8] is about one third of the number expected from the standard sun model[9];

b) there are new results from the Caltech-SIN-TUM Collaboration working at the Goesgen reactor. They found no indication for neutrino oscillations[10];

c) results from the group working at the Bugey reactor have been circulated[11]. In this experiment indication was found at 3 standard deviations level for neutrino disappearance;

d) several new experiments performed at accelerators and dedicated to the neutrino oscillations have been or are being completed. Their results do not show any indications for neutrino oscillations;

e) a few hundred atmospheric neutrino interaction events in the deep underground proton decay detectors are now available[12]: they do not show evidence for neutrino oscillations.

Two experiments searching for neutrino oscillations have been performed recently at Goesgen and Bugey reactors with quite similar detectors and experimental methods.

Nuclear reactors are pure source of isotropically emitted $\bar{\nu}_e$'s. Neutrinos originate from the beta decay of fission products, have energies up to $\sim$10 MeV, and fluxes

$$\phi(\bar{\nu}_e \ cm^{-2}sec^{-1}) \simeq 1.5 \cdot 10^{12} PL^{-2}$$

where P is the reactor thermal power in MW and L the source-detector distance in meters. The Goesgen and Bugey reactors have both high power ($\sim$2800 MW) and large cylindrically shaped core.

In both experiments neutrinos are detected through the reaction $\bar{\nu}_e + p \rightarrow e^+ + n$, the detectors consisting of a sandwich of scintillation target cells and ^{3}He proportional chambers. The targets act as positron calorimeter and neutron moderators. The neutrons, once thermalized, are captured in the ^{3}He chambers. The signature for a neutrino candidate event is a delayed coincidence between a positron and a neutron.

Systematic uncertainties on the ratio R = $N(\bar{\nu}_e)$observed/$N(\bar{\nu}_e)$expected derive from the knowledge of neutrino flux and spectrum, detection efficiency, reaction cross-section etc., and amount to $\sim$(5÷10)%. These incertitudes, may be reduced to the level of a few per cent comparing the rates of $\bar{\nu}_e$ interactions in two positions: deviations from L^2 dependence will indicate neutrino oscillations.

In the Geosgen experiment data were taken with the detector at 37.9 m and 45.9 m from the reactor core[9].

The detector at Bugey was placed at 13.6 and 18.3 meters from the neutrino source[6]. Control tests were done of the detector stability and of the effective reactor mean power.

The Goesgen and Bugey experimental results are shown in Fig. 2. The ratio of the integrated experimental yields in the two positions R_{12} for the Goesgen experiment

$$R_{12} = 1.01 \pm 0.03 \pm 0.02$$

is consistent with the absence of oscillations; for the Bugey experiment

$$R_{12} = 1.102 \pm 0.014 \pm 0.028$$

indicates a neutrino disappearance at 3 σ level.

These results together with those derived from the event rate in

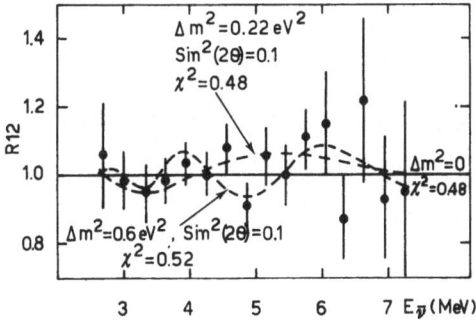

Fig. 2a – The ratio of $\bar{\nu}_e$ events measured at the two positions at Goesgen. Expected curves for different parameters are shown.

Fig. 2b – The ratio of $\bar{\nu}_e$ events measured at the two positions at Bugey. The full line shows the fuel burn-up correction.

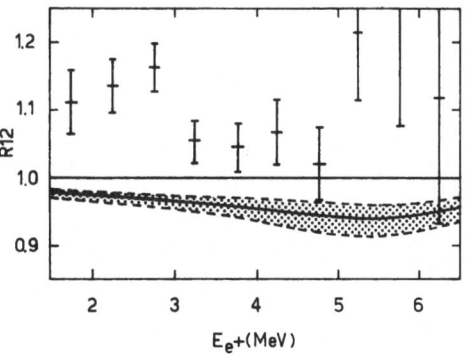

the two detectors as a function of $E_{\bar{\nu}_e}$, are displayed in the Δm^2-$\sin^2(2\theta)$ plot, assuming the 2-neutrino mixing hypothesis, Fig. 3. While the Goesgen result is represented by a continuous line which is the contour of the forbidden region, the Bugey experiment gives two possible oscillation regions, the most probable value $\Delta m^2 = 0.2$ and $\sin^2(2\theta)$ =0.25 is not excluded by the Goesgen experiment.

It appears very important to confirm these oscillation effects, and in order to do that systematic incertitudes have to be reduced mainly in the reactor neutrino spectrum and detection efficiency. Moreover, following the Bugey result, in order to maximize the oscillation effects, measurements should be done at distances from the reactor core such that $\frac{L}{E} \simeq 6$. For larger L smearing effects will hide the $\frac{L}{E}$ dependence.

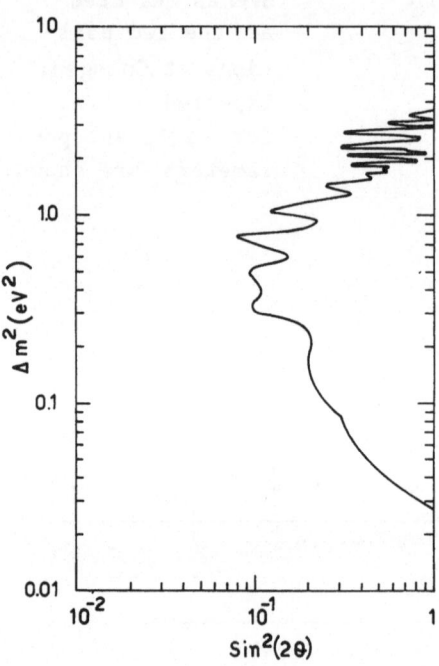

Fig. 3a - Limits on Δm^2 and $\sin^2(2\theta)$ at Goesgen.

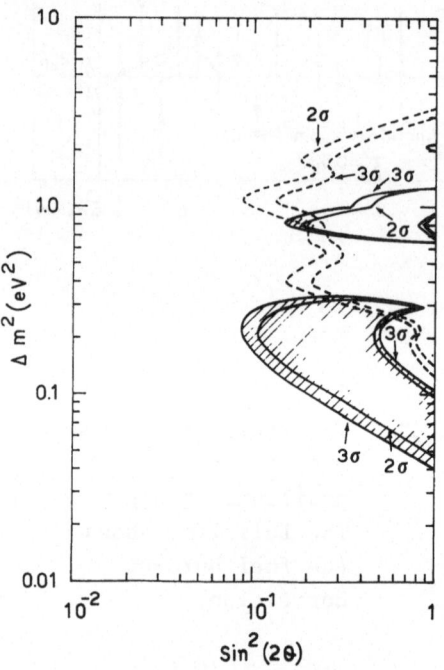

Fig. 3b - Limits on Δm^2 and $\sin^2(2\theta)$ at Bugey.

Several dedicated experiments have been performed in the last period at accelerators, most of them at CERN where a new low energy neutrino beam from the PS was installed pointing towards the existing BEBC, CDHS and CHARM detectors as shown in Fig. 4.

Neutrinos from accelerators have energies in the range $10^2 \lesssim E_\nu \lesssim 10^5$ MeV, the beams are quite well collimated and the ν fluxes are quite well known, so that both appearance and disappearance measurements as possible.

For experiments of the appearance type the lower limit on the sensitivity will be set by the number of events due to ν_e or, may be, ν_τ initially present in the ν_μ beam and by the number of events simulating electron or τ neutrino interactions; these events in dedicated experiments are expected to be less than 0.5 per cent of ν_μ events in wide band beams and less than 10^{-4} ν_μ in narrow band beams.

For experiments of the disappearance type, the sensitivity to neu-

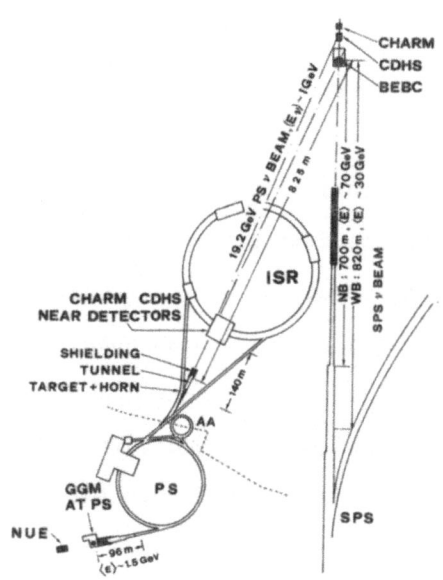

Fig. 4 - Layout of the CERN neutrino beam.

trino oscillations depends mostly on the uncertainty on the far to near events ratio. The corresponding systematic error has a number of contributions from: hadron production energy and transverse momentum spectra, beam geometry, detection efficiency, background, etc. The overall systematic uncertainty in the event ratio can be evaluated to be a few per cent.

Experiments of "disappearance type" have been performed with the new CERN PS low energy neutrino beam by the CDHS[13] and CHARM[14] collaborations, and at Fermilab by the CCFR[15] collaboration with a dichromatic neutrino beam.

In order not to be dependent on the absolute neutrino flux a second detector of the same type of the main one was placed near to the neutrino source. Data were taken simultaneously by the near and the far detector.

The ratio of event rates in the two detectors was evaluated for each experiment and a Monte Carlo simulation was used to correct for differences in the ν spectrum and in the detector geometry.

In addition to the integrate event rates, the ratio of the rates was considered as a function of neutrino energy or muon projected length in the detector.

The experimental results which are compatible with the absence of muon neutrino oscillations, may be expressed in terms of Δm^2 and $\sin^2 2\theta$ (Fig. 5) under the assumption of two neutrino oscillations.

The CHARM collaboration at CERN was able to do also an appearance type experiment selecting events characterized by an electromagnetic shower, compatible with a quasi-elastic ν_e induced charged current event and comparing the ratios "ν_e"/ν_μ in the near and in the far detectors as a function of energy.

Under the hypothesis of two neutrino mixing, the limit shown in

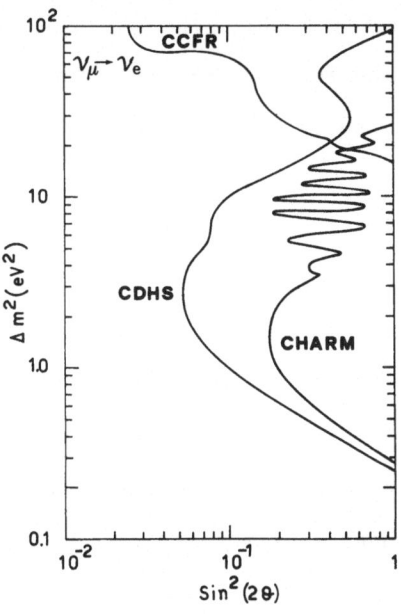

Fig. 5 – Limits on Δm^2 and $\sin^2(2\theta)$ for ν_μ disappearance experiments.

Fig. 6 have been obtained.

    Moreover, an experiment to study $\nu_\mu \rightarrow \nu_e$ transition is under way at CERN using BEBC as detector[16]. In this experiment a magnetic horn is used to focus the beam with an increase by almost one order of magni-

Fig. 6 – Limits on Δm^2 and $\sin^2(2\theta)$ for appearance $\nu_\mu \rightarrow \nu_e$ experiments.

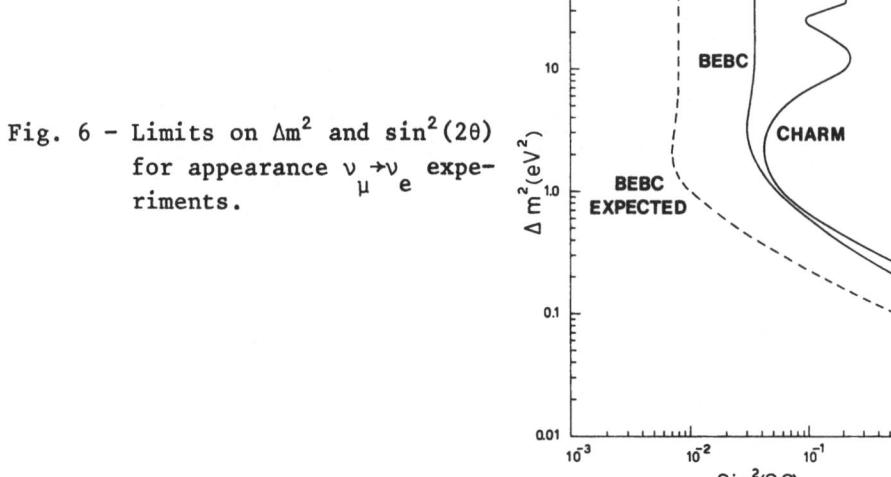

tude in the neutrino flux. The bubble chamber is filled with a mixture of 70% neon in hydrogen, the electron neutrino interactions are identified with an average efficiency of 90% and their energy measured with ~15% resolution.

A preliminary result obtained with $\sim\frac{1}{4}$ of the final statistics is shown in Fig. 6.

Details of the experiments are listed in Table I.

From the previous discussion it appears that in order to improve the experimental knowledge on neutrino oscillations larger $\frac{L}{E}$ values are needed and, consequently, an increase in the neutrino event rate by a significant factor (≥ 10). To this end high flux accelerators at medium energies appear more suitable than high energy machine since, because of the dependence on the $\frac{L}{E}$ factor, in the latter case the distance to make competitive contributions would be inconveniently large. Finally it is worth noting that for small $\sin^2(2\theta)$ values appearance experiments have better sensitivities.

Table 1

Accelerator Experiments	L (meters)	$<E>$ (GeV)	Limits on Δm^2 for maximum mixing (eV)^2	$\sin^2 2\theta$ minimum value	Statistics	Systematic errors
$\nu_\mu \to \nu_x$						
CDHS	130 880	3	$0.26 \geq \Delta m^2 \geq 90$	0.053	22.000 3.300 events	2.5%
CHARM	125 900	1.5	$0.3 \geq \Delta m^2 \geq 25$	0.19	2.043 270 events	5.0%
CCFR	715 1116	40–230	$30 \geq \Delta m^2 \geq 1000$	0.02	33.700 32.400 events	1.5%
$\nu_\mu \to \nu_e$						
CHARM	125 900	1.5	≤ 0.2	0.04	$R_1 = "\nu_e"/\nu_\mu = (9.2\pm1.1\pm1.1)\%$ $R_2 = "\nu_e"/\nu_\mu = (8.9\pm2.0\pm1.0)\%$	
BEBC	825	1.5	≤ 0.16	0.03	$N(\nu_e)$ 0 events $N(\nu_\mu)$ 130 events	$<5.10^{-3}$
BEBC*	825	1.5	$\sim 0.08*$	$\sim 0.007*$		

*extrapolated limits to the full statistics.

Most of the cosmic neutrinos are generated in the earth's atmosphere by primary cosmic rays, extraterrestrial sources are believed to be much weaker. Their flux, calculated from observed muons, peaks at low energy and contains twice as many muon neutrinos as electron neutrinos[17].

Solar neutrinos are produced in fusion processes in the solar core[18]. In both cases neutrino fluxes are low and neutrino detection difficult.

Solar neutrinos have been detected[8] in a 4×10^5 litre tank of C_2Cl_4, located in the Homestake gold mine, using the reaction $\nu_e + {}^{37}Cl \rightarrow {}^{37}A + e^-$ and measuring the subsequent ${}^{37}A$ decay. The average experimental result, expressed in ${}^{37}A$ atoms per day and the latest calculations for the expected event rate[9] show a large discrepancy

$$R = \frac{0.37 \pm 0.06}{1.44 \pm 0.62} = 0.26 \pm 0.12$$

If neutrino mixing and oscillations are considered as a possible explanation for this solar neutrino deficiency, the ratio

$$R = P(\nu_e \rightarrow \nu_e) \cong \sum_i |U_{ei}|^4 \cong 0.26 \pm 0.12$$

implies large neutrino mixing angles and $\Delta m^2 \gtrsim 10^{-11}$ eV^2.

Results have been reported on atmospheric neutrino interactions in the deep underground detectors constructed with the major aim of detecting proton decay events. In the Kolar Gold Field[19], NUSEX[20], IMB[21] and Kamiokande[22] experiments the observed rate of events attributable to neutrino interactions, their energy spectrum, the relative number of upwards and downwards neutrinos, as well as the relative numbers of ν_μ's and ν_e's are in accord with the predictions based on calculated fluxes[12].

However, large systematic uncertainties on the calculations and experimental difficulties in the neutrino direction reconstruction would allow the detection of neutrino oscillation effects only for very large mixing angles.

A very massive detector characterized by a high energy and space resolution would be needed for an oscillation sensitive search with solar and atmospheric neutrinos independent of flux normalization.

It will consist in detecting and measuring the highly directional solar neutrino-electron elastic scattering, where both neutral and charged current interaction amplitudes contribute: the shape of e^- recoil spectrum will depend on the composition in ν_e and others ν type of the solar neutrino flux. For cosmic neutrinos ν_e and ν_μ fluxes have to be measured as well as their up-down asymmetry which depends on the mixing parameters and on the number of neutrino flavours[23].

CONCLUSIONS

The presently known experimental results can be summarized as follows:

a) results from Bugey reactor indicate $\Delta m^2 \sim 0.2$ eV^2 and $\sin^2(2\theta) \sim 0.25$;

b) $\bar\nu_e$ mass, as measured by the ITEP group, turns out to be $m_{\bar\nu_e} > 20$ eV;

c) the solar neutrino deficiency may be explained as the result of neutrino oscillations with $\Delta m^2 \gtrsim 10^{-11}$ eV^2 and large mixing angles;

d) measurements on double beta decay may be interpreted as an indication that $m_{\nu_e} \lesssim$ few eV.

Experimental confirmation of these results appears very important. Experiments are already in progress for measuring the neutrino mass in many different ways and results will be compared with the ITEP ones in the near future.

Moreover, in a short time from now the BEBC experiment will be completed and its results compared with the Bugey results.

To progress further:

1) a detector very massive with very high resolution should be used to perform a measurement of the solar neutrino flux composition. In a medium-low Z detector $\sim$1000 neutrino-electron scattering with the electron energy larger than 5 MeV are produced per Kiloton per year: several thousend events are needed for a sensitive analysis;

2) systematic incertitudes in reactor experiments may be reduced for example going back to the ILL reactor with an improved detector to compensate for the smaller neutrino intensity;

3) high intensity, medium energy neutrino beams at accelerators and high resolution detectors should be developed. $\nu_\mu \rightarrow \nu_e$ transitions may in this way be explored with high accuracy, and $\nu_\mu \rightarrow \nu_\tau$ oscillations may be detected measuring with high accuracy the neutral to charged current event ratio at different positions.

REFERENCES

1) V.A. Lubimov, Proceed. International Conference on High Energy Physics, J. Guy and C. Costain eds., Brighton (1983), 386.

2) E. Fiorini, Fifth Workshop on Grand Unification, Providence RI (April 1984).

3) H. Primakoff, S.P. Rosen, Ann. Rev. Nucl. Part. Sci. 31:145 (1981).

4) L. Wolfenstein, CMU-HEP 84-8 and reference therein.

5) See for example: D. Fargion and R. Mignani, Roma-Preprint 370 (1983).

6) J.F. Cavaignac, A. Hoummada, D.H. Koang, B. Vignon, Y. Declais,

H. de Kerret, H. Pessard, J.M. Thenard, LAPP–EXP–84–03–INS–84–11.

7) S.M. Bilenky and B. Pontecorvo, Phys. Rep. 41:225 (1978).

8) R. Davis jr., B.T. Cleveland, J. Rowley, Proceed. Telemark Neutrino Mass Miniconference (1980), 38 – Univ. of Wisconsin Report 186.

9) J. Bahcall et al., Rev. Mod. Phys. 54:767 (1982).
B.W. Filippone, Phys. Rev. Lett. 50:412 (1983).

10) K. Gabathuler et al., Phys. Lett. 138B:449 (1984).

11) See ref. 6).

12) See D.H. Perkins, CERN/EP 84-7.

13) F. Dydak et al., Phys. Lett. 134B:281 (1984).

14) F. Bergsma et al., CERN-EP/84-36.

15) I.E. Stockdale et al., Phys. Rev. Lett. 52:1384 (1984).

16) PS(180) – Athens-Padova-Pisa-Wisconsin Collaboration, Results presented by M. Baldo-Ceolin at CERN-PSC Meeting, March 1984.

17) D.S. Ayres et al., Phys. Rev. D29: 902 (1984).

18) See ref. 9).

19) M.R. Krishnaswamy et al., Proceed. FWOGW, Philadelphia (April 1983).

20) G. Battistoni et al., Phys. Lett. 133B: 454 (1983).

21) R.M. Bionta et al., Phys. Rev. Lett. 51:27 (1983).

22) M. Koshiba, Proceed. ICOBAN '84, Park City, Utah (1984).

23) P.H. Frampton, S.L. Glashow, Phys. Rev. D25:1982 (1982).

SEARCHES FOR MIXED HEAVY NEUTRINOS

IN MESON DECAYS AND IN MUON CAPTURE

R. Prieels

Institut de Physique
Université Catholique de Louvain
2, Chemin du Cyclotron
B - 1348 Louvain-la-Neuve, Belgium

1. INTRODUCTION

We have seen in the preceeding lectures that the standard electroweak theory gives satisfaction in describing one generation of fermions.

Starting from massless fermions, Yukawa couplings to scalar Higgs fields and spontaneous symmetry breaking are able to give mass to the vector bosons and to the fermions.

The origin of several generation stays however unsolved, and their integration into the theory gives rise to a mass matrix and, after diagonalisation, to a matrix relation between fields of definite mass and fields with definite properties under the weak interaction. The quarks are inter-related through the Kobayashi-Maskawa mixing matrix, whilest similarly, in the lepton sector, the U matrix connects the left neutrino fields and the V matrix the right neutrino fields.

Of course, these mixings between neutrinos exists only if their masses are non degenerate. In the standard theory where no right handed neutrino is provided, a neutrino can acquire mass if he is of Majorana type. It is the most economical solution but imposes a lepton number violation of two units ($\Delta L = 2$). If right handed neutrinos are introduced into the theory, then Dirac type neutrinos with "Dirac masses" different from zero can exist.

Concerning their mass values a variety of theories like GUT's, SU(5), SO(10)... (the standard model may well not be the final theory) predict masses between the 10^{-6} eV up to 120 MeV.

Cosmological constrains tell us that if the universe is to be bound due to stable neutrinos, their mass for all generation together should not exceed 100 eV or should be greater than 2 GeV. Between these two bounds neutrinos should decay.

On the experimental side, many approaches are on their way to search for these masses and generation mixing. They investigate : beta decay end points, meson decays in flight or at rest, charged lepton decays, internal bremsstrahlung in electron capture, nuclear recoil in electron or muon capture, heavy neutrino decays, oscillation experiments.

I will discuss here partly only three of them which have meaning only in the assumption of both mixing and masses different from zero i.e. oscillation measurements, two lepton meson decay and nuclear recoil.

The assumption of mixed neutrinos mass eigenstates ν_i in the flavor-states ν_ℓ is the key condition of eventual neutrino oscillations[1]. The mixing enters through the unitarity U matrix as :

$$|\nu_\ell\rangle = U_{\ell i}|\nu_i\rangle$$

It was the merit of Shrock[2] to have called attention to a class of experiments which would allow the observation of eventual neutrino-mixing induced by non-diagonal matrix-elements as small as 10^{-6}. This method consists essentially in a precision spectroscopy of the charged particle emitted in conjunction with the neutrino of a given flavour-state. A mixture of different mass eigenstates would be revealed by peaks of "anomalous" momentum in the charged-particle spectrum ; their positions being related to the mass of the neutrino eigenstates and their intensities to the square of the non-diagonal matrix-element. In the same spirit nuclear recoil spectroscopy, in muon capture for example, may also reveal abnormal peaks induced by massive neutrino components in the emitted neutrino.

The region of neutrino-mass and mixing investigated in these decay or capture experiments is complementary to the region sampled by the oscillation experiments in their appearance or disappearance mode. This is shown in Fig. 1 where we clearly see the "micromixing" reached by the spectroscopic method to be compared to the "millimixing" sensitivity of the oscillation experiments. The limitation however of the former method is the poor minal mass (1 MeV) that can be observed due to the finite energy resolution of the measuring devices.

The sensitivity to the mixing in the meson decay is related to the suppression of the V-A constraint as the neutrino mass increases. The intensities of the charged lepton lines in the momentum spectrum are equal to

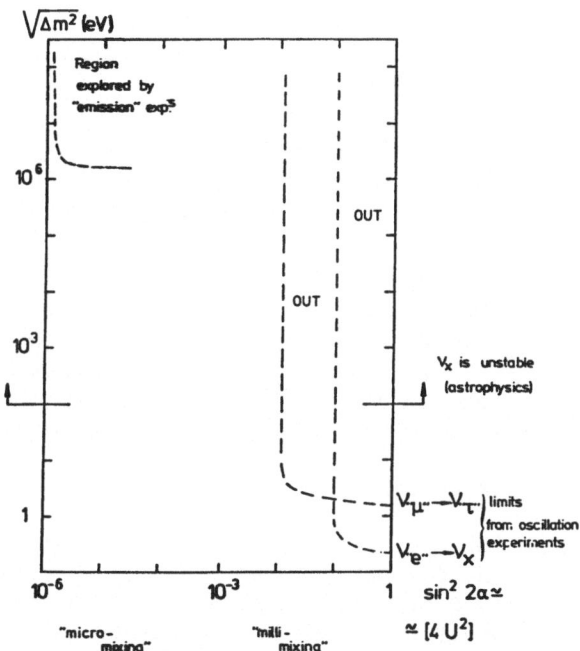

Fig. 1. Qualitative comparison of the sensitivity-region of the neutrino oscillation experiments and the "emission-type" ones.

$$|U_{\ell i}|^2 \rho m_{\nu_i}$$

where ρ contains phase space and V-A suppression effects and is shown in Fig. 2.

Table 1 gives examples of decays with the maximum neutrino mass that can be reached. Let us precise here that those decays give informations on different non diagonal elements of the U matrix.

Table 1

matrix elements	meson decay or reaction products	maximum ν-mass	ref.
$U_{\mu i}$	$\pi^+ \rightarrow \mu^+ + \nu_\mu$	30 MeV	3,4)
	$K^+ \rightarrow \mu^+ + \nu_\mu$	390 MeV	5)
	$K^- \rightarrow \mu^+ + \nu_\mu + \nu + \bar{\nu}$	450 MeV	6)
	$\mu^- + {}^3He \rightarrow {}^3H + \nu_\mu$	100 MeV	7,8)
U_{ei}	$K^+ \rightarrow e^+ + \nu_e$	490 MeV	9)
	$\pi^+ \rightarrow e^+ + \nu_e$	130 MeV	10,11)
	$e^- + A \rightarrow A' + \nu_e$	2-3 MeV	

2. MUON CAPTURE IN HELIUM 3

A research for massive neutrinos in muon-capture is developed by the LOUVAIN-SIN collaboration at SIN. It was noted[7] that pion and kaon decay are practically unable to observe with good sensibility possible heavy neutrinos in the rather extended mass range of 30-90 MeV and that a measurement of the triton recoil spectrum from the reaction $\mu^- + {}^3He \rightarrow {}^3H + \nu_\mu$ should allow to bridge this gap. In the same paper[7], an analysis of an old precursor muon-capture experiment[12] was used to set new limits for the muon-flavor mixings in the 30-90 MeV mass range. Fig. 3 shows this experimental result onto which bounds were calculated. The aim of the currently developed experiment is to lower these limits by an order of magnitude. This aim is shown with the already known results in Fig. 4.

Improvements of the pioneering experiment can be done in two ways : first by the use of a beam with better focusing which is expected to decrease by a factor of ten the noise in the spectrum region of interest and secondly in improving the resolution to ~ 1 % by building a new type of detector. The tool chosen for the determination of the kinetic energy of the triton is a gas scintillation proportional chamber (GSPC) filled with ${}^3He$[13-16].

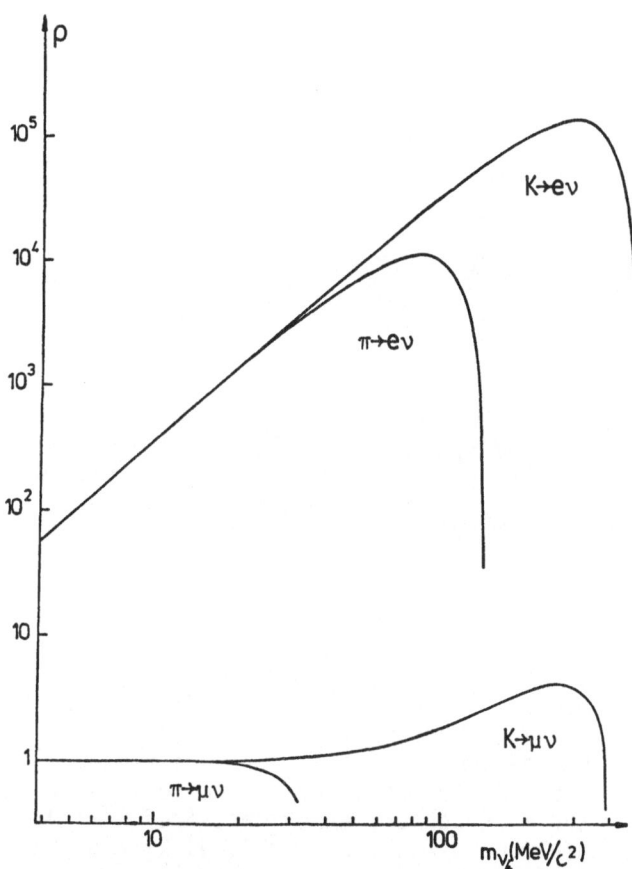

Fig. 2. Sensitivity enhancement factor for observing "anomalous" peaks, corresponding to components with neutrino masses different from zero in the decay of kaons and pions.

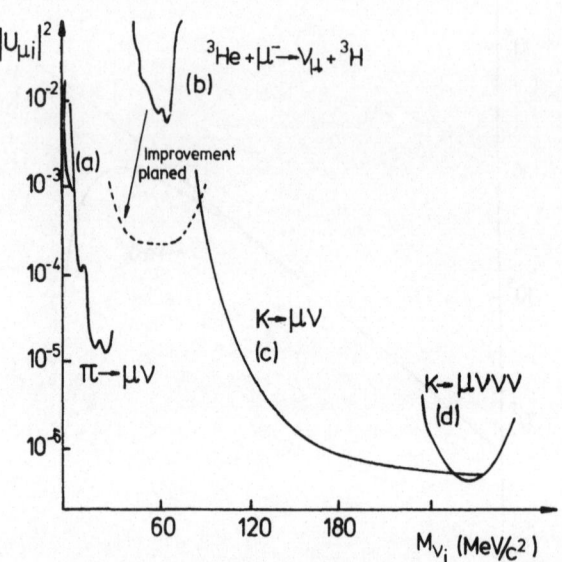

Fig. 3. Bounds of $|U_{\mu i}|^2$ vs neutrino mass. The goal limit of the running experiment is indicated by a dashed line. Other bounds shown in the figure are from : a) $\pi \to \mu\nu$ heavy neutrino searches (ref.[3] and ref.[4]), b) an analysis of the muon-capture experiment (ref.[7,12]), c) K $\to \mu\nu$ heavy neutrino search (ref.[5]), d) K$^+ \to \mu\nu\bar{\nu}\nu$ decay experiment (ref.[6]).

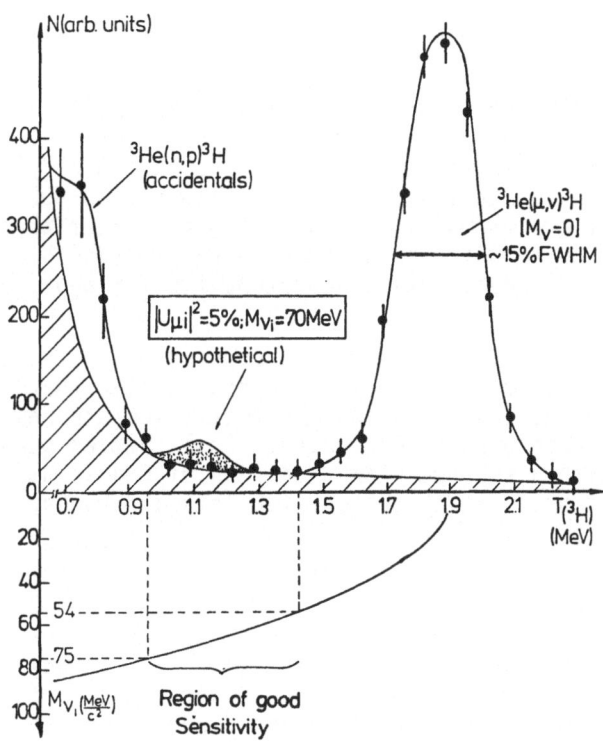

Fig. 4.   Triton-recoil-energy spectrum observed by the authors of
ref. [12] and relationship between this energy and the mass
of a corresponding neutrino. The peak drawn as a dotted
area illustrates the effect of a hypothetical heavy neutrino of $m_{\nu_i} = 70$ MeV/c^2 emitted with a mixing strength
of $|U_{\mu i}|^2 = 0.05$.

If a particle is stopped in a gas, it looses its energy either by ionizing or exciting the gas atoms. In collision processes, isomeric molecular excited states He_2^* are formed and emit their typical far-ultraviolet radiation. The photons produced by atoms excited during the stopping process are called the primary light. If the primary electrons produced by ionization are drifted into a region of high enough electrid field, new excited atoms will be formed. This means that an amount of radiation (secondary light) proportional the number of primary electrons and thus to the kinetic energy of the stopped particle will be produced.

Two problems were encountered in building this helium-GSPC detector.

The first one is, in contrast to a Xenon-GSPC type detector, that there is a mismatch of the emission spectrum (600-1000 Angstrom) with the sensitive region of the photomultiplier. We observed that a small addition of nitrogen (100-5000 ppm) can play the role of the normally used wavelength shifter through a transfer of excitation to nitrogen which have an emission band of (3000-3800 Angstrom) adapted with a photomultiplier equipped with a quartz window.

Secondly, with his quenching properties, nitrogen prevents the UV-light to reach the walls of the counter where these hard photons from He_2^* can cause photoeffect, liberate new electrons and initiate a long discharge process. Still better quenching can be obtained with a supplementary addition of CH4 in small amount. This allows us to reach better stability and higher rate.

A schematic drawing of this detector is shown in Fig. 5. Is also shown a time sequence indicating two sharp pulses correlated to the primary light emitted during the muon stop and the triton recoil. The time structure of the occurence of the second pulse refers to the muon lifetime in the gas. Those peaks are followed after a drift time period by two broader structures whose form depend strongly on the location and direction of the muon-stop and the triton-recoil. The second structure integrated contains the energy signature of the recoil. The "parasitic" first one can be handle as a supplementary information, or rejected if a reserved field is applied some time after the muon-stop, resulting in a controlled dead time period. The definitive method is not yet chosen.

As resolution is concerned, tests have been done both with nitrogen or with xenon admixture. The actual achieved resolution is of 4 % and improvements are still under way[17]. Noise and performance study need still to be done before the definition of the ultimate design.

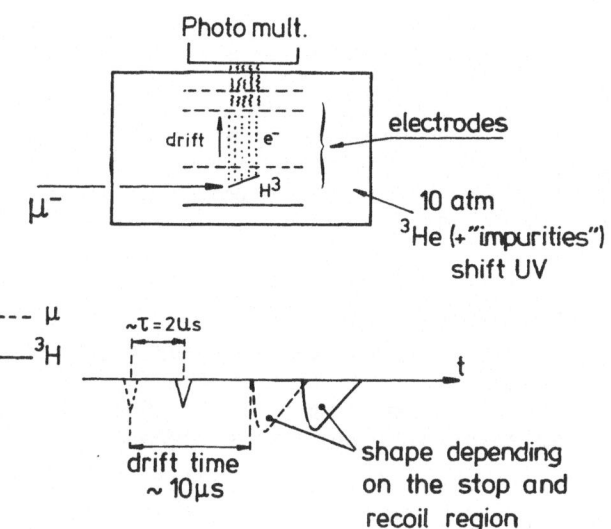

Fig. 5. Sketch of the gas detector developed to measure the
recoil energy of ^{3}H in the reaction $\mu + {}^3He \rightarrow {}^3H + \nu$.
Below : time structure of one event. The two first
pulses are correlated to the primary light emitted during
the muon-stop and the triton recoil. The two later ones
correspond to the light emitted by the excited atoms
"near" the photomultiplier after the electron drift
period.

3. PION MESON DECAY INTO ELECTRON AND NEUTRINO

Neutrino mass limits from this meson decay already exist from a by-product of a TRIUMF branching ratio measurement[18] of the $\pi^+ \rightarrow e^+\nu$ and $\pi^+ \rightarrow \mu^+\nu$ decays. A dedicated experiment is under way performed by a LOUVAIN-ZURICH-LAUSANNE collaboration at SIN. Fig. 6 shows the already published result and the goal limits of our experiment. Let us note here that the limit "b" from the branching ratio measurement is model dependent as it is calculated with the assumption of no pseudoscalar contribution. I will report here on the status of our experiment.

Let us remind here that the method consists in detecting an eventual abnormal peak in the momentum spectrum of the positron emitted in the $\pi^+ \rightarrow e^+\nu$ decay. This peak would occur at a momentum lower than 70 MeV/c and eventually even lower than 50 MeV/c. Two points has then to be considered in order to achieve the best result. First a good energy resolution : the better it is, the more sensitive we will be on the mixing parameter of an eventual heavy neutrino. Secondly, to detect with great sensitivity a positron peak below 50 MeV/c, it is crucial to reduce the 10000 times more abundant positron background coming from pion muon decay chain building up, as time goes on, the known Michel energy-spectrum.

The different life times of the pion and the muon already help if one restricts the observation time of the first 30 nanoseconds. This is shown in Fig. 7 taken from ref. [10]. But this is not enough and a supplementary rejection is needed.

If the pions are stopped in a scintillator we can perform an analysis of the resulting pulse. If one integrates the charge of the rising part of the pulse and also, separately, the falling part of the same signal, the ratio of both areas will be constant whatever the energy release is during the stopping process. If, however, the pion decays whilest in the scintillator and if the produced muon looses some energy inside the scintillator too, then the electronic pulse shape will be deformed and the previous ratio will change depending on the instant of the decay. This phenomenon is shown in Fig. 8 where the ridge at the left side of the figure corresponds to single particle events and the other one to double particle events the second being produced later than the rise time of the first pulse. If on the other hand the thickness of the scintillator is such that the energy release of the decay electron remains always small, this effect does not affect the pulse-distribution. Clearly, selecting "single events", a separation of about one nanosecond can be achieved[19].

To get a good energy resolution we choosed to use the magnetic pair-spectrometer of the LAUSANNE-ZURICH-MUNICH collaboration as a single particle momentum analyser. Fig. 9 shows a drawing of our setup where "T" is our scintillator target. The energy resolution obtained from a part of our data by a retracing method

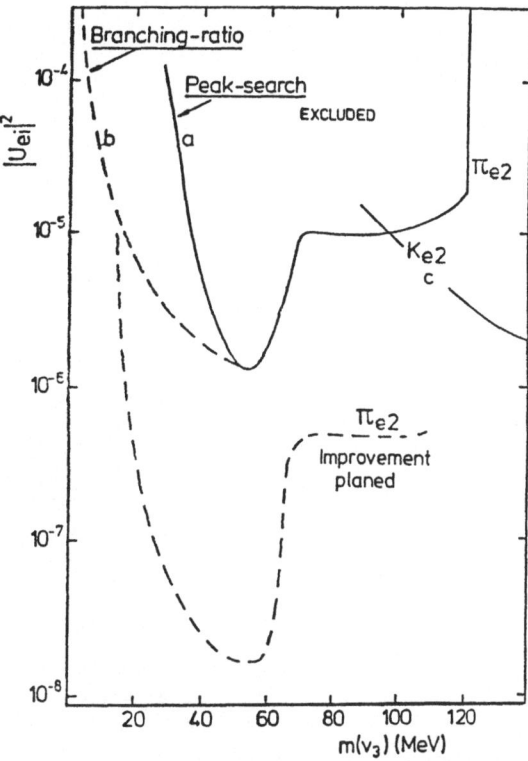

Fig. 6. Bounds on $|U_{ei}|^2$ vs neutrino mass. The goal limit of the described experiment is shown at the bottom of the figure. Are also shown, the bounds from : a) the neutrino peak search from $\pi \to e\nu$ decay (ref. [10]), b) the branching ratio measurement $\pi \to e\nu/\pi \to \mu\nu$ (ref. [18]), c) the Ke2 results (ref. [9]).

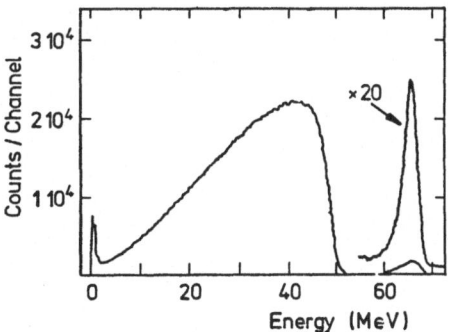

Fig. 7. Positron energy spectrum from $\pi \to e\nu$ decay obtained at Triumf in an observation window (3-28 ns) after stopping of the pion.

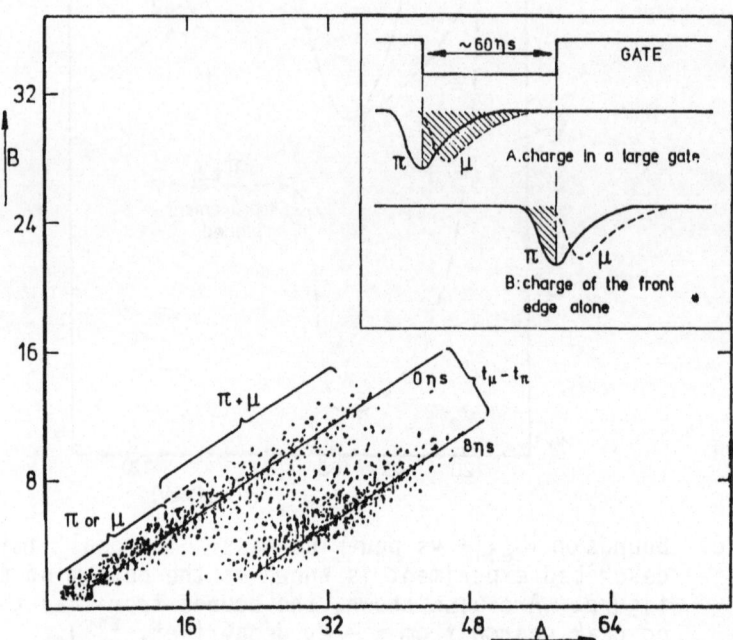

Fig. 8. Biparametric plot of the two samplings from the target
detector. The left edge corresponds to "single particle",
whereas the right one corresponds to double particles
events separated by more than the rise time of a single
pulse. A separation of two pulses within one nanosecond
can be achieved.
The insert shows the gating scheme for a typical π-μ sequence. The shaded areas indicates the part of the
signal integrated.

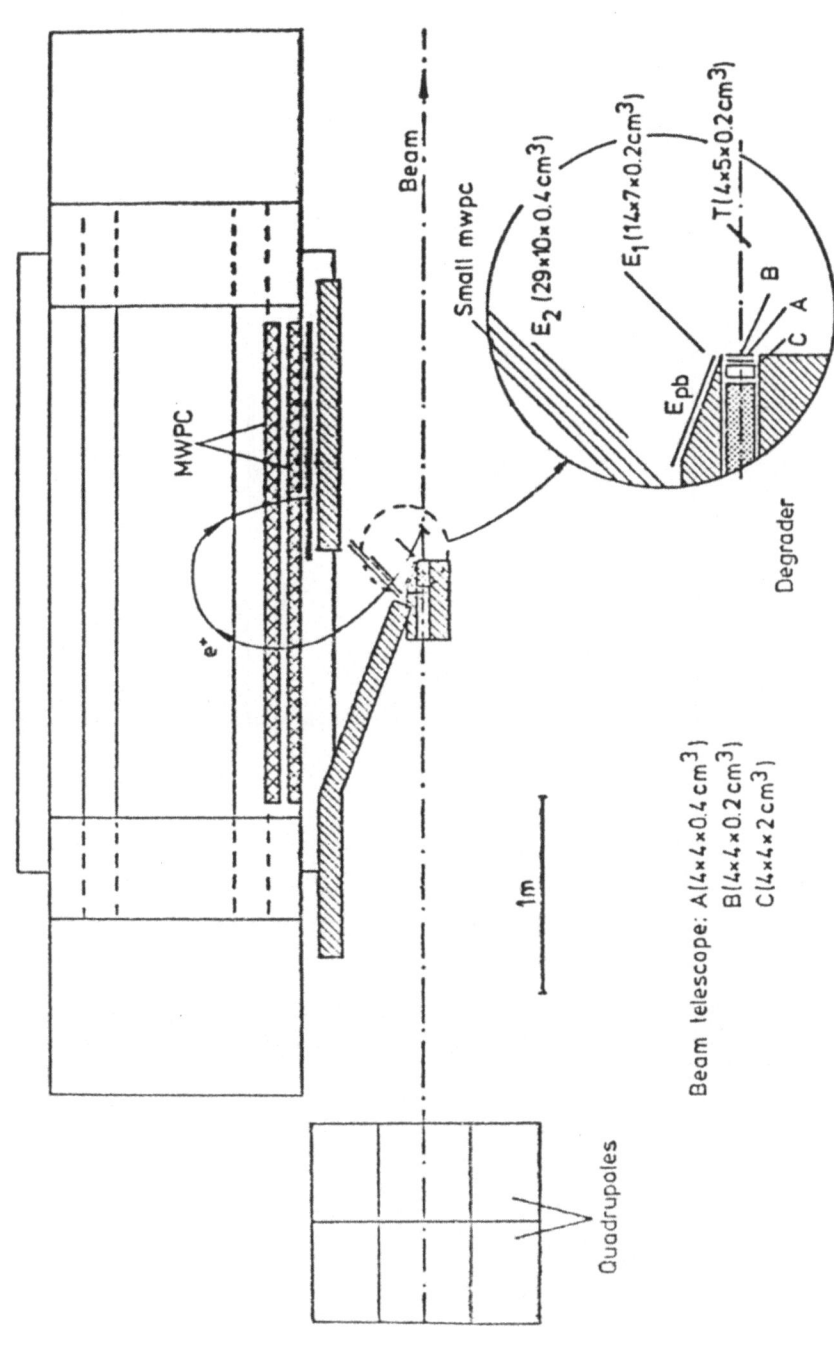

Fig. 9. Setup of the $\pi \to e\nu$ experiment of the LOUVAIN-ZURICH-LAUSANNE collaboration. Pions are stopped in the scintillator target "T" and the emitted positrons are momentum analysed by a transformed pair-spectrometer.

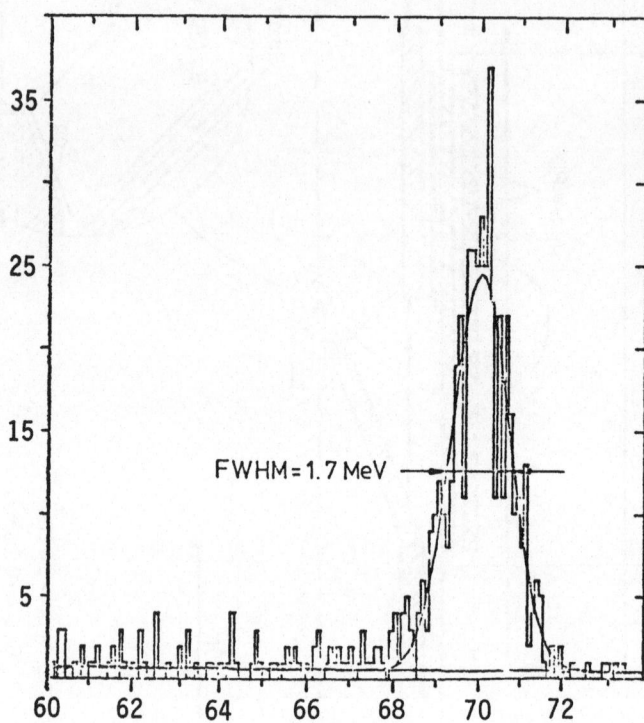

Fig. 10.   Positron momentum spectrum of π → eν decay between 60 and 75 MeV/c.   The momentum is computed by a retracing method.

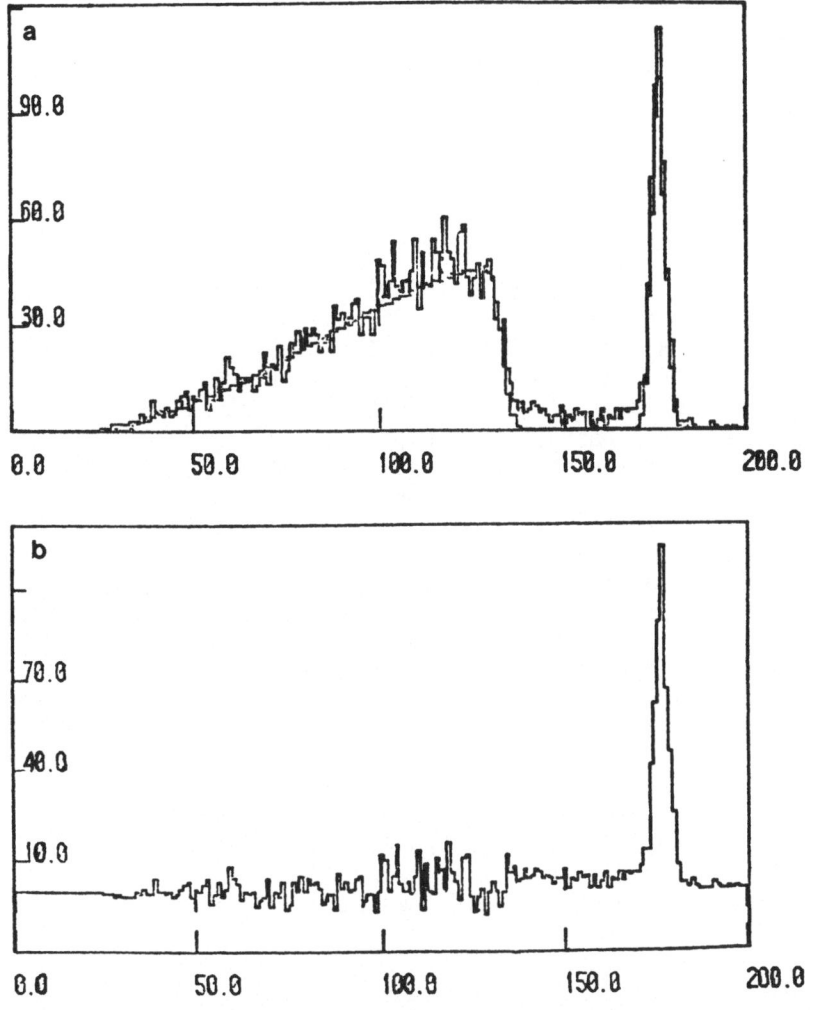

Fig. 11. a) Total positron energy spectrum computed by a retracing method after cuts to reduce the π-μ-e cascade
contribution. These positrons are emitted within
the 30 first nsec. after the pion stop. "Michel"
to "π → eν" ratio is 6.4 ; total π → eν events are
441 with a resolution of 1.7 MeV/c FWHM.

b) Experimental spectrum obtained after substraction of
a fitted "Michel" spectrum folded with an approximate
response function.

is illustrated in Fig. 10. The combined effects of both energy resolution improvements and Michel spectrum suppression is illustrated in the spectrum shown in Fig. 11a to be compare to Fig. 7. From our data shown in Fig. 11a we subtracted an adjusted Michel spectrum deformed according to our acceptance ; the resulting spectrum is shown in Fig. 11b.

Data are still under analysis ; the overall amount of $\pi \rightarrow e\nu$ events recorded to date is about 6000.

We are indebted to Dr. J.P. Deutsch for many useful discussions.

4. REFERENCES

1. B. Pontecorvo, ZhETH (USSR) 53:1717 (1967).
   A.K. Mann and H. Primakoff, Phys. Rev. D15:655 (1977).
   J.N. Ng, Nucl. Phys. B191:125 (1981).
2. R.E. Shrock, Phys. Lett. 96B:159 (1980).
3. R. Abela et al., Phys. Lett. 105B:263 (1981).
4. R.C. Minehart et al., preprint 1984.
5. R.S. Hayano et al., Phys. Rev. Lett. 49:1305 (1982).
6. C.Y. Pang et al., Phys. Rev. D8:1989 (1973) quoted in ref. 5.
7. J.P. Deutsch et al., Phys. Rev. D27:1644 (1983).
8. J.P. Deutsch et al., SIN proposal R-82-14.
9. J. Heintze et al., Nucl. Phys. B149:365 (1979).
10. D.A. Bryman et al., Phys. Rev. Lett. 50:1546 (1983).
11. R. Prieels et al., SIN proposal R-81-09.
12. D.R. Clay et al., Phys. Rev. 140:B586 (1965).
13. C.A.N. Conde and A.J.P.L. Policarpo, Nucl. Instr. and Meth. 53:7 (1967).
14. W. Herold et al., Nucl. Instr. and Meth. 217:277 (1983).
15. W. Herold, ETH Thesis Nr. 7133.
16. R. Ferreira Marques, ETH Thesis Nr. 7111.
17. J. Egger et al., SIN Newsletters (1984).
18. D.A. Bryman et al., preprint 1983.
19. C. Amsler et al., SIN Newsletters 15:NL28 (1983).

THE EXOTIC ATOMS OF QCD: GLUEBALLS, HYBRIDS AND BARYONIA

T. Barnes

Rutherford Appleton Laboratory
Theory Division
Chilton, Didcot, Oxon. OX11 0QX
England

INTRODUCTION

In these three lectures I review the theoretical basis underlying
the expected "exotic" states in QCD, the theory of quarks and
gluons.   The first lecture is an historical introduction to QCD,
meant to motivate it to the non-specialist.   The second lecture
is a critical review of the two most important models used by
theorists to try to understand unusual quark and gluon states –
the MIT bag model and QCD on a lattice.   Recent developments in
both models of possible experimental relevance are particularly
emphasized.   Finally, in the third lecture I discuss the status
of three candidate "exotic" states seen in ψ radiative decays, the
$\iota(1440)$, $\theta(1700)$ and $\xi(2220)$, and review theorists' suggestions of
what these states might be.

I.   QCD: an historical introduction

QCD, the theory which we now believe describes the
interactions of quarks and gluons, is a special case of a type of
theory known as "non-Abelian gauge theories".   The prototype
non-Abelian gauge theory was invented in 1954 by two theorists [1]
who were "messing around" with an interesting theoretical idea,
namely local gauge invariance.   It was known that one could
invent the highly successful theory of electron-photon
interactions, QED (quantum electrodynamics), by requiring that the
theory be invariant under local gauge transformations.   Let us
first recall how this was done, and then we shall see what the two
theorists, Yang and Mills, were up to with non-Abelian theories.

Suppose we have only free Dirac electrons.   This particle and
its antiparticle can be described by the four-component complex

spinor field $\psi(x)$, with the Lagrangian density [2].

$$\mathcal{L}_0 = \psi^+(x)\gamma_0 \, (i\slashed{\partial}-m)\psi(x) \qquad (1.1)$$

First, we note that the Lagrangian field ψ is invariant under a constant change of phase in the electron field ψ.

$$\psi \to \psi' = e^{i e\theta}\psi \qquad (1.2)$$

$$\mathcal{L}_0 \to \mathcal{L}_0' = e^{-i e\theta}\overline{\psi}' \, (i\slashed{\partial}-m)\psi e^{i e\theta} = \mathcal{L}_0 \qquad (1.3)$$

This change of ψ by a constant phase is also called a constant gauge transformation.

The importance of looking for transformations which leave the Lagrangian $\mathcal{L}$ invariant is that they lead to conservation laws, which may be derived from the transformation through Noether's theorem [3]. In the case of a constant gauge transformation (1.2), the conserved quantity which results is the total electric charge

$$Q = \int \psi^+\psi \, d^3x \qquad (1.4)$$

Now suppose we require that the theory (Lagrangian) be invariant under a gauge transformation which depends on the spatial coordinate, a so-called "local" gauge transformation, rather than the everywhere constant "global" transformation (1.2). This local change of phase in ψ is

$$\psi(x) \to \psi'(x) = e^{\, i e\theta(x)} \, \psi(x) \qquad (1.5)$$

This time, the free electron theory (1.1) is <u>not</u> invariant under the local gauge transformation (1.5), but instead changes by

$$\mathcal{L}_0(\psi') - \mathcal{L}_0(\psi) = - \, (e\overline{\psi}\gamma_\mu\psi)(\partial_\mu\theta) \qquad (1.6)$$

How can we modify so as to eliminate this violation of local gauge invariance? One way is to introduce an additional vector field A_μ, which couples to the charged Fermi field ψ through an interaction term

$$\mathcal{L}_I = -(e\overline{\psi}\gamma_\mu\psi)A_\mu \qquad (1.7)$$

and which changes under a local gauge transformation as

$$A_\mu(x) \to A'_\mu(x) = A_\mu(x) - \partial_\mu\theta(x) \qquad (1.8)$$

The sum of the free electron and electron-photon Lagrangians is then invariant under a local gauge transformation.

$$\mathcal{L}_0(\psi') + \mathcal{L}_I^! = \bar{\psi}'(i\not{\partial}-m)\psi' - e\,\bar{\psi}'\gamma_\mu\psi'A_\mu'$$

$$= \bar{\psi}(i\not{\partial}-m)\psi - e(\bar{\psi}\gamma_\mu\psi)(\partial_\mu\theta) - e\bar{\psi}\gamma_\mu\psi A_\mu + e(\bar{\psi}\gamma_\mu\psi)(\partial_\mu\theta)$$

$$= \mathcal{L}_0(\psi) + \mathcal{L}_I \qquad (1.9)$$

We may now take an imaginative leap and assume particles associated with the vector field A_μ we have invented really exist, and so add to the lagrangian a term $\mathcal{L}_0(A_\mu)$ which allows the vector A_μ to produce physical, massless vector quanta.

$$\mathcal{L}_{total} = \mathcal{L}_0(\psi) + \mathcal{L}_I - \tfrac{1}{4}\,F_{\mu\nu}^2 \qquad (1.10)$$

$$\mathcal{L}_0(A) = -\tfrac{1}{4}\,F_{\mu\nu}^2 = \tfrac{1}{2}(\vec{E}^2-\vec{B}^2) \qquad (1.11)$$

The tensor $F_{\mu\nu}$ is just the electric and magnetic field strengths,

$$F_{\mu\nu} = \partial_\mu A_\nu - \partial_\nu A_\mu \qquad (1.12)$$

with explicit components like

$$F_{01} = E_1 \qquad (1.13)$$

and

$$F_{12} = B_3 \qquad (1.14)$$

The lagrangian we have invented (1.10) by imposing local gauge invariance on a massive Dirac electron is the lagrangian of QED. The interaction between the massive fermion ψ and the massless vector field A_μ, which appears as $\mathcal{L}_I$ in (1.7), gives the familiar electron-photon vertex in QED perturbation theory

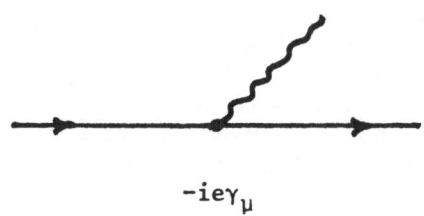

$$-ie\gamma_\mu$$

Fig. 1. The electron-photon vertex in QCD.

This theory, with gauge transformations $e^{ie\theta(x)}$ that depend only on the single parameter function $\theta(x)$ is known as an Abelian, U(1) gauge theory. "Abelian" here means that any two gauge transformations commute, and U(1) means that the gauge transformation "matrix" $e^{ie\theta}$ which acts on the basic Fermi field is a unitary ($U^+=U^{-1}$), one-dimensional "matrix".

Evidently this principle of local gauge invariance may have underlying physical significance, in that it has led us from simple electric charge conservation to the theory of electron-photon interactions. It is interesting to see what happens when we apply this same invariance principle to other global conservation laws.

This is exactly the idea Yang and Mills explored in their 1954 paper, which introduced non-Abelian gauge theories. The global conservation law they investigated was isospin, which was known to be conserved to a good approximation in the strong interaction through experimental results on pion-nucleon scattering. So, in "Conservation of Isotopic Spin and Isotopic Gauge Invariance", Yang and Mills asked what theory results when we impose local isospin invariance on a nucleon field. The basic fermion (nucleon) field ψ is now a doublet, with proton and neutron components

$$\psi = \begin{bmatrix} P \\ N \end{bmatrix} \tag{1.15}$$

and global isospin invariance means that the theory is invariant under the constant gauge transformation G,

$$\psi \rightarrow \psi' = G\psi \tag{1.16}$$

$$G(\theta) = e^{ig\sum_{a=1}^{3} \frac{1}{2}\sigma^a \theta^a} \tag{1.17}$$

The fundamental Fermi field $\psi = \begin{bmatrix} P \\ N \end{bmatrix}$ has n=2 degrees of freedom, and the unitary transformation (1.17) which generalizes the QED U(1) $e^{ie\theta}$ change of phase is a 2x2 unitary matrix. A theory which is unchanged by such a transformation is said to be invariant under global SU(2) (2x2 unitary) gauge transformations. The S in SU(2) is an abbreviation for "special", and means that the transformation matrix G has determinant +1. The Pauli matrices $\frac{1}{2}\sigma^a$ which determine the direction of infinitesimal gauge transformations are said to be the generators of the gauge transformation. The total number of generators, and hence parameters $\{\theta^a\}$, in an SU(n) gauge theory is n^2-1 in general, $2^2-1=3$ here.

Imposing local SU(2) gauge invariance requires the introduction of a new vector field A_μ^a, just as in QED local gauge invariance required the photon field A_μ. The difference here is that the vector field carries <u>isospin</u> (more generally, whatever charge the gauge transformation operates on), so it would have the same quantum numbers as the rho meson. The complete locally gauge invariant lagrangian Yang and Mills found for SU(2) was

$$\mathcal{L}_{total} = \mathcal{L}_0(\psi) - g(\overline{\psi}\gamma_\mu \frac{\sigma^a}{2}\psi)\underbrace{A_\mu^a} \qquad - \frac{1}{4}F_{\mu\nu}^{a\,2} \qquad (1.18)$$

isospin 1 vector

In Feynman diagram language, the second (Fermi-Bose coupling) term gives anF interaction vertex very much like the familiar QED vertex, except for the appearance of the SU(2) generators (isospin matrices) $\frac{1}{2}\sigma^a$;

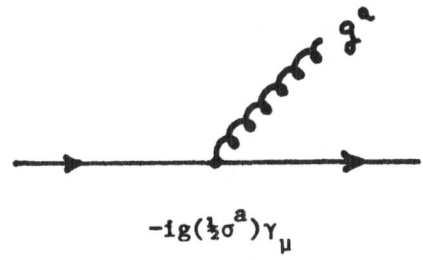

$$-ig(\tfrac{1}{2}\sigma^a)\gamma_\mu$$

Fig. 2.   The quark-gluon vertex in SU(2) Yang-Mills.

The surprise comes in the last term $\frac{-1}{4}F_{\mu\nu}^{a\,2}$, which in QED was just the free photon lagrangian.   In non-Abelian theories there is an additional boson-boson interaction in $F_{\mu\nu}^a$ required by local gauge invariance

$$F_{\mu\nu}^a = \partial_\mu A_\nu^a - \partial_\nu A_\mu^a + \underline{gf^{abc}A_\mu^b A_\nu^c} \qquad (1.19)$$

The f^{abc} are group theoretic structure coefficients, which arise from the commutation of two generators. In SU(2),

$$[\frac{\sigma^a}{2}, \frac{\sigma^b}{2}] = i\epsilon^{abc}\;\frac{\sigma^c}{2}, \qquad (1.20)$$

so f^{abc} (SU(2)) = ϵ^{abc}.   Obviously, in QED, with a single parameter, commutators never appear and no photon-photon interaction arises.   This quadratic A^2 term in $F_{\mu\nu}$ gives rise to rather complicated three- and four-boson couplings in perturbation theory [4]

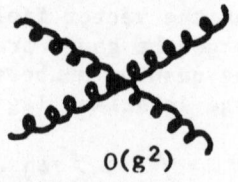

O(g)   O(g²)

Fig. 3.   Three-and four-gluon couplings in QCD.

An interesting question is the mass of these gauge bosons.
Yang and Mills are very careful in their remarks on this question
in their 1954 paper; they first note that one might argue that
such a boson should be massless, because there is no explicit mass
term in the Lagrangian. They then note that the above
argument is suspect because of divergences in the theory, so
"dimensional arguments are not satisfactory", and hence that "We
have therefore not been able to conclude anything about the mass
of the ... quantum". As present day indications are that it may
be impossible to isolate a gauge boson in an unbroken gauge
theory, the note of caution evident in these remarks appears
justified.

Having reviewed the origin of Yang-Mills theory in 1954, as a
model of isospin interactions between nucleons and hypothetical
I=1 vector bosons with a local gauge invariance under SU(2), it is
of interest to see how it became today's theory of the interaction
of quarks and gluons, based on color SU(3);

1954	1973
Yang-Mills	QCD
n=2 isospin: SU(2)	n=3 color: SU(3)
$\frac{1}{2}\sigma^a$, $\begin{bmatrix} \text{Proton} \\ \text{Neutron} \end{bmatrix}$	$\frac{1}{2}\lambda^a$, $\begin{bmatrix} \text{red quark} \\ \text{green quark} \\ \text{blue quark} \end{bmatrix}$
incorrect	correct

QCD was postulated to be the correct theory of quark-gluon
interactions in 1973 by Gross and Wilczek [5] and Politzer [6]
primarily because it exhibited a property known as asymptotic
freedom.   In 1969, studies of deep inelastic electron-proton
scattering at SLAC [7] had shown, surprisingly, that the
underlying physical scattering amplitudes depended only on the
dimensionless quantity $x=q^2/2M_p\nu$.

196

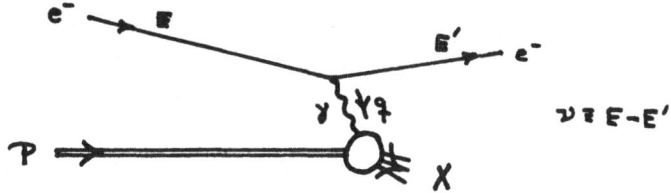

Fig. 4.   Deep-inelastic scattering conventions.

This dependence of the physics on only the dimensionless
quantity x, known as <u>scaling</u>, indicated that there were no
intrinsic mass scales inside the proton beyond the energies of the
scattering - a few GeV.   This implied that the quarks inside the
proton were not heavy, very tightly bound objects (one way to
explain why they had not been seen in isolation), but were rather
quite light objects, which behaved as though they were essentially
free inside the proton, if struck hard enough by the photon.
This experimental result led several theorists to search candidate
theories of quarks and their binding particles (gluons) for models
in which the binding grew weaker as the scale of momenta (of the
photon in D.I.S., for example) grew larger.

How do theorists look for such behavior in a field theory?
One approach is to ask how a particular scattering amplitude, for
example a quark-quark-gluon vertex, changes when we multiply all
the momenta by the same scale λ.

Fig. 5.   A renormalization group transformation.

The change in $g_R(\lambda)$ due to an infinitesimal change in scale
can be written as a differential equation, known as the
Callen-Symanzik equation [8].

$$\frac{dg_R(t)}{dt} = \beta(g_R) \qquad (1.21)$$

where

$$t = \ln(\lambda) \qquad (1.22)$$

The behavior of the coupling strength $g_R(t)$ is determined qualitatively by the "beta function" $\beta(g)$, which may be calculated in perturbation theory. If $\beta(g) > 0$, going to larger momenta (increasing t) gives a larger coupling strength. Conversely, negative $\beta(g)$ gives a decreasing coupling strength with increasing momenta, which is just the desired "asymptotic freedom" behavior seen in deep inelastic scattering. So, we test theories for scaling by looking for a negative beta function $\beta(g)$.

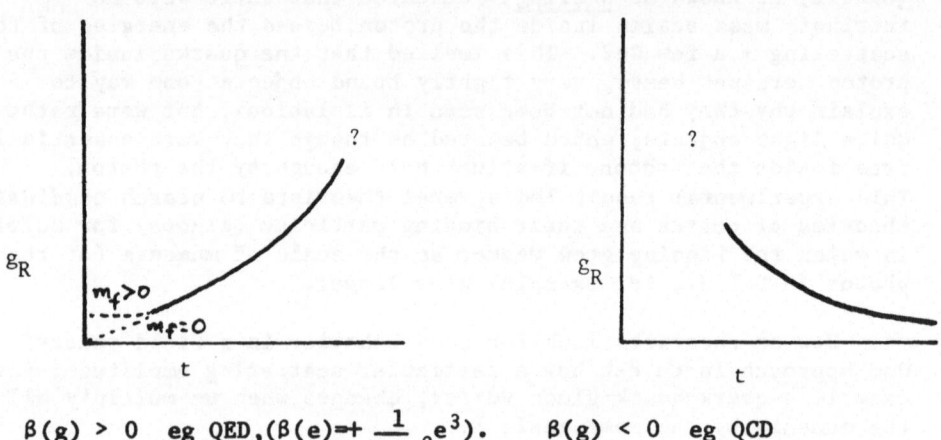

$\beta(g) > 0$ eg QED, $(\beta(e)=+\frac{1}{12\pi^2}e^3)$. $\qquad$ $\beta(g) < 0$ eg QCD

Fig. 6. Positive and negative beta functions and their running coupling constant. (Note in QED the static charge $e_R(t=0)$ is non-zero because the charge stops running at $q^2 \sim m_f^2$).

Theories with the required negative beta function are, it develops, rather hard to find. The only such theories which Gross and Wilczek, and independently Politzer, found were non-Abelian gauge theories, of the type introduced by Yang and Mills in 1954. For such a gauge theory based on the color gauge group SU(3) and with N_F independent types of fermions (each type a color triplet), the beta function $\beta(g)$ is given by [4].

$$\beta(g) = -\frac{1}{24\pi^2} (\frac{33}{2} - N_F) g^3 + O(g^5) \qquad (1.23)$$

As long as g is sufficiently small and the number of types (<u>flavors</u>, not colors) of quarks with masses smaller than our momentum scale is less than 33/2 (certainly true at present), we have $\beta(g) < 0$ and hence scaling at sufficiently high energies in this SU(3) gauge theory.

The linking of scaling (experimental) and asymptotic freedom (theoretical) was the principal reason for the initial proposal that a Yang-Mills theory was the correct theory of the strong interaction between quarks and gluons. There were of course several other problems that QCD could account for; in particular we note the following.

1.   A statistics problem in the quark model of baryons arose regarding the overall symmetry of certain states. We expect states of fermions to be totally antisymmetric, according to the Pauli principle.   The puzzle was that baryon states seemed to be totally symmetric, for example the Δ^{++} state of three "up" quarks u was

$$\left|\Delta^{++},\ S_z=3/2\ >\ =\ \left|u\uparrow u\uparrow u\uparrow>\right.\right. \tag{1.24}$$

This is explained by QCD if we assume that there is an additional "color" label (r,g,b) for each quark, and that the overall state including the color label is actually antisymmetric

$$\left|\Delta^{++}\ >\ =\ \tfrac{1}{\sqrt{6}}(\left|u_r\uparrow u_g\uparrow u_b\uparrow>\ -\ \left|u_r\uparrow u_b\uparrow u_g\uparrow>\ +\ ...)\right.\right. \tag{1.25}$$

2.   Zweig's rule [9] that quark-antiquark pairs do not annihilate inside a hadron, devised to explain selection rules like $\phi \not\rightarrow \rho\pi$, is understood because this is a high-order process in QCD, requiring three intermediate gluons

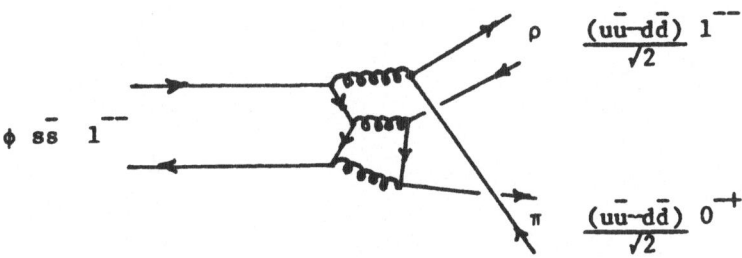

Fig. 7.   A Zweig-forbidden process in QCD.

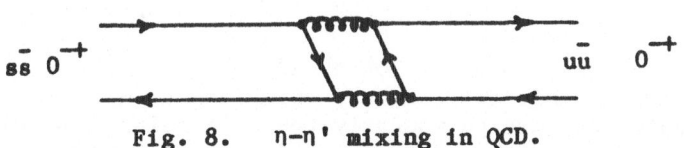

Fig. 8. $\eta-\eta'$ mixing in QCD.

In the $\eta-\eta'$ system, where $s\bar{s}-u\bar{u}-d\bar{d}$ mixing does occur with large amplitude, only two intermediate gluons are required for the coupling.

Note that this fails to explain the unmixed nature of 2^{++} tensor mesons f-f', which two gluons can connect! Zweig's rule is still an area of much theoretical discussion and controversy, particularly as regards evidence for the hypothetical "glueballs", hadrons composed mostly of gluons [10].

3.   The Adler anomaly calculation of the rate for $\pi^0 \to \gamma\gamma$ [11] was published in 1969, and with laudable candor its author noted that his result for the decay rate was about 9x smaller than the experimental number.

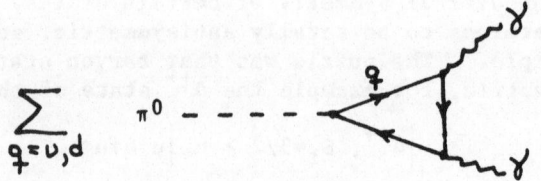

Fig. 9.   The Adler anomaly calculation of $\Gamma(\pi^0 \to \gamma\gamma)$.

Lo and behold, the introduction of QCD and the additional 3 colors of fermions running around the loop brought the predicted rate for $\pi^0 \to \gamma\gamma$ into agreement with experiment.

4.   A great puzzle in the quark model was the lack of experimental evidence for any states except $q\bar{q}$ and qqq (and $\overline{qqq}$) in the hadron spectrum.   We begin to understand this in QCD if we consider what representations (what strengths of color charge) we çan make with various combinations of quarks (color 3) and antiquarks (color $\bar{3}$).   Note that color 1, a color singlet, has no net color charge.

State	Color State
q	3
q̄	3̄
q q̄	3x3̄ = 8 + 1
qq	3x3 = 6 + 3̄
q̄q̄	3̄x3̄ = 6̄ + 3
qqq	3x3x3 = 10 + 8 + 8 + 1

Only the experimentally observed states occur in colorless combinations.   This has led many theorists to suggest that colored particles are permanently bound in color singlet combinations by a strong long-range force, an effect known as "confinement".   This possibility has not been proven, but we shall see that lattice gauge theory calculations of the interquark potential energy certainly make it look plausible.

So, we see that the theory suggested by Yang and Mills as a model of nucleon interactions is an unqualified disaster. As proposed, it would predict permanently bound neutrons and protons, and we would only see I=0 nuclei. (This is assuming confinement). As a theory of strong interactions at a deeper level, between quarks and gluons, it is paradoxically an unqualified success.

II. Bag and Lattice Models of the "Exotic Atoms" of QCD.

In this lecture I review two of the theoretical tools which are currently used to model the "exotic" states of gluons or quarks and gluons which are expected in QCD.

I should begin with a word of warning about nomenclature. In these lectures I will use "exotic" in quotes to refer to any non-$q\bar{q}$ meson or non-qqq baryon. For historical reasons, the word exotic in the context of QCD has come to mean any state which cannot be made from a $q\bar{q}$ pair (if a meson) or from qqq (if a baryon). In the meson case, this might be a doubly-charged meson m^{++} or a meson with spin, parity and charge-conjugation parity forbidden to $q\bar{q}$ states, for example $J^{PC} = 1^{-+}$. Unconventional states which have conventional quantum numbers, for example a $J^{PC} = 0^{++}$, I=0 meson which is secretly composed mostly of $q^2\bar{q}^2$, have come to be known as cryptoexotics [12]. As there may be significant mixing between cryptoexotic basis states and conventional $q\bar{q}$ states, we can see why identification of unusual but non-exotic states has become a matter of some controversy in QCD.

The first model I will review is the MIT bag model; after an introduction I will concentrate on the spectrum of states containing large gluonic components which have been predicted using the bag model. In particular we will look at gg "glueballs", $q\bar{q}g$ "hybrid mesons" and briefly at q^3g "hybrid baryons".

Second, I will review important recent developments in calculations of the masses of cryptoexotic gg "glueball" states using Monte-Carlo methods on the lattice. I discuss in particular the recent controversy over the glueball mass scale on the lattice - the "string tension crisis"-, and finally will briefly mention other interesting results related to cryptoexotic and exotic states which lattices may soon produce.

a)   The MIT bag model

A version of the bag model was introduced by P.N. Bogoliubov

[13] in 1968, as a relativistic model of free quarks confined by an unspecified mechanism to a spherical region of radius ~ 1 fermi. Bogoliubov at this early date noted that baryon magnetic moments, semileptonic weak decays (g_A/g_V) and charge radii could be explained by such a model.

The real impetus for developing this model in the early 70's was the introduction of QCD in 1973, together with the scaling observed at SLAC in 1969, which indicated that quarks at short distances might feel relatively weak QCD forces. This gave the picture of free quarks permanently confined inside a hadron qualitative theoretical and experimental support, and led to the development of the bag model in 1974 [14] and 1975 [15]. In the simple 1974 "zeroth-order" bag model of A. Chodos, one visualizes a hadron as a region in space in which a fixed number of quarks (or gluons) are enclosed. The model is made Lorentz invariant by the addition of a surface-pressure (or external energy density) term B_0 to the lagrangian, which is the principal improvement on Bogoliubov. No colored fields (quarks or gluons) are allowed outside the bag.

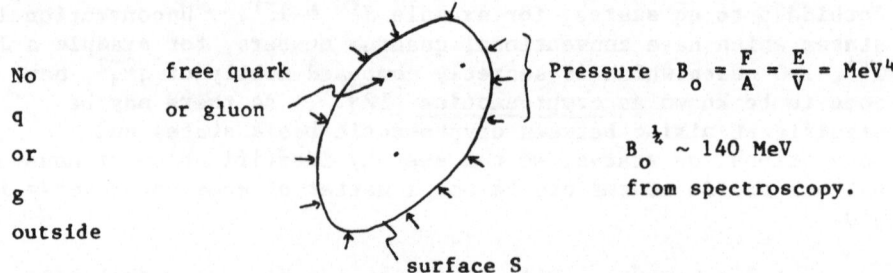

No   free quark   Pressure $B_0 = \dfrac{F}{A} = \dfrac{E}{V} = MeV^4$

q   or gluon

or $B_0^{\frac{1}{2}} \sim 140$ MeV

g   from spectroscopy.

outside

surface S

Fig. 10. "Zeroth-order" bag model.

Inside the bag the quark field ψ and the gluon field $\tilde{A}^a$ obey the free Dirac equation and the free Klein-Gordon equation, respectively

$$(\not{\partial} - m)\,\psi = 0 \qquad \text{inside S}$$
$$\psi = 0 \qquad \text{outside S} \qquad (2.1)$$
$$\Box^2 \tilde{A}^a = 0 \qquad \text{inside S}$$
$$\tilde{A}^a = 0 \qquad \text{outside S} \qquad (2.2)$$

There are two boundary conditions which are imposed on these fields in the bag model. The "first" boundary condition is equivalent to requiring that no color current flows through the bag surface S anywhere. This effectively imposes confinement - no net color charge enclosed - on bag model states ab initio.

First boundary condition

quarks:
$$\{ i \not{n} \psi = \psi \} \big|_S \qquad (2.3)$$
$$(n_\mu = \text{unit 4-normal to S, } n \text{ at rest}).$$

gluons:
$$n_\mu F^a_{\mu\nu} \big|_S = 0 \qquad (2.4)$$
"antimetallic b.c."
$$(n \cdot \vec{B}^a \big|_S = 0 = n \times \vec{E}^a \big|_S \quad \text{at } r).$$

These are eigenmode equations which determine the energies of quark and gluon modes in the bag.

The second boundary condition is a pressure balance equation which determines the bag surface S. For quarks alone, this is

Second boundary condition

$$< \text{bag} \big| \, n \cdot \partial (\bar{\psi}\psi) \, \big| \text{bag} > \big|_S = 2B_0 \qquad (2.5)$$

In practice, one usually assumes that the bag is actually spherical (true for the $S_{\frac{1}{2}}$ quark modes) and satisfies this equation in the mean. Nonspherical bags are very complicated mathematical physics problems, and have only been studied in a few special cases.

Now suppose we consider the massless quark modes in the bag model, assuming the bag is a sphere of radius R. One may solve the eigenmode equation implicit in (2.3) to find an infinite sequence of quark modes, a few of which are given below

Massless Quark Modes

$$\chi = kR = \omega R$$

J^P	N=1	N=2	N=3	
$\frac{1}{2}^+$	2.043	5.396	8.578	...
$\frac{1}{2}^-$	3.812	7.002		
$\frac{3}{2}^+$	3.204			
$\frac{3}{2}^-$	5.123			
...				

Most studies of hadrons in the bag model use the $S_{\frac{1}{2}}$ $\chi=2.043$ quark (and antiquark) mode. This is because only radially and orbitally excited states require the other modes to zeroth-order, and the bag model has special problems with these states.

The quark density in the $S_{\frac{1}{2}}$ mode is shown below

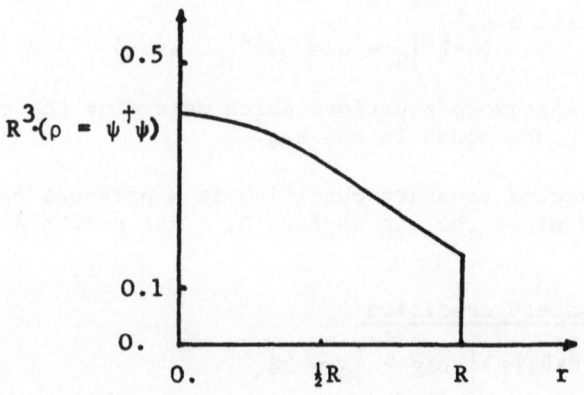

Fig. 11.   Quark density in the lowest $S_{\frac{1}{2}}$ mode ($m_q=0$).

and is a smooth, monotonically decreasing function of r from r=0
to r=R, where it discontinuously drops to zero.

The mass of a hadron in the zeroth-order bag model is simply
the sum of the quark (or gluon) constituent energies and the bag
pressure energy.   Taking the light S-wave $q\bar{q}$ mesons as a specific
example, we have for massless quarks ($\chi=2.043$),

$$E(q\bar{q}) = \underbrace{2}_{\substack{q \text{ and } \bar{q}}} \cdot \underbrace{\frac{\chi(=KR)}{R}}_{\substack{\text{one} \\ \text{quark} \\ \text{energy}}} + \underbrace{\frac{4\pi}{3} R^3}_{\substack{\text{Volume} \\ \text{of bag}}} \underbrace{B_o}_{\substack{\text{bag} \\ \text{energy} \\ \text{density}}} \qquad (2.6)$$

We must satisfy the second boundary condition (2.5) to
determine the bag radius.   This is equivalent to minimizing the
energy E as a function of the surface S, and assuming S is a
sphere (in this case it is), we find the optimum radius R_o as
follows

$$\frac{\partial E}{\partial K} \Big|_{R_o} = 0 \qquad (2.7)$$

$$\therefore R_o = \left(\frac{\chi}{2\pi B_o}\right)^{\frac{1}{4}} \qquad (2.8)$$

The total energy, given this R_o, is

$$E = 2 \cdot \underset{\substack{\uparrow \\ q \\ \text{and} \\ \bar{q}}}{\frac{\chi}{R_o}} \cdot (1 + \underset{\substack{\uparrow \\ \text{bag}}}{\frac{1}{3}})$$

That is, the bag contributes ⅓ of the total hadron energy.   This
is always true in the zeroth-order bag model with massless
constituents if R_0 is determined as in (2.6-7).

Suppose we see what radii and masses actually result.   In
1975, the MIT bag model group [15] suggested a bag strength of
$$B_0^{\frac{1}{4}} = 146 \text{ MeV} \tag{2.9}$$
which gives a plausible meson bag radius,
$$R_0 = 1.02 \text{ fm} \tag{2.10}$$
but a disturbingly high mass.

$$E = \frac{8\chi}{3R_0} = 1.05 \text{ GeV} = E_\rho = E_\omega = E_\pi \tag{2.11}$$

We can imagine lowering the mean mass scale in (2.11) by just
using a smaller bag pressure $B^{\frac{1}{4}}$, as E is proportional to it here.

What we <u>cannot</u> do is to separate E_ρ (experimentally 770
MeV) from E_π (140 MeV).   These states differ primarily in quark
spin orientation ($S_{q\bar{q}}$ (ρ)=1, $S_{q\bar{q}}$ (π)=0), and one assumes in the
zeroth-order bag model that there is no interaction between the
quarks.   This is evidently not a very good approximation in S-wave
meson spectroscopy!   The required improvement in the bag is the
introduction of the spin-spin forces due to QCD, in particular
transverse gluon exchange.

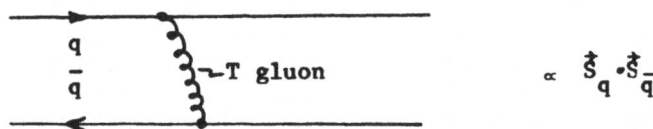

Fig. 12.   The QCD spin-spin force from gluon exchange.

There are two difficulties in the zeroth-order bag model
which we note in passing.

The first problem is seen when we calculate the mass of a
hadron composed of 3n quarks in the lowest mode;
$$E_{3n} = n^{\frac{3}{4}}E_3 \tag{2.12}$$

If true, this would mean that a single hadron of 6 light
quarks is only 0.84 times as massive as two nucleons; hence the

deuteron would be a six quark object rather than a loosely bound
P-N pair. This obviously incorrect result may be rectified by
gluon exchange corrections; this is a matter of current study in
attempts to derive the nucleon-nucleon interaction from the bag
model [16].

The second problem is the existence of spurious states in the
bag due to excitation of the CM coordinate of the quarks, which is
a well-known problem in the shell model in nuclear physics. For
example, from $S_{\frac{1}{2}}$ and $P_{\frac{1}{2}}$ modes we can make a conventional 1^{++} $q\bar{q}$
meson

$$\left| q\bar{q},\ 1^{++} \right) = \frac{1}{\sqrt{2}} (\left| S_{\frac{1}{2}}\bar{P}_{\frac{1}{2}} \right> + \left| P_{\frac{1}{2}}\bar{S}_{\frac{1}{2}} \right>) \qquad (2.13)$$

as well as the exotic J^{PC} spurious state

$$\left| q\bar{q},\ J^{PC} = 1^{-+} \right> = \frac{1}{\sqrt{2}} (\left| S_{\frac{1}{2}}\bar{P}_{\frac{1}{2}} \right> - \left| P_{\frac{1}{2}}\bar{S}_{\frac{1}{2}} \right>) \qquad (2.14)$$

The latter state does not exist unless the bag is a real,
physical object - a picture which few people would accept today.
If the bag is instead merely a way of truncating quark and gluon
wavefunctions, and has no physical existence, then we must take
care to distinguish physical states like (2.13) from spurious CM
excitations like (2.14).

Gluon Modes

Now we return to the problem of putting gluons in the bag
model and calculating spin-dependent forces between gluons. We
solve the free wave equation (2.2) for the gluon field $\vec{A}^a(\vec{x})$
inside the bag, subject to the "antimetallic" first boundary
condition (2.4), assuming once again that the bag surface, S is a
sphere of radius R. The solutions in a spherical cavity,
presumably to the delight of any electrical engineers present,
turn out to be TE and TM cavity resonator modes [17]. The E and
M are color electric $\vec{E}^a$ and magnetic $\vec{B}^a$ fields, but the color
index a=1-8 at this level just comes along for the ride. The
sequence of modes and their eigenenergies are

	$(X_g = k_g R = \omega_g R)$		
TE	N=1	N=2	...
1^+	2.744	6.117	
2^- ...	3.870	7.443	
TM			
1^-	4.493	7.725	
2^+ ...	5.763	9.905	

In color SU(3), single gluon states are color octets. We can make finite energy color singlet states - "glueballs" - by combining two gluons, and most bag model studies of glueballs consider states which can be made by combining the lowest TE(χ_g=2.744) and TM(χ_g=4.493) modes.

$$gg \qquad (TE)^2 \; = \; 0^{++}, \; 2^{++}$$

$$(TE)(TM) \; = \; 0^{-+}, \; 2^{-+}$$

The $1^{++}(TE)^2$ state is forbidden by Bose statistics, and the $1^{-+}(TE)(TM)$ state is a spurious CM excitation, rather like the spurious 1^{-+} $q\bar{q}$ state discussed earlier. The absence of a J=1 gg glueball is not surprising, because Yang's theorem [18] says that we cannot make a J=1 state from two transverse photons (or gluons).

What masses do we find for these glueballs? Assuming that the bag pressure B_o is the same as for quarks, we find

$$E_{gg} \; = \; \begin{matrix} 1.25E_{q\bar{q}} \\ 1.53E_{q\bar{q}} \end{matrix} \qquad \begin{matrix} 0^{++}, \; 2^{++}(TE)^2 \\ 0^{-+}, \; 2^{-+}(TE)(TM) \end{matrix} \qquad (2.15)$$

We may assemble other color singlet combinations from quarks and gluons, notably $q\bar{q}g$ "hybrid" mesons [19] and $q^2\bar{q}^2$ "baryonium" [12]. Using m_ρ as a typical E_{gg}, we find a zeroth-order bag model spectrum for these "exotic atoms of QCD" which looks like this

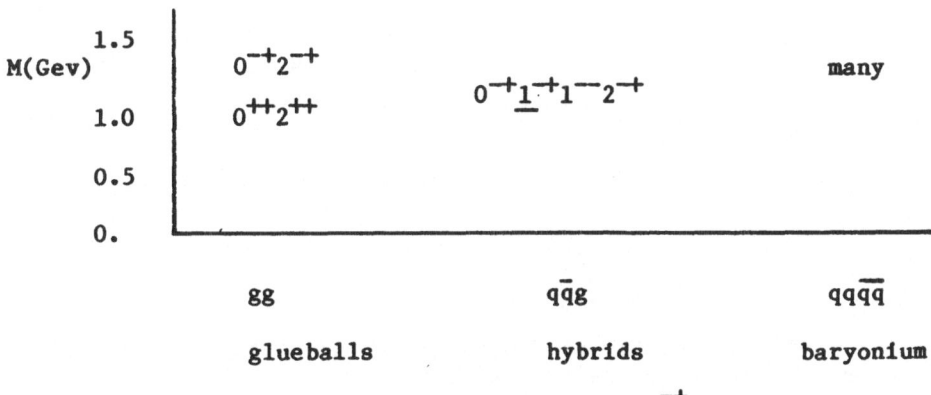

Fig. 13. Zeroth-order bag model spectrum of exotics and crypto-exotics.

Many unusual states are evidently expected from about 1 GeV upwards in the bag model. A more detailed estimate of spectroscopy, and a prediction of which exotics or cryptoexotics are expected to be lightest, obviously requires inclusion of spin-dependent gluon exchange effects, which we shall now consider.

The Hamiltonian of QCD has a quark-gluon coupling term rather like the $\vec{j}\cdot\vec{A}$ of electrodynamics.

$$H_I = \sum_{a=1}^{8} \vec{j}^a \cdot \vec{A}^a \qquad (2.16)$$

This Hamiltonian in the bag model connects pure quark states to mixed quark and glue states, and diagonalizing it produces states (the π, for example) which are superpositions of each

$$|\pi> = 0(1) \mid q\bar{q}> + 0(g) \mid q\bar{q}g > + \ldots$$

In the bag model one may evaluate the wavefunction overlap integrals implicit in the interaction (2.16), and reduce the vertex to simple spin and color factors [20].

$$q_i \qquad\qquad q_j \quad = \quad \underbrace{ib_1}_{\text{overlap}} \quad \underbrace{\vec{S}_q \cdot \vec{\epsilon}_g^*}_{\text{spin}} \quad \underbrace{\lambda_{ji}^a}_{\text{color}} \quad gR^{-1} \qquad (2.17)$$

$$b_1 = 0.1389 \; , \chi_q = 2.073, \chi_g(TE) = 2.744$$

The shift in energy due to (TE) gluon exchange in the ρ/π system for example is then easily calculable;

$$\delta E \qquad\qquad\qquad = \frac{128\pi}{3} \; \frac{b_1^2}{\chi_g} \; \vec{S}_q \cdot \vec{S}_{\bar{q}} \; \alpha_s R^{-1} \qquad (2.18)$$

$$E_\rho - E_\pi = 0.94 \; \alpha_s R^{-1}$$

using the MIT mean ρ/π radius of 0.79 fm [15], the α_s needed to fit $E_\rho - E_\pi$ is

$$\alpha_s = \frac{E_\rho - E_\pi}{0.94R^{-1}} \quad = 2.6 \qquad (2.19)$$

A fit to a number of S-wave $q\bar{q}$ and q^3 states led the MIT group to a slightly smaller value of

$$\alpha_s = 2.2 \qquad (2.20)$$

The resulting fit, as we can see in the figure shown below, is generally quite impressive, and was followed by studies of the unconventional states with gluon constituents as well as $q^2\bar{q}^2$ states.

Evidence that there might be some trouble with the bag model parameters (an overestimated α_s) has come from studies of heavy quark systems [21] and from glueball states [22]. The dependence of the gg state masses shown earlier on α_s (through gluon exchange effects) is shown below

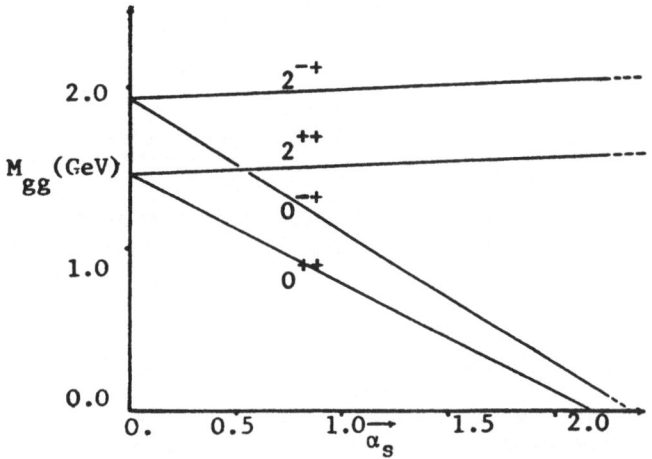

Fig. 14.   Low-lying gg glueball spectrum in the bag model [22]. (R = 1 fm for illustration, $\alpha_s \sim 0.7$ preferred).

really be of order $\alpha_s \approx 1$. [22,23], and that $q\bar{q}$ and qqq hyperfine splittings were strongly dependent on the short-distance part of the bag model wavefunction, which the bag model underestimates –

hence requiring an incorrect large $\alpha_s \sim 2.2$ in the 1975 fit [15].
The value $\alpha_s = 2.2$ gives a light 0^{-+} glueball, which no one wants.
The value $\alpha_s = 0.7$ makes it possible to account for the $\iota(1440)$ and
$\theta(1700)$ as glueballs and predicts a 0^{++} glueball at ~ 1.0 GeV
[22]. This is so like the lattice results of Ishikawa et al [24]

$$M_{2^{++}} = 1620 + 100 \text{ MeV}$$
$$M_{0^{-+}} = 1420 + 240 \text{ MeV}$$
$$\qquad\qquad\qquad - 170 \qquad\qquad\qquad (2.21)$$
$$M_{0^{++}} = 740 \pm 40 \text{ MeV}$$

that one might believe that there is general theoretical agreement
regarding the expected spectrum of gluonic hadrons. As we shall
see in discussing lattices, evidence has arisen in the last year
that physics may be more complicated than the bag model indicates,
in that the overall mass scale of lattice glueballs used above may
be an underestimate.

Let us take a devil's advocate viewpoint for a moment and ask just
how reliable the bag model predictions for an unusual system of
hadrons - such as glueballs - are likely to be. The problem is
primarily the choice of parameter values in the model; we
distinguish three different sources of uncertainty.

1) The bag radius R or bag pressure $B_0^{\frac{1}{4}}$ typically varies by $\sim$
30% from calculation to calculation, depending on what is used as
input. This gives an $\sim 30\%$ uncertainty in the mean mass of the
multiplet, E_0. Unfortunately there is evidence that $B_0^{\frac{1}{4}}$ is not a
universal hadronic constant [25], so this freedom of choice is
likely to remain.

2) There are various spin-independent effects in the bag model
which are not well understood, and which are often guessed or
included as additional free parameters. These include the
Casimir effect in the bag, quark and gluon self-energies and mass
shifts due to pion couplings. An additional uncertainty of $\sim 30\%$
in E_0 arises from this source.

3) As seen in comparing $q\bar{q}$ and gg spectroscopy, there is some
controversy regarding the effective value of the quark-gluon
coupling α_s for light hadrons, and extreme estimates typically
range over $\alpha_s = 0.7 \rightarrow 2.2$. Multiplet splittings proportional to
α_s are thus uncertain by an overall factor of about 3.

To summarize, a mythical bag model prediction of the masses
of gg glueballs which reads

$$M(0^{++}gg) = 660 \text{ MeV}$$
$$M(2^{++}gg) = 1320 \text{ MeV}$$

$$\alpha_s = 2.2 \qquad\qquad (2.22)$$

should at present be interpreted by the experimentalist as

$$E_0(gg) \ (= \tfrac{5}{6}M(2^{++}) + \tfrac{1}{6}M(0^{++})) \ \sim 1200 \pm 600 \text{ MeV}$$

$$M_{2^{++}} - M_{0^{++}} \sim 200 \to 700 \text{ MeV} \qquad\qquad (2.23)$$

Level <u>orderings</u> tend to be on much firmer ground than the detailed expectations for level positions in a given bag model calculation.

The source of such uncertainty is the lack of solid experimental candidates for gg and other unusual states as opposed to $q\bar{q}$ and $q\bar{q}q$. It is easy to fit a spectrum of known states ($q\bar{q}$, q^3) with a reasonably correct model, but it is quite another matter to predict a new spectroscopy such as gg, with rather uncertain parameters and perhaps exhibiting new physical effects.
Having reviewed predictions for gg states and issued the above disclaimer for detailed predictions of level masses, we shall now see what other gluonic hadrons we might expect to find below 2 GeV if the bag model is correct and there are indeed light glueballs.
We may envision combined states of quarks and gluons which form the required color singlet combinations, just as do $q\bar{q}$, q^3 and gg;

$$q\bar{q} = 3\times\bar{3} = \underline{1} + 8 \qquad\qquad qqq = 3\times3\times3 = \underline{1} + 8^2 + 10$$

$$\text{gluons: } g = 8 \ \ldots \qquad gg = 8\times8 = \underline{1} + 8^2 + 10 + \overline{10} + 27$$

$$q\bar{q}g \supset 8\times8 = \underline{1} + \ldots \qquad qqqg \supset 8\times8 = \underline{1} + \ldots$$

The term "hybrid" is now used to describe a product state of quarks and gluons, and is the compromise outcome of a long terminology war between "hermaphrodite" [19] and "meikton" [26]. The latter two suggestions represent independent attempts to restore the classical Greek idiom to particle physics. "Hybrid" (from the Latin "hybrida", meaning the offspring of a tame sow and wild boar) was proposed by Fishbane, Horn, Karl and Meshkov [27].

The spectrum of hybrid $q\bar{q}g$ mesons in the bag model has been calculated by several groups [19,20,26], and the detailed masses predicted depend on theoretical systematics as described above. One such result [20] is shown below.

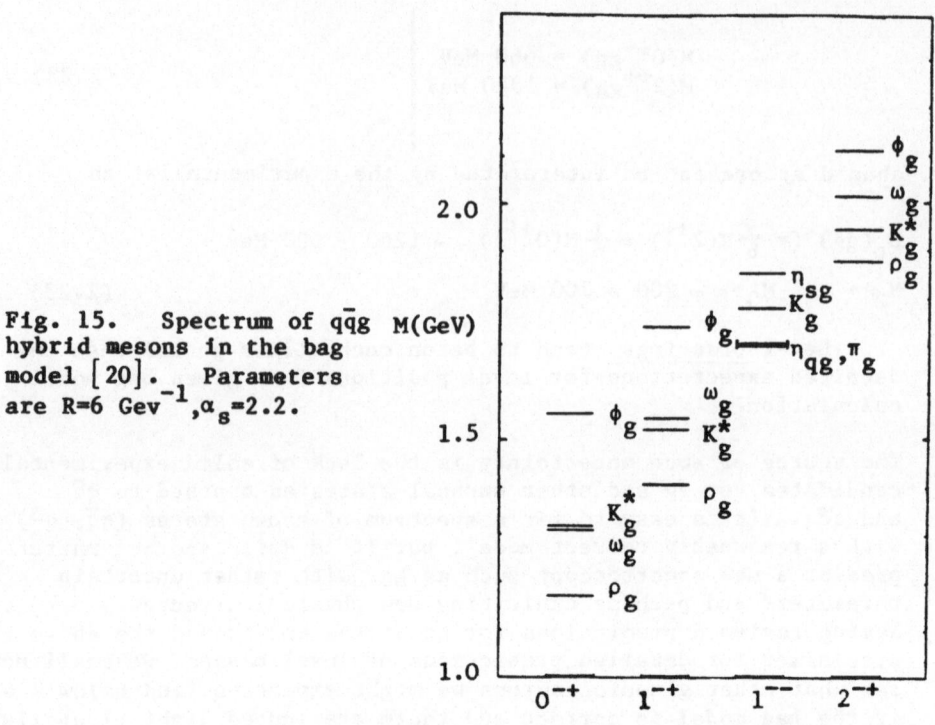

Fig. 15. Spectrum of $q\bar{q}g$ hybrid mesons in the bag model [20]. Parameters are $R=6$ Gev^{-1}, $\alpha_g=2.2$.

The level ordering $0^{-+} < 1^{-+} < 1^{--} < 2^{-+}$ and the true J^{PC} exotic 1^{-+} are the most interesting parameter-independent features. Barnes and Close [28] have suggested that the light $I=0$ 0^{-+} state $\omega_g(\sim 1.3)$ might be the $\iota(1440)$; we shall discuss this in the final lecture. Another interesting observation is the existence of an extra bump in the $(f\pi)$ 2^{-+} system at ~ 1850 MeV [29], too light to be a radial excitation of the $A_3(1660)$. Chanowitz and Sharpe [26] suggest that this might be the $I=1$ $\rho_g(2^{-+})$ $q\bar{q}g$ state. Finally, we note that some P-wave signal has been seen experimentally by Apel et al [31] in the $\eta\pi$ system from ~ 1.4 GeV upwards, as would be expected if the $I=1$ $q\bar{q}g$ exotic $\rho_g(1^{-+})$ were being produced [30]. Distinctive features which should help confirm or refute candidate $q\bar{q}g$ states are the predicted exotic 1^{-+} states, couplings to photons which are usually rather smaller than quark model $q\bar{q}$ predictions, and the very characteristic $m(\omega_g)$ $- m(\rho_g) \sim 100$ MeV splitting expected for non $- 1^{--}$ $q\bar{q}g$ states due to the process $q\bar{q}g \rightarrow gg \rightarrow q\bar{q}g$ [20,26].

Along with hybrid mesons come hybrid baryons, q^3g. The most promising area to find these is in the light quark baryon spectrum of N's and Δ's, and bag model calculations of the expected

212

spectrum have been carried out [33,34]. The lightest q^3g state is expected to be an $I=\frac{1}{2}$, $J^P=\frac{1}{2}^+$ state rather like the famous Roper resonance N(1440). There are arguments for [33] and against [33,34] Roper=q^3g in the literature.

This ends the discussion of the bag model as a tool for the study of conventional and gluonic hadrons. It is evidently a very useful and admirably simple model, although one must be aware of its problems (spurious states, "systematic" theoretical uncertainties) when comparing its predictions with experiment.

b)   Lattice QCD (Monte - Carlo)

Rather than give a detailed review of lattice QCD as I have done for the bag model, I will instead be rather schematic about details and quickly get to results and recent developments which are of relevance to experimentalists.

A few theoretical points should be covered initially. In lattice calculations, one replaces the continuum gluon lagrangian (1.18-19) by a nonunique discrete generalization. One frequent choice of the action
$S = + \int d^4x\, L$ is the "fundamental action" due to Wilson [35].

$$S = \sum_{\square} S_{\square} = -\frac{\beta}{6} \sum_{n,\mu<\nu} Tr\left[U_{(\mu)}U_{(\nu)}U^+_{(\mu)}U^+_{(\nu)} + h.c. \right] \qquad (2.24)$$

$Tr(UUU^+U^+) \sim \vec{E}^2 - \vec{B}^2 =$ gauge invariant

$U_{(\mu)} = \exp(iga\lambda^a A^a_{(\mu)}) = $ SU(3) matrix "link"

Fig. 16.   An elementary plaquette on the lattice [35]. The two free parameters are $\beta=6/g^2$ and a.

The details of S don't really concern us here; all we require is that it approach the correct continuum theory as a→0. (The question of which discrete S approaches the continuum action most rapidly is actually a matter of much theoretical interest at present.)

How are masses evaluated in Monte-Carlo calculations?
First, one generates complete sets of the SU(3) matrices $\{U(n_x,$
$n_y, n_z, n_\tau)\}$ (each set is called a "configuration"), with a weight
proportional to

$$W(\{U\}) = e^{-S(\{U\})} \qquad (2.25)$$

This is the weight expected in the true quantum field theory
vacuum (in Euclidean time) as formulated in the language of path
integrals.   One then evaluates correlation functions of an
operator $\mathcal{O}$ which creates the state of interest from the vacuum.
These correlation functions decay exponentially in Euclidean time,
and the slope is the particle's mass

$$\lim_{\substack{T \to \infty \\ (T=n_\tau a)}} \sum_{\vec{n}} < \mathcal{O}^+_{(\vec{n}, n_T)}(U)\, \mathcal{O}_{(\vec{0}, 0)}(U) > = ce^{-M_{\mathcal{O}} T} \qquad (2.26)$$

$M_{\mathcal{O}}$ = mass of lightest state $\mathcal{O}$ creates
If we are studying glueballs, the operators $\{\mathcal{O}\}$ can be taken as
various combinations of Wilson loops (traces of U's around closed
paths).   The expectation value of two Wilson loops at different
separations tells us the mass M of the lightest particle
(glueball) which one such loop can create.

A predicted mass (of the lightest 0^{++} glueball, for example),
depends on the two parameters $\beta = \frac{6}{g^2}$ and a.   For dimensional
reasons it is of the form

$$M(0^{++}gg) = n(\beta)a^{-1}(\beta) \qquad (2.27)$$

M is a physical constant, say $M(0^{++} gg) = 1.1$ GeV.   The inverse
lattice spacing $a^{-1}(\beta)$ and the coefficient $n(\beta)$ are individually
not physical constants.   If we choose to work first at $g^2 = 5.7$
($\beta = 1.05$) and later at $g^2 = 6.0$ ($\beta = 1.00$), consistency demands that
M = 1.1 GeV be found at either coupling β.   <u>The lattice spacing</u>
$a^{-1}(\beta)$ <u>is however not constant,</u> but varies with $\beta = \frac{6}{g^2}$ in a
calculable fashion.   This "running scale"
may seem surprising, but is really just the familiar running
coupling constant

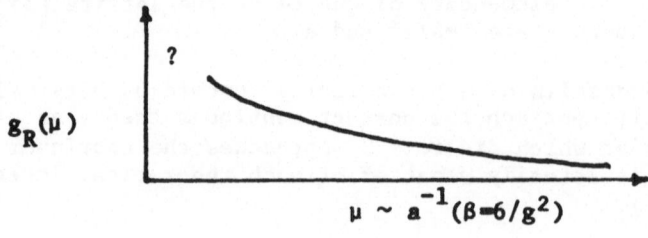

in another guise. In the continuum, the arbitrary
renormalization point $\mu (\propto \Lambda_{MS})$ must be chosen to set the
overall mass scale; given $g(\mu)$, the value of $g(\mu')$ at some other
μ' is known through the renormalization group. On the lattice, we
use not μ but a^{-1} to set the mass scale, so the values of a^{-1} at
different g's are determined by the same renormalization group
equations;

$$(a^{-1} \frac{\partial}{\partial a^{-1}} + \beta(g) \frac{\partial}{\partial g}) \, g_R(a^{-1}) = 0 \qquad (2.28)$$

The two-loop $\beta(g)$ function gives $\left[\beta \equiv \frac{6}{g^2} \text{ is } \underline{\text{not}} \text{ the function} \right.$
$\beta(g) \left. \right]$.

$$a(\beta \equiv \frac{6}{g^2})^{-1} = \Lambda_0 c_1 \beta \, \frac{51}{121} \, e^{c_2 \beta} \quad (c_1, c_2 \text{ known}) \qquad (2.29)$$

It is an obvious consistency check to see if a physical
quantity like $M(0^{++} gg)$ really is constant, assuming $a^{-1}(\beta)$
scales as shown above. If the physical quantity measured really
looks constant in $\beta = \frac{6}{g^2}$, it is said to exhibit "scaling". A
nice example is the physical string tension $\sqrt{K}$ (the force between
two heavy q and $\bar{q}$ sources at large distance), determined by
measuring large Wilson loops. The results of Barkai, Creutz and
Moriarty [36] on a 12^4 lattice are shown below.

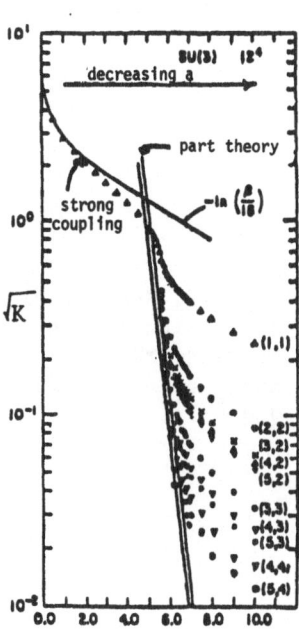

Fig. 17. Monte-Carlo evaluation
of the lattice string tension [36].
(An example of scaling).

For $\beta > 6.$, $g^2 < 1.$, where weak-coupling perturbation theory begins to be valid, we can see that $\sqrt{K}$ doesn't look constant unless we assume that the units it is measured in, a^{-1}, have strong β dependence. The expected dependence $a^{-1}(\beta)$ (2.29) for two values of the physical scale Λ_0 are shown in this figure, and the Monte-Carlo and theoretical results track rather well. The string tension between infinitely heavy q and $\bar{q}$ sources can be determined by measuring Wilson loops because they are simply related to the static interquark potential. A rectangular Wilson loop of sides R x T, fixed T >> R, is related to V(R) by

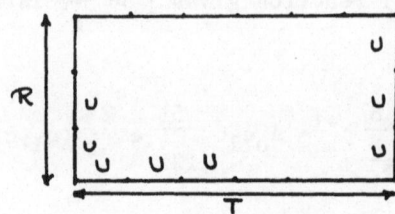

$$W(T,R) \equiv < \frac{1}{3} \; \mathrm{Re} \; \mathrm{Tr} \left\{ \underset{\mathrm{path}}{\Pi \; U} \right\} > \; \rightarrow \; ce^{-TV(R)} \qquad (2.30)$$

V(R) may then be determined by taking logs of ratios of different length W's.
A recent example of the Monte-Carlo evaluation of V(R) in SU(3) by Stack [37] using this same technique is shown below. The short-range Coulombic
part and the (probable) linear confining part of V(R) are clearly in evidence.

Fig. 18. Monte-Carlo evaluation of the heavy quark-anti-quark potential in QCD [37].

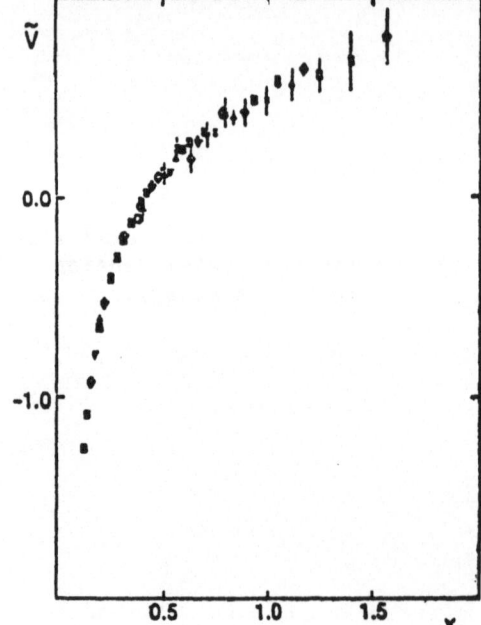

The determination of the string tension $\sqrt{K}$, the slope of the linear part of the potential $V(R)$, has become a matter of some controversy of late, for reasons relating to the glueball spectrum predicted by Monte-Carlo QCD. Ishakawa, Sato, Schierholz and Teper [24] predicted a rather light glueball mass using Monte-Carlo techniques;

J^{PC}	M_{gg} MeV)	quoted statistical error
0^{++}	740	± 40
0^{-+}	1420	$+ 240$ $- 170$
2^{++}	1620	± 100
1^{-+} ggg exotic	1730	± 220

Other J^{PC} exotic ggg states are found to have masses > 3 GeV.

These masses could nicely account for the i(1440) and $\Theta(1700)$ if there is a gg scalar - coupled to $\pi\pi$ - near 740 MeV. An experimental search at the ISR found no evidence for such a state below 1150 MeV [38]. What has gone wrong?

As we have seen, Monte-Carlo calculations give masses in terms of the lattice spacing a^{-1}, which must itself be determined by fitting some known quantity. What Schierholz and Teper [39] actually find is

$$M_{gg}(0^{++}) \sim 1.1\ a^{-1}(\beta=5.7) \qquad (2.31)$$
$$M_{gg}(2^{++}) \sim 2\tfrac{1}{2}\ a^{-1}(\beta=5.7)$$

and so forth. The physical value of $a^{-1}(\beta=5.7)$ was taken from a measurement of the string tension

$$\sqrt{K} = \lim_{R \to \infty} \frac{dV(R)}{dR} \qquad (2.32)$$

due to Creutz and Moriarty [40], which gives a lattice spacing

$$a^{-1}(\beta=5.7) = 0.77 \pm 0.12\ \text{Gev} \qquad (2.33)$$

It is assumed (from potential model fits) that $\sqrt{K} \simeq 420$ MeV, and this gives one a value for the physical constant Λ_o (lambda lattice) (2.29). The Λ_o which gave the a^{-1} ($\beta=5.7$) quoted in (2.33) is

$$\Lambda_o = (6\pm1) \cdot 10^{-3}\ \sqrt{K} \qquad (2.34)$$

The problem with using the string tension as input is that one must work at quite large R to see the asymptotic slope in (2.32). The value of Λ_o quoted above was not measured at large enough R, so $dV(R)/dR$ was still decreasing significantly. The more recent measurement of Stack [37] gives

$$\Lambda_o = (11\pm3) \cdot 10^{-3}\ \sqrt{K} \qquad (2.35)$$

Alternately, one may use quarkonium masses (usually m_ρ) rather than $\sqrt{K}$ to set the scale of $a^{-1}(\beta=5.7)$ for M_{gg}. Two groups which use m_ρ as input [41, 42] find

$$\Lambda_o \simeq (12\pm1) \cdot 10^{-3} \sqrt{K} \qquad (2.36)$$

As Ishakawa et al use $\Lambda_o = 6.10^{-3}\sqrt{K}$, it is clear that their values of M_{gg} may have to be scaled up by a factor of two! This "string tension crisis" feeds through to the bag model as well, which previously gave glueball masses rather similar to the Monte-Carlo results
of Ishikawa et al.

If the glueball mass scale on the lattice was indeed too small by a factor of two, it probably means that the glueball bag radius R_{gg} was previously too large by a factor of two, as bag model glueball masses are proportional to
R_{gg}^{-1}. If so, the obvious conclusion for the candidate glueball states ι and Θ is

$$\iota(1440) \neq gg \; 0^{-+} \qquad (2.37)$$

$$\Theta(1700) \neq gg \; 2^{++}$$

because the experimental masses are a factor of two below the revised lattice masses. It also implies that the bag constant B_o which implies $R_{q\bar{q}} \sim R_{gg}$, may require two very different values to describe $q\bar{q}$ and gg spectroscopy.

Even if the new, larger lattice parameter Λ_o is confirmed in high accuracy string tension measurements, it may yet prove impossible to accurately extract glueball masses, because one must work at very small separations to measure the relevant Wilson loop correlation functions. Typically one measures glueball masses using Wilson loops separated by one or two lattice spacings, whereas it is known that $q\bar{q}$ correlation functions don't settle down to their asymptotic behaviour (2.26) before ~ 10 lattice spacings [43]. Glueball correlation functions are unfortunately very "noisy" and probably cannot be measured at such large separations.

Finally, I will close with a "commercial" for three lattice calculations which should soon produce interesting results.

A high accuracy, large lattice measurement of the lattice string tension is being carried out by J. Stack, which will probably confirm the larger glueball mass scale implicit in (2.35-36).

Monte-Carlo calculations of the $q\bar{q}g$ 1^{-+} exotic meson mass and the light baryonium state $q^2\bar{q}^2$ 0^{++} mass are being carried out in Edinburgh and should soon produce results [44].

means that it may soon be possible to calculate strong decay
widths of unusual states like glueballs and hybrid mesons.
Narrow widths have often been suggested as characteristic features
of these states [27,46], but no one has been able to calculate the
widths of such "exotic" states reliably.   This $\rho-\pi\pi$ coupling
constant evaluation means that Monte-Carlo techniques may put
strong decay width calculations for light quark and gluon states
within the reach of theorists at last.

III. ϕ Radiative Decays

In this final lecture I will review the experimental and
theoretical status of states seen in ϕ radiative decays, $\phi \to X\gamma$.
A review of this largely experimental subject is not out of place
here because much of the theoretical work on the "exotic" systems
gg and $q\bar{q}g$ was stimulated by the unexpected results seen in $\phi \to$
$X\gamma$.   Here I shall concentrate in particular on the three states
$\iota(1440)$ 0^{-+}, $\theta(1700)$ 2^{++} and $F(2220)$ $(2n)^{++}$.

ϕ radiative decays are a natural place to look for glueballs
because in perturbation theory these decays proceed through a
two-gluon intermediate state with variable invariant mass, $0 < M_{gg}$
$< M_\phi$.

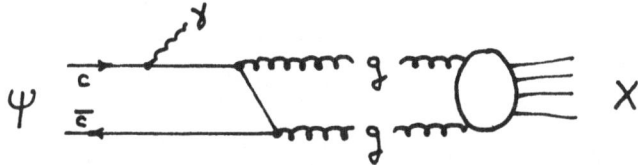

Fig. 19. Tree diagram process $\phi \to \gamma gg$. (gg $\to$ hadrons, X).

One may look at inclusive or exclusive hadrons X, and search
for resonances; any bump in the M_X distribution which is not a
well-established $q\bar{q}$ state is a natural glueball candidate.

What J^{PC} do we expect to produce?   Lacaze and Navelet [47] did a J^{PC} analysis of the two transverse gluon final state in the tree diagram $\psi \to \gamma gg$, and found the following production rate for various $J^{PC} gg$ combinations

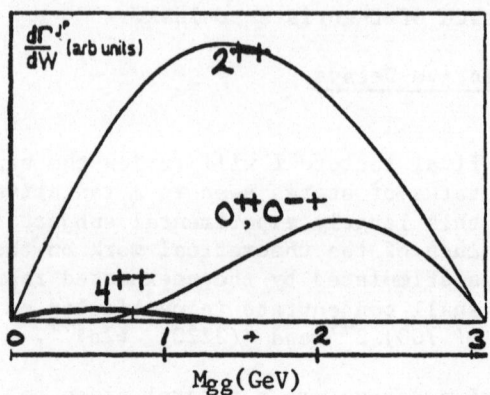

Fig. 20.   Tree diagram production rate of gg states in $\psi \to \gamma gg$ [47].

If $\psi \to X\gamma$ is indeed dominated by two transverse gluon production, we expect the hadronic final state X to consist almost exclusively of $J^{PC} = 0^{++}, 0^{-+}$ and 2^{++} combinations.   The 0^{++} and 0^{-+} signals peak at rather large invariant mass and are small for $M_{gg} < 1$ GeV.   Other J^{PC} are essentially absent.

The first experimental results from the Mark II group [48] were measurements of the branching fraction $\psi \to X\gamma/\psi \to$ all for various X.   The results showed, surprisingly, that the largest signal seen was $\psi \to$ "E(1420)"γ.

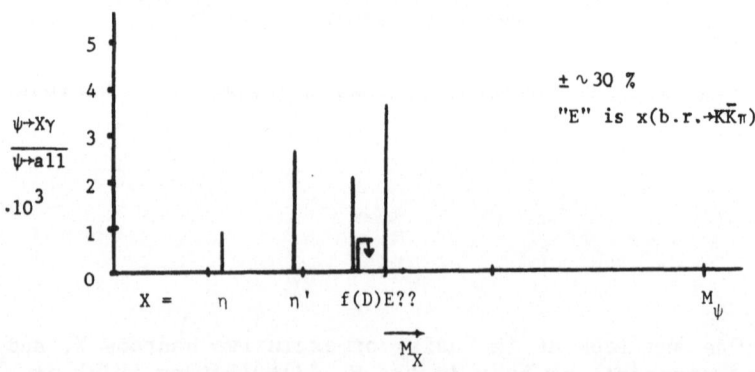

Fig. 21. $\psi \to X\gamma/\psi \to$ all branching fractions at Mark II [48].

The "E", a 1^{++} s$\bar{s}$ state, was expected to be a very weak signal, because two transverse gluons cannot make a 1^{++} state. Chanowitz [49] suggested that the $\psi \rightarrow$ "E"γ signal was actually due to the production of a pseudoscalar glueball, and that a spin-parity measurement of this state would give $J^{PC} = 0^{-+}$, not 1^{++}. When this was indeed shown to be a $J^{PC} = 0^{-+}$ state, it was given the name iota, i(1440), to distinguish it from the 1^{++} s$\bar{s}$ E.

A recent measurement of the i(1440) signal at the Mark III detector in $\pi^0 K^+ K^-$ states is shown below.

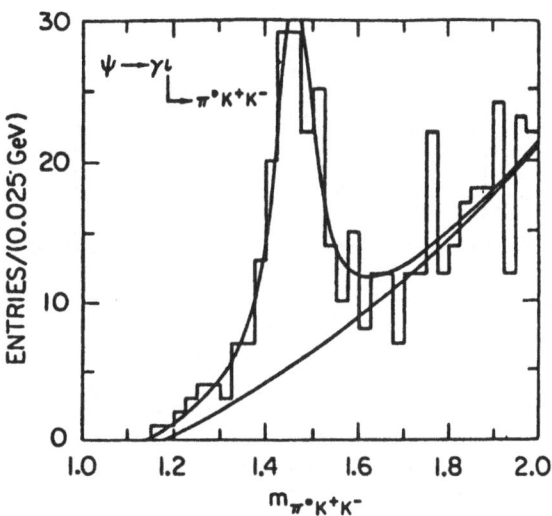

Fig. 22.   The i(1440) in $\pi^0 K^+ K^-$ at Mark III [53].

It is an amusing accident of history that the original detection of the "E" – in P$\overline{P}$ annihilation at rest, by Armenteros et al in 1963 [50] – may have actually been an early detection of the i(1440). A later analysis of P$\overline{P}$ annihilation data [51] found a mixture of 0^{-+}(i?) and 1^{++}(E?) components, with the former being dominant at rest. The original "E" bump in P$\overline{P} \rightarrow (KK\pi)\pi\pi$ is reproduced below.

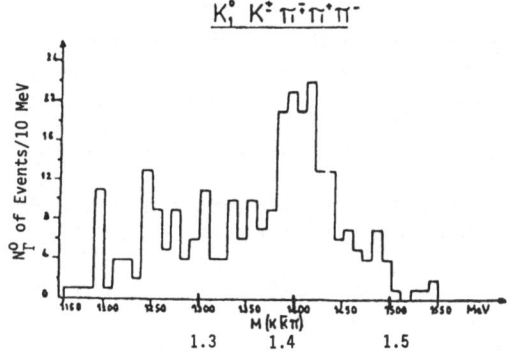

Fig. 23.   Possible detection of the i(1440) in $K_S K^{\pm}\pi$ in PP annihilation at rest in 1963 [50].

If one assumes that the ι(1440) is a gg 0^{-+} glueball, one may calculate which additional glueball and hybrid (qq̄g) mesons one might expect to find. Various bag model estimates of the masses of these states gave similar spectra

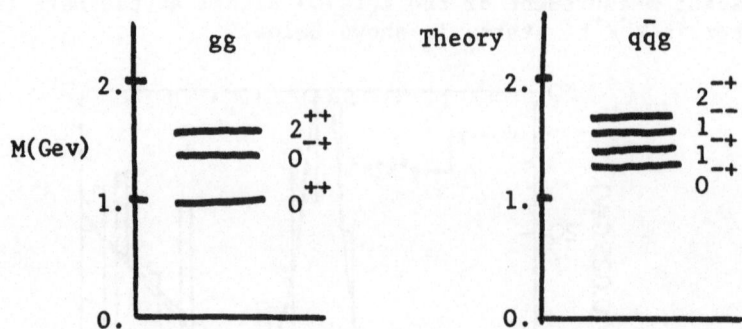

Fig. 24.   Typical bag model spectra of gg and qq̄g states.

to ~ 100 MeV [46]. Some references noted that, as quark-gluon couplings are independent of flavor, glueballs might be expected to decay to all flavor channels with approximately equal amplitude, for example

$$\Gamma(\text{gg } 2^{++} \rightarrow \pi\pi) \sim \Gamma(\text{gg } 2^{++} \rightarrow K\bar{K}) \qquad (3.1)$$

The status of the ι(1440) has been somewhat obfuscated by the search for it in the ηππ channel. The preference of the decay ι → KK̄π for low mass KK̄ pairs led to the suggestion that this was a quasi two-body decay ι → δπ → (KK̄)π.   As the δ decays to both KK̄ and ηπ, the Crystal Ball collaboration looked for evidence of the ι(1440) in ηπ$^+$π$^-$.   They found only a possible hint of the ι(1440), as a shoulder on a very large signal at ~ 1.7 GeV [52]. The Mark III group has also looked at ηπ$^+$π$^-$, and find that structure does exist in "δ"π → ηππ at about 1.4 GeV, but that the mass of the enhancement they see is significantly below 1.4 GeV; in one fit a mass of ~ 1.38 GeV is quoted [53].   The absence of the ι(1440) in ηππ and the presence of a different signal nearby in mass is a surprising result, and we certainly would like to have independent confirmation and a J^{PC} measurement if this new state is not the ι(1440).

Theorists can probably learn more from electromagnetic decays of the ι(1440) than from its hadronic decay modes.   This is because radiative decays such as (qq̄)* → (qq̄)γ can be calculated to perhaps 30% accuracy in various quark models.   For this reason I believe that the way to resolve possible assignments for the

i(1440) (and other unusual states) is through their
electromagnetic couplings, for example the recent Mark III and
Crystal Ball measurements of i → ρψ through ψ → γ(γρ) [53,54].
The γρ° mass distribution seen at Mark III is shown below.

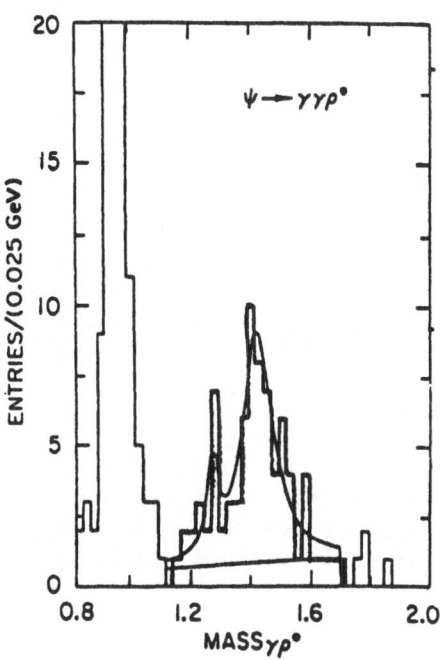

Fig. 25.   Possible evidence for i(1440) → γρ at Mark III [53].

Various assignments we might imagine for the i(1440) and the
corresponding rate Γ(i → ρψ) are

i(1440) assignment	Γ(i → ρψ)
n_R [55] (assumes ≠ n_R(1275))	0.9±0.5 MeV [56]
qq̄g ω_g(0^{-+}) [28]	~1.5MeV [57]
gg [49] (Predicts (γφ) (1440)).	~ 0.3 MeV [58] (wide range)

The experimental result from Mark III is

$$\Gamma(i \to \rho\gamma) = 1.4 \pm 0.6 \text{ MeV } [53] \qquad (3.2)$$

The glueball assignment for the i(1440) may already be in
trouble in many models, and an improved measurement of i → ρψ
together with more careful theoretical work will probably make it
untenable.   The hybrid qq̄g assignment for the iota is in

223

reasonable agreement with $i \to \rho\gamma$. Taken literally, it predicts a $\psi \to \gamma(\gamma\phi)$ signal with $\sim 2/9$ the strength of $\psi \to \gamma(\gamma\rho)$ at a $(\gamma\phi)$ invariant mass of $m(\phi_g \, 0^{-+} \, q\bar{q}g) \sim 1.7$ GeV if $i = q\bar{q}g$, due to $\psi \to \gamma\phi_g$, $\phi_g \to \gamma\phi$. There should also be a $\gamma(\gamma\omega)$ signal at ~ 1.3 GeV due to $\psi \to \gamma \, i(0^{-+})$, $i \to \gamma\omega$, but with only $1/9$ the strength of $\psi \to \gamma i$, $i \to \rho\gamma$. A competing assignment is $i = \eta_R(1440)$ assuming the mass of the first radial I=0 $q\bar{q}$ state η_R is at 1440 MeV. The rate expected for $\eta_R(1440) \to \rho\gamma$ by Godfrey and Isgur [56] is 0.9 ± 0.5 MeV, again consistent with the Mark III measurement. This assignment can only be correct if the previously announced $\eta_R(1275)$ is not a radial eta state. In the Godfrey-Isgur model the i(1440) cannot be the η_R'; the latter is found to be almost pure $s\bar{s}$, and would not couple to $\gamma\rho$ significantly. The above discussion shows how the accurately predictable $q\bar{q}$ electromagnetic couplings, together with accurate experimental data, may make it possible to distinguish the options $q\bar{q}/q\bar{q}g/gg$ in the non-exotic sectors, even in the very complicated I=0 pseudoscalar sector. To finally understand the i(1440) we should be able to model $i \to \rho\gamma$ and $i \to \gamma\gamma$ in terms of its quark and gluon content to $\sim 30\%$.

If pressed to speculate given the data available today, I would say that the i(1440) is most likely a radially excited eta, with some $s\bar{s}$ component ($m_i > m_\pi*$). The puzzle $\psi \to \gamma i / \psi \to \gamma\eta'$ is understandable as a reflection of the rapid increase in the strength of the $0^{-+}gg$ source which drives this channel with $M_x = 1. \to 1.5$ GeV, as found by Lacaze and Navelet [47], Fig. 20.

$\Theta(1700)$

Next I will briefly discuss the $\Theta(1700)$ 2^{++} meson, which certainly has no place in the $q\bar{q}$ meson spectrum. This state was first seen by the Crystal Ball [59] in $\psi \to (\eta\eta)\gamma$. It has since been seen in $K_S^0 K_S^0$ and K^+K^- final states at Mark III [53]. The mass and width determined by Mark III are

$$M_\Theta = 1719 \pm 6 \text{ MeV} \qquad \Gamma_\Theta = 117 \pm 23 \text{ MeV} \qquad (3.3)$$

The prominence of the $\Theta(1700)$ in $K\bar{K}$ final states makes a radial f(1270) $(u\bar{u} + d\bar{d})/\sqrt{2}$ 2^{++} unlikely, and it is far too light to be a radial excitation of the f'(1516) $s\bar{s}$ 2^{++}. There have been many suggestions that the Θ might be a tensor glueball. If so, flavor blindness of glueball decays would suggest

$$\Gamma(2^{++}gg \to \pi\pi) \sim \Gamma(2^{++}gg \to K\bar{K}). \qquad (3.3)$$

(See however reference [62].) Similarly, $\Theta = f_R$ would give

$$\Gamma(f_R \to \pi\pi) \gg \Gamma(f_R \to K\bar{K}). \qquad (3.4)$$

Experimentally, no $\Theta \to \pi\pi$ has yet been seen!

$$\phi \to \gamma\Theta \cdot \Theta \to \pi\pi \qquad\qquad < 3.10^{-4} \qquad\qquad (3.5)$$

compared with

$$\phi \to \gamma f(1270) \cdot f \to \pi\pi \quad \sim \quad 2.10^{-3} \qquad\qquad (3.6)$$

A possible clue to the Θ comes from the large $\eta\eta$ signal

$$\phi \to \gamma\Theta \cdot \Theta \to \eta\eta > 10. \quad \phi \to \gamma f \cdot f \to \eta\eta. \qquad (3.7)$$

These results can be understood if there is hidden strangeness – an $s\bar{s}$ pair – in the Θ. This is certainly consistent with the large $K\bar{K}$ branching fractions, for example

$$\phi \to \gamma\Theta \cdot \Theta \to K^+K^- \qquad \sim 4.8.10^{-4} \qquad (3.8)$$

compared with the $s\bar{s}$ 2^{++} meson branching fraction

$$\phi \to \gamma f'(1516) \cdot f' \to K^+K^- \sim 1.6.10^{-4} \qquad (3.9)$$

This evidence for hidden strangeness in the Θ led Chanowitz [60] to suggest that the Θ might be one of the $q^2\bar{q}^2$ states predicted by Jaffe [12] using the MIT bag model, in particular an I=0 2^{++} state with hidden strangeness.

$$\Theta \stackrel{?}{=} \frac{u\bar{u} + d\bar{d}}{\sqrt{2}} \, s\bar{s} \qquad\qquad (3.10)$$

This assignment explains $\Theta \to K\bar{K}, \eta\eta \gg \pi\pi$, but immediately raises two additional problems if we accept it:

1) Why is Γ_Θ only ~ 100 MeV? The "folklore" regarding $q^2\bar{q}^2$ states with fall-apart decays

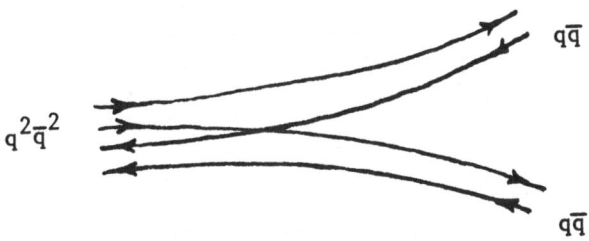

Fig. 26. Fall-apart decay of a $q^2\bar{q}^2$ state.

is that they are extremely broad, so that they do not show up as resonance bumps.

2) If for some reason $q^2\bar{q}^2$ states have ~ 100 MeV hadronic widths after all, why has only the Θ been seen? There should be a literal forest of such states, some visible in (experimentally)

featureless exotic channels like $\pi^+\pi^+$.

The $\Theta(1700)$ is evidently something of a mystery. Perhaps measurement of its electromagnetic couplings, such as $\Theta \rightarrow \gamma\gamma$, can help us test $\Theta = q^2\bar{q}^2$. There already exist theoretical estimates of this coupling [61], which is predicted to be very large, unlike the option $\Theta = gg$. As for the absence of $\Theta \rightarrow \pi\pi$, theorists have begun to back away from the flavor-blind picture of glueball decays [62], probably in response to the absence of any unusual states in $\phi \rightarrow \gamma\pi\pi$ [53].

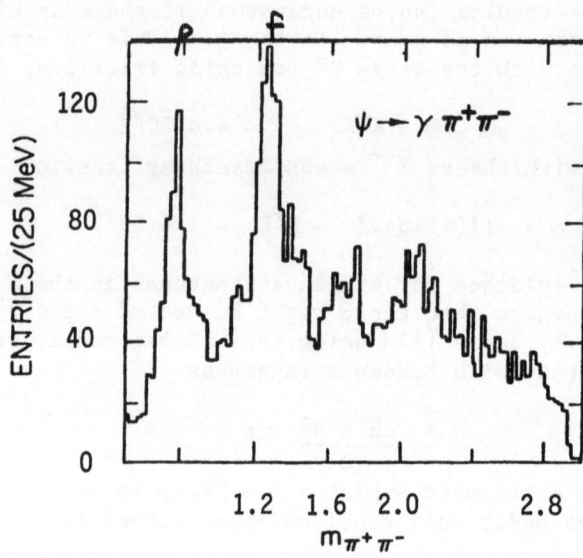

Fig. 27.   Absence of $\phi \rightarrow \gamma\Theta, \Theta \rightarrow \pi^+\pi^-$ at Mark III [53].

F(2220)

Finally, as if that wasn't enough, at the Brighton Conference the Mark III group announced the discovery of a <u>very narrow</u> state in $\phi \rightarrow \gamma K\bar{K}$, called the F(2220) [63]. The reported mass and width are

$$M_F = 2.22 \pm .03 \text{ GeV} \tag{3.11}$$

$$\Gamma_F = 30 \pm 15 \pm 20 \text{ MeV} \tag{3.12}$$

This amazingly narrow width is consistent with zero, given the experimental resolution! Most people would expect a hadron with $M \sim 2$ GeV to have a width measured in 100's of MeV's.

226

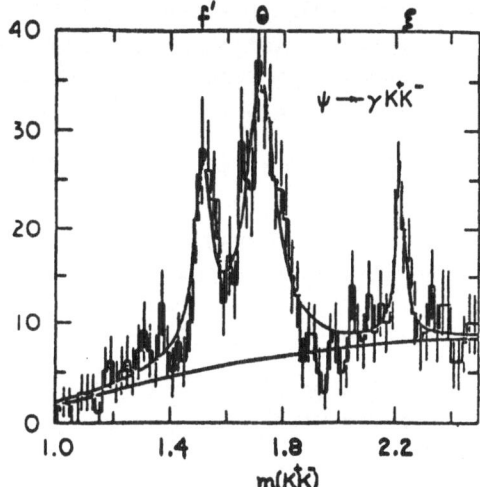

Fig. 28. The ꟻ(2220) meson
in $\psi \to \gamma\xi$, $\xi \to K^+K^-$
at Mark III [53].

There are two options for the ꟻ(2220), based on what the width
turns out to be.

$$\Gamma_{ꟻ} > 1 \text{ MeV} \to \text{Very narrow hadron.} \qquad (3.13)$$
$$2^{++}\text{gg?}$$
$$^3F_2 \; 2^{++} \; s\bar{s} \; \lceil 64 \rceil?$$

$$\Gamma_{ꟻ} < 1 \text{ KeV} \to \text{Non-hadronic; Higgs } (0^{++})? \; \lceil 65 \rceil \qquad (3.14)$$

The tensor glueball suggestion could account for the
anomalously small width, since theorists cannot yet calculate
glueball strong decay widths with any certainty.

The Godfrey-Isgur relativistic $q\bar{q}$ model predicts surprisingly
narrow widths for some orbitally excited $q\bar{q}$ states, in particular

$$\Gamma(^3F_2 \; s\bar{s}) = 45\text{--}60 \text{ MeV} \qquad (3.15)$$
and the expected mass and production cross section are also in

satisfactory agreement with experiment. Characteristic features of this $s\bar{s}$ assignment include $F \not\rightarrow \pi\pi$, $F \rightarrow (K \bar{K}^* + c.c.) \sim F \rightarrow K\bar{K}$. A convincing test of this $s\bar{s}$ assignment for the F is the predicted production of a corresponding $2^{++}\omega^* \, ^3F_2$ state at $\sim$ 2.0 GeV in $\phi \rightarrow \gamma\pi\pi$. The width expected for this $\omega^* \, 2^{++}$ is

$$\Gamma(^3F_2 \, \omega^*) \cong 150 \text{ Mev } \lfloor 56 \rfloor. \tag{3.16}$$

If $\Gamma_F < 1$ KeV, then the $F(2220)$ is evidently not a hadron. One alternate suggestion is that the F might be a Higgs 0^{++} boson. In the simplest model, with a single Higgs doublet, the experimental rate for $\phi \rightarrow F\gamma$ is already too large $\lceil 65,66 \rceil$,

$$\phi \rightarrow \gamma F(2220) \cdot F \rightarrow K^+K^- \sim 8.10^{-5} \tag{3.17}$$

compared with

$$\phi \rightarrow \gamma H^o \cdot H^o \rightarrow K^+K^- \sim 0.5.10^{-5} \tag{3.18}$$

One would also expect to see a significant $\mu^+\mu^-$ signal, if the Higgs couplings accounts for both quark and lepton masses;

$$H^o \rightarrow \mu^+\mu^-/K^+K^- = \frac{m_\mu^2}{3m_s^2} \sim 10\% \tag{3.19}$$

With more than one Higgs doublet, one can arrange mixing angles so that the problems with $\phi \rightarrow \gamma H^o$ do not occur $\lceil 65 \rceil$.

It should be clear that the measurement of the $F(2220)$ width on an MeV scale and then of its J^{PC} are matters of crucial importance for model builders in QCD and now perhaps in other fields as well.

Acknowledgements

It is a pleasure to thank the organizers of the Erice school on Exotic Atoms, Professors P. Dalpiaz, G. Fiorentini and G. Torelli, for their kind invitation to give these lectures. I would also like to express my gratitude to my fellow participants for many interesting and enjoyable conversations. Finally, the assistance I have received from F.E. Close, S. Godfrey and N. Isgur in completing this manuscript is gratefully acknowledged.

References

Lecture I

[1] C.N. Yang and R.L. Mills, Phys. Rev. 96 (1954) 191.
[2] J.D. Bjorken and S. Drell, Relativistic Quantum Fields (McGraw-Hill, 1964).
[3] E. Noether, Nachr. kgl. Ges.Wiss. Gottingen (1918) 235.
[4] W. Marciano and H. Pagels, Phys. Rep. 36C (1978) 137.
[5] D. Gross and F. Wilczek, Phys. Rev. Lett. 30 (1973) 1343.
[6] H.D. Politzer, Phys. Rev. Lett. 30 (1973) 1346.
[7] W.H.K. Panofsky, in Proc. of the 14th Intl. Conf. on High-Energy Physics, Vienna, 1968 (CERN).
[8] C.G. Callan, Phys. Rev. D12 (1970) 1541.
     K. Symanzik, Comm. Math. Phys. 18(1970) 227.
[9] G.Zweig, CERN preprint 8419/TH412 (1964).
[10] H.J. Lipkin, Phys. Lett. 124B (1983) 509.
     S.J. Lindenbaum, Brookhaven preprint BNL 33286.
[11] S.L. Adler, Phys. Rev. 177 (1969) 2426.

Lecture II

[12] R.L. Jaffe, Phys. Rev. D15 (1977) 267.
[13] P.N. Bogoliubov, Ann. Inst. Henri Poincaré, 8(1968) 163.
[14] A. Chodos, R.L. Jaffe, K. Johnson, C.B. Thorn and V.F. Weisskopf, Phys. Rev. D9 (1974) 3471.
[15] T. de Grand, R.L. Jaffe, K. Johnson and J. Kiskis, Phys. Rev. D12 (1975) 2060.
[16] S.A. Williams, F.J. Margetan, P.D. Morley and D.L. Pursey, Phys. Rev. Lett. 49 (1982) 771. [17] T. Barnes, F.E. Close and S. Monaghan, Nucl. Phys. B198 (1982) 380.
[18] C.N. Yang, Phys. Rev. 77 (1980) 242.
[19] T. Barnes and F.E. Close, Phys. Lett. 116B (1982) 365.
[20] T. Barnes, F.E. Close and F. de Viron, Nucl. Phys. B224 (1983) 241.
[21] R.L. Jaffe and J. Kiskis, Phys. Rev. D13 (1975) 1355.
[22] T. Barnes, F.E. Close and S. Monaghan, Phys. Lett. 110B (1981) 159.
[23] T. Barnes, Phys. Rev. D30 (1984) 1961.
[24] M. Teper, in Proc. of the Intl. Europhysics Conf. on High-Energy Physics, Brighton, 1983 (Rutherford Appleton Laboratory).

[25] The most recent results for glueball masses on the lattice would require a larger B_0 in the bag model than is required for qq and q^3 spectroscopy. Similar problems occur for cc states.

[26] M. Chanowitz and S. Sharpe, Nucl. Phys. B222 (1983) 211.

[27] P.M. Fishbane, D. Horn, G. Karl and S. Meshkov NBS-81-0896 (1981).

[28] This suggestion has appeared in [20] and [30].

[29] C. Daum et al., Phys. Lett. 89B (1980) 285.

[30] T. Barnes, Proc. of the XVIII Moriond Conf. (1983), 525.

[31] W.D. Apel et al., Nucl. Phys. B193 (1981) 269.

[32] T. Barnes and F.E. Close, Phys. Lett. 123B (1983) 89.

[33] E. Golowich, E. Haqq and G. Karl, Phys. Rev. D28 (1983) 160.

[34] T. Barnes and F.E. Close, Phys. Lett. 128B (1983) 277.

[35] K. Wilson, Phys. Rev. D10 (1974) 2445.
M. Bander, Phys. Rep. 75C (1981) 205.

[36] D. Barkai, M. Creutz and K.J.M. Moriarty, Phys. Rev. D29 (1984)1207.

[37] J.D. Stack, Phys. Rev. D29 (1984) 1213.

[38] T. Akesson et al., Phys. Lett. 133B (1983) 268.
D. Morgan and M.R. Pennington, RL-83-126 (1983).

[39] G. Schierholz and M. Teper, Phys. Lett. 136B (1984) 64.

[40] M. Creutz and K.J.M. Moriarty, Phys. Rev. D26 (1982) 2166.

[41] G. Martinelli, G. Parisi, R. Petronzio and F. Rapuano, Phys Lett. 122B (1983) 283.
G. Parisi, R. Petronzio and F. Rapuano, Phys. Lett. 128B (1983) 418.

[42] K.C. Bowler, in Proc. of the Intl. Europhysics Conf. on High-Energy Physics, Brighton, 1983 (Rutherford Appleton Laboratory) 15.

[43] K.C. Bowler, G.S. Pawley, D.J. Wallace, E. Marinari and F. Rapuano, Nucl. Phys. B220 (1983) 137.

[44] T. Barnes and A. Thomson, to be published.

[45] S. Gottlieb, P.B. MacKenzie, H.B. Thacker and D. Weingarten, Phys. Lett. 134B (1984) 346.

[46] C.E. Carlson, J.J. Coyne, P.M. Fishbane, F. Gross and S. Meshkov, Phys. Lett. 99B (1981) 353.
K. Senba and M. Tanimoto, Phys. Lett. 106B (1981) 215.

Lecture III

[47] R. Lacaze and H. Navelet, Nucl. Phys. B186 (1981) 247.

[48] D.L. Scharre et al., Phys. Lett. 97B (1980) 329.

[49] M. Chanowitz, Phys. Rev. Lett. 46 (1981) 981.

[50] R. Armenteros, in Proc. of the Sienna Intl. Conf. on Elem. Part., 1(1963) 287.

[51] P. Baillon et al., Nuo. Cim. 50A (1967) 393.

[52] C.E. Newman-Holmes et al., SLAC-PUB-2971 (1982). C. Edwards et al., SLAC-PUB-3111 (1983).

[53] W. Toki, SLAC-PUB-3262 (1983).

[54] D. Hitlin, CALT-68-1071 (1983).

[55] S. Godfrey and N. Isgur, Mesons with Chromodynamics, U. Toronto preprint (1984).

[56] S. Godfrey and N. Isgur, personal communication.

[57] T. Barnes and F.E. Close, RAL-84-055 (revised).

[58] S. Iwao, Lett. Nuo. Cim. 35 (1982) 209.

[59] C. Edwards et al., Phys. Rev. Lett. 48(1982) 458.

[60] M. Chanowitz, LBL-13398 (1981). More recently Chanowitz has suggested a glueball interpretation for the $\theta(1700)$, in LBL-166530 (1983).

[61] F.M. Renard, SLAC-PUB-3126 (1983).

[62] M.S. Chanowitz and S.R. Sharpe, LBL preprint LBL-16489 (1983).

[63] K. Einsweiler, in Proc. of the Intl. Europhysics Conf. on High-Energy Physics, Brighton, 1983 (Rutherford Appleton Laboratory)348.

[64] S. Godfrey, R. Kokoski and N. Isgur, $F(2.22$: an L=3 s$\bar{s}$ Meson? U. Toronto preprint (1984).

[65] H.E. Haber and G.L. Kane, SLAC-PUB-3209 (1983).

[66] F. Wilczek, Phys. Rev. Lett. 39 (1977) 1304.

QUARKONIUM SPECTROSCOPY

W. Buchmüller

CERN, Geneva, Switzerland

1. INTRODUCTION

Over the past ten years heavy quarkonia have been extensively studied[1),2)], both experimentally and theoretically. They occur as families of narrow resonances and are believed to be bound states of heavy quark-antiquark pairs. So far, two sets of resonances have been discovered: the ψ- and Υ- spectroscopies, which are related to the c- and b-quark species.

Heavy quarkonia provide direct evidence for the quark structure of hadrons, and the interest in the ψ- and Υ- families has always been stimulated by the hope that these systems may be the "hydrogen atoms" of strong interactions. It is the theme of these lectures to discuss to what extent this hope has been fulfilled and to point out where the quarkonium spectroscopy is still incomplete.

The bound state structure of quarkonia is very simple. As the constituents are heavy, a non-relativistic treatment based on the Schrödinger equation and a static potential is sufficient to a good approximation. The phenomenological success of potential models, as well as theoretical limitations of this approach, are discussed in Section 2. Similar to atomic physics the non-relativistic potential model can be extended to account for electromagnetic and hadronic transitions and spin-dependent forces, as we will see in Sections 3 and 4. Because of their simplicity, quarkonia are promising systems with respect to quantitative QCD tests; the present status as well as some theoretical expectations are discussed in Section 5. Due to the gluonic degrees of

233

freedom bound states in addition to those predicted by potential models are expected. Such "extra" states are the subject of Section 6. Section 7 contains a brief summary.

Before we proceed to a discussion of potential models, let us briefly recall a few basic experimental facts[3]. Most of our knowledge about the ψ- and Υ-spectroscopies has been obtained at electron-positron storage rings where quarkonium resonances are seen in the reaction (cf. Fig. 1)

$$e^+ e^- \rightarrow \text{hadrons}$$

At centre-of-mass energies between 3.0 GeV and 4.5 GeV, the ψ-family (J/ψ, ψ',...) was discovered and the Υ-resonances (Υ, Υ',...) were found in the range from 9.4 GeV to 11 GeV (cf. Fig. 2). Further states have been discovered by studying the decays of the resonances produced in $e^+ e^-$ collisions. The analysis of the cascade decays

$$\Upsilon' \rightarrow \gamma \gamma \, \Upsilon$$
$$ \hookrightarrow e^+ e^-$$

and the inclusive photon spectrum

$$\Upsilon' \rightarrow \gamma + \text{anything}$$

led to the discovery of the charmonium χ states as well as the η_c and η_c'. More recently, analogous states in the Υ-spectroscopy, χ_b and χ_b', have been found. In addition to photon transitions, hadronic transitions such as

$$\Upsilon' \rightarrow \Upsilon \, \bar{\pi} \bar{\pi} \quad , \quad \Upsilon' \rightarrow \Upsilon \, \bar{\pi} \bar{\pi} \quad ,$$

have been observed.

An impressive amount of experimental information about the ψ and Υ systems has been accumulated as a few numbers may illustrate: the Mark III group[4] has analyzed 2.7×10^6 ψ decays, the latest Crystal Ball analysis[5] of the inclusive γ spectrum from ψ' involves 1.8×10^6 hadronic ψ' decays, and the Υ mass is known up to $\Delta M/M \sim 10^{-6}$[2],

$$M_\Upsilon = 9460.0 \pm 0.1 \quad \text{MeV} \quad ,$$

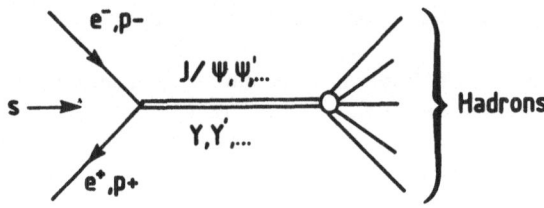

Fig. 2   Cross-section for $e^+e^- \rightarrow$ hadrons, showing the Υ resonances
[CUSB data, from Ref. 3)].

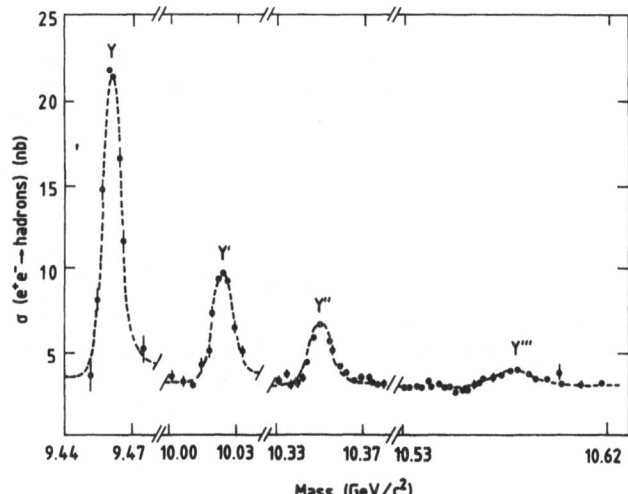

Fig. 1   Quarkonium resonances in the reaction $e^+e^- \rightarrow$ hadrons.
$p_-(p_+)$ is the electron (positron) four-momentum;
$s = (p_- - p_+)^2$ is the total centre-of-mass energy squared.

which corresponds to QED accuracy in hadron spectroscopy! As we
will see, however, there are still important aspects in the theory
of quarkonia which have not yet been tested and further experimental work at electron-positron as well as hadron colliders is
needed to complete our understanding of heavy quark-antiquark
bound states.

2. HEAVY QUARKS AND THE ($Q\bar{Q}$) POTENTIAL

The theory of hadron structure is quantum chromodynamics (QCD)[6]. The fundamental building blocks are the colour triplet spin-$\frac{1}{2}$ u-, d-, s-, c-, b-, (t-) quarks and a colour octet of spin-1 gluons. The resulting quark-antiquark attraction is described by a Coulomb-like potential with a "running coupling" whose strength depends on the quark-antiquark separation r,

$$\alpha_s(r) \equiv \frac{g_s^2(r)}{4\pi} = \frac{12\pi}{33-2n_f} \frac{1}{\ln \frac{1}{r^2 \Lambda^2}} \quad , \tag{2.1}$$

where n_f is the number of "effective" flavours and Λ the QCD scale parameter, i.e., $\Lambda \sim 300$ MeV. Due to the increase of $\alpha_s(r)$ with r, a quark-antiquark pair cannot be separated. For the "confinement" radius, i.e., the typical hadronic size, one obtains from (2.1):

$$r_c \sim \frac{1}{\Lambda} \sim 0.7 \text{ fm} \quad . \tag{2.2}$$

Light quarks are quarks whose intrinsic ("current") mass is small compared to the QCD scale, i.e., $m_Q \ll \Lambda$. Examples are the u- and d-quark masses: $m_u \sim 5$ MeV, $m_d \sim 9$ MeV. The approximate chiral symmetry associated with light quarks is spontaneously broken through the formation of vacuum condensates,

$$\langle 0|\bar{u}u|0 \rangle \sim \langle 0|\bar{d}d|0 \rangle \sim \Lambda^3 \quad . \tag{2.3}$$

This generates an effective ("constituent") mass $\tilde{m}_Q$ for light quarks,

$$\tilde{m}_Q \sim \Lambda \sim \frac{1}{2} m_\rho \sim \frac{1}{3} m_N \quad , \tag{2.4}$$

and leads to the additive quark model as an approximate description of ordinary hadrons.

Heavy quarks are quarks whose current mass is large compared to the scale parameter Λ. So far we know two heavy quark species: the c-quark with $m_c \sim 1.5$ GeV and the b-quark with $m_b \sim 4.8$ GeV. Bound states of heavy quarks can be treated as non-relativistic two-body systems to good approximation. Thus with respect to their J^{PC} quantum numbers, these ($Q\bar{Q}$) bound states form a positronium-like spectrum. The total spin $\vec{S} = \vec{s}_Q + \vec{s}_{\bar{Q}}$ can take the values 0 and 1. The angular momentum $\vec{L}$ and $\vec{S}$ form the total angular momentum $\vec{J} = \vec{L} + \vec{S}$. One distinguishes

singlet states: S = 0,   J = L,
and
triplet states: S = 1,   J = $\begin{cases} 1 \\ L-1, L, L+1 \end{cases}$; $\begin{array}{l} L = 0 \\ L \neq 0 \end{array}$

with parity and C-parity

$$P = (-)^{L+1} \quad , \quad C = (-)^{L+S} \quad . \tag{2.5}$$

Of particular interest are the triplet states with L = 0. They
carry the quantum numbers of the photon, $J^{PC} = 1^{--}$, and are
therefore dominantly produced in e^+e^- collisions (cf. Fig. 3).
Figure 4 shows the narrow $(b\bar{b})$ resonances of the Υ spectroscopy as
well as the various observed hadronic and photon transitions.

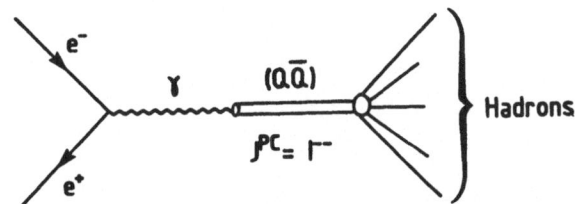

Fig. 3 Mixing between γ and $(Q\bar{Q})$ resonances, the dominant
production mechanism for hadrons in the final state.

The mass spectrum of a quarkonium family is computed by means
of the Schrödinger equation:

$$M_n(Q\bar{Q}) = 2m_Q + E_n(m_Q, V) \quad ,$$
$$\left[-\frac{1}{m_Q} \Delta + V(r) \right] \varphi_n(r) = E_n \varphi_n(r) \quad . \tag{2.6}$$

Obviously, the main problem of quantitative spectroscopy is the
choice of the correct $(Q\bar{Q})$ potential V(r). So far the potential
has not yet been computed from first principles, although the
progress in lattice-QCD[7] appears promising. At present, one
therefore has to rely on simple models.

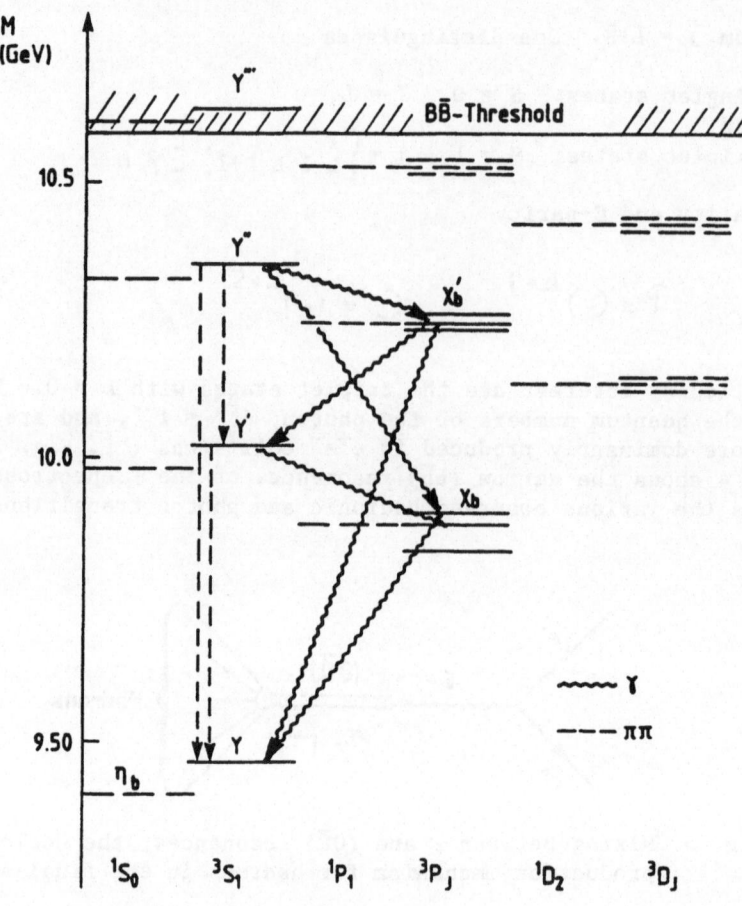

Fig. 4 T-spectroscopy. The quantum numbers of the various states
are given by $n^{2S+1}L_J$, L = S, P, D, ... (see text). Full
(dashed) lines correspond to observed (predicted) states.

(i) QCD-like models

In this class of models, the known asymptotic behaviour of
the static $(Q\bar{Q})$ potential in QCD is incorporated. One has

$$V(r) \underset{r \ll \frac{1}{\Lambda}}{\sim} - \frac{4}{3} \frac{d_c(r)}{r} \quad , \qquad (2.7a)$$

$$V(r) \underset{r \gg \frac{1}{\Lambda}}{\sim} k r \quad , \qquad (2.7b)$$

where k is the "string tension". Equation (2.7a) corresponds to a static Coulomb field at short distances whereas (2.7b) reflects the linear increase of the binding energy with distance for a colour-electric flux tube (cf. Fig. 5). In QCD-like models the potential at intermediate distances is chosen ad hoc.

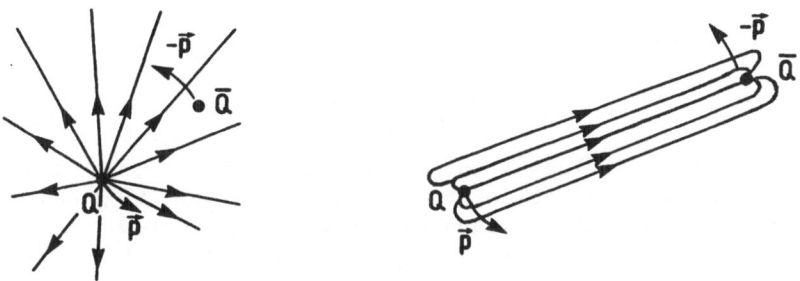

Fig. 5 Field configurations which determine the $(Q\bar{Q})$ potential at small and large distances.

The simplest choice is the Cornell model[8] for which the
first extensive study of quarkonia was carried out:

$$V(r) = -\frac{4}{3}\frac{\alpha_s}{r} + kr \quad , \quad \alpha_s = const. \qquad (2.8)$$

Already for ($b\bar{b}$) bound states, the vacuum polarization corrections
to α_s are important quantitatively. Models with "running
coupling" $\alpha_s(r)$ have been studied by many authors. A particularly
simple and successful example was suggested by Richardson[9]. The
Fourier transform

$$\tilde{V}(q) = \int d^3x \, e^{-i\vec{q}\vec{x}} \, V(r) \quad , \quad r = |\vec{x}| \, , \quad q = |\vec{q}| \quad , \qquad (2.9)$$

has the asymptotic behaviour

$$\tilde{V}(q) \underset{q \gg \Lambda}{\sim} -\frac{4}{3}\frac{48\pi^2}{33-2n_f}\frac{1}{q^2 \ln\frac{q^2}{\Lambda^2}} \quad , \qquad (2.10a)$$

$$\tilde{V}(q) \underset{q \ll \Lambda}{\sim} -8\pi k \frac{1}{q^2} \quad , \qquad (2.10b)$$

where k is the string tension introduced in (2.7b). Richardson's
interpolating formula for (2.10a) and (2.10b) reads

$$\tilde{V}(q) = -\frac{4}{3}\frac{48\pi^2}{33-2n_f}\frac{1}{q^2 \ln(\frac{q^2}{\Lambda^2}+1)} \quad . \qquad (2.11)$$

Equation (2.11) defines a potential which depends only on a single
scale parameter Λ as it is the case for the true QCD potential.
Consequently, the string tension k and Λ are related.
Richardson's fit uses Λ = 398 MeV which yields k = 0.147 GeV^2.

The potential (2.11) describes the ψ and Υ spectroscopies
remarkably well. Yet, in order to relate Λ to a well-defined QCD
scale parameter, e.g. $\Lambda_{\overline{MS}}$, one has to proceed to a next level of
sophistication and incorporate next-to-leading order perturbative
QCD corrections for the ($Q\bar{Q}$) potential. This has been done[10]
and the running coupling of the potential, $\alpha_p(q) = (-3/16\pi)q^2\tilde{V}(q)$,
has been studied in terms of its β-function. For Υ spectroscopy,
however, this development is phenomenologically rather unimportant
and we will therefore not discuss it further in these lectures.

(ii) Empirical models

Theoretically unbiased one may also attempt to describe quarkonium spectroscopy using simple empirical potentials. Examples are the logarithmic[11] and small power potentials. Such models check the QCD-content of QCD-like models. A successful example has been suggested by Martin[12]

$$V(r) = A + B \, r^{0.1} \quad . \tag{2.12}$$

How well do potential models work? Two QCD-like models and Martin's model are compared with experiment in Table 1. All of them work very well, in fact better than one should expect. Within the theoretical uncertainties, the three models cannot be distinguished. Figure 6 shows various potentials with the mean square radii of the two quarkonium families. The ψ and Υ spectroscopies probe distances between 0.1 fm and 1.0 fm. In this range the phenomenological potential appears to be essentially logarithmic and there is evidence neither for a Coulombic nor a linearly rising part.

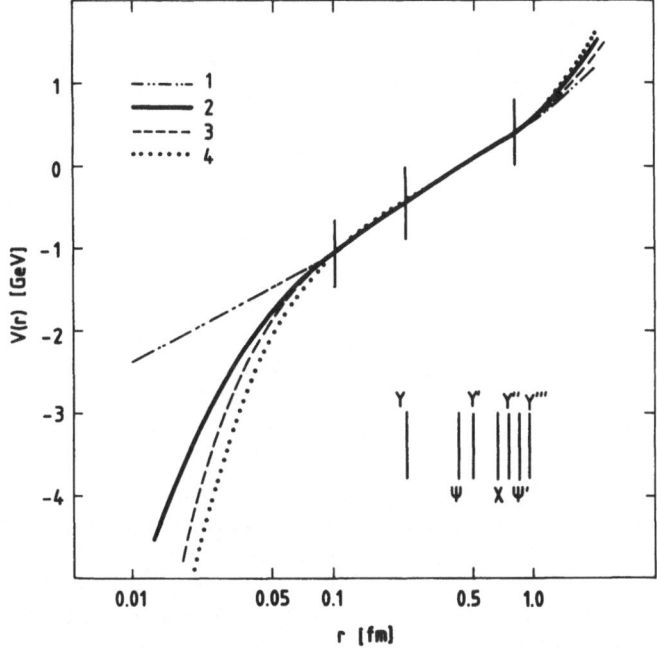

Fig. 6 Different $(Q\bar{Q})$ potentials [from Ref. 13)]: (1) Ref. 12), (2) Ref. 10), (3) Ref. 14), (4) Ref. 8).

Table 1   Predictions of three potential models compared with
          experiment. Excitation energies for different (c$\bar{c}$) and
          (b$\bar{b}$) states are given in GeV. The ratios of leptonic
          widths with respect to the ground state are listed in
          brackets. The asterisk denotes input to fix the
          parameters of the potential.

	State	Experiment Refs. 2), 3)	Cornell Ref. 8)	"QCD" Ref. 10)	Martin Ref. 12)
(c$\bar{c}$)	2S	0.589 (0.46±0.06)	0.589* (0.44)	0.60 (0.46)	0.592* (0.40)
	1P	0.425	0.427*	0.42	0.407
(b$\bar{b}$)	2S	0.561 (0.44±0.03)	0.560* (0.39)	0.56* (0.44)	0.560* (0.41)
	3S	0.890 (0.32±0.03)	0.898 (0.27)	0.89 (0.32)	0.900 (0.35)
	1P	0.440 (±1)	0.463	0.43	0.401
	2P	0.796 (±5)	0.811	0.79	0.782

There are, however, two hints that the potential at
distances below 0.1 fm may be as predicted by perturbative QCD.
The leptonic width of the Υ resonance, $\Gamma_{ee}(\Upsilon) \equiv \Gamma(\Upsilon \rightarrow e^+e^-)$, is
proportional to the square of the wave function at the origin,

$$\Gamma_{ee}^{(o)}(\Upsilon) = \frac{16\pi e_b^2 d_{EM}^2}{M_\Upsilon^2} |\varphi_\Upsilon(o)|^2 \quad , \tag{2.13}$$

and thereby particularly sensitive to the short distance part of
the potential. Contrary to other models QCD-like models with
running coupling and QCD corrections to $\Gamma_{ee}^{(o)}$ taken into account
yield the correct value for the Υ leptonic width[13]. Theory and
experiment are compared in Table 2. The second hint comes from
the recently observed χ_b states in the Υ spectroscopy (cf.
Fig. 4). Next to Υ, the χ_b resonances are the quarkonia of
smallest size[13]

242

$$\langle r^2 \rangle^{\frac{1}{2}}_{\Upsilon} \quad < \quad \langle r^2 \rangle^{\frac{1}{2}}_{\chi_b} \quad < \quad \langle r^2 \rangle^{\frac{1}{2}}_{J/\psi} \quad < \quad \cdots \qquad (2.14)$$

$$0.23 \text{ fm} \qquad 0.39 \text{ fm} \qquad 0.42 \text{ fm}$$

The sensitivity of their centre-of-gravity (c.o.g.) position with respect to the short distance behaviour of the $(Q\bar{Q})$ potential is shown in Table 3: the more singular the potential, the larger the mass difference $M(\chi_b) - M(\Upsilon)$. In the pure Coulombic case, one has $M(\chi_b) - M(\Upsilon) = M(\Upsilon') - M(\Upsilon)$.

Table 2 Υ leptonic width of a "QCD"- model[13] compared with experiment[2].

	"QCD"- model	Experiment
$\Gamma_{ee}(\Upsilon)$ (keV)	1.07±0.24	1.24±0.05

Table 3 χ_b-Υ mass differences for different potentials compared with experiment.

	Martin Ref. 12)	Richardson Ref. 9)	Cornell Ref. 8)	Experiment Ref. 2)
$M_{c.o.g.}(\chi_b)-M(\Upsilon)$ [MeV]	401	436	463	440±1

Given the empirical fact that potential models describe quarkonium energy levels with an accuracy up to a few MeV, one may conclude from the success of Richardson's potential that the regime below 0.1 fm is governed by perturbative QCD. Obviously, for the final evidence we have to wait for the ζ-spectroscopy of the t-quark.

Voloshin has made a prediction for the χ_b-Υ mass difference based on QCD sum rules[15]

$$M(\chi_b) - M(\Upsilon) = 0.37 \pm 0.03 \text{ GeV} \quad , \qquad (2.15)$$

and it has been emphasized[16], even after[17] the experimental discovery of the χ_b states, that the χ_b–Υ mass difference should distinguish between potential models and QCD sum rules. It is conceivable[18], however, that the discrepancy between (2.15) and the experimental result (cf. Table 3) is a problem of the specific calculation in Ref. 15) and not of the QCD sum rule approach in general.

What is the theoretical significance of the phenomenological success of potential models? Consistency of the non-relativistic potential model demands that corrections are not too large. A complete treatment of relativistic corections[19] has not been achieved and appears impossible without a deeper understanding of the confinement problem. Yet the order of magnitude of relativistic corrections to mass differences is easily estimated,

$$\Delta_{REL}\left(M_{\Upsilon'}-M_{\Upsilon}\right) \sim \left\langle \frac{v^2}{c^2}\right\rangle_{\Upsilon} \left(M_{\Upsilon'}-M_{\Upsilon}\right) \sim 45 \text{ MeV} \, . \quad (2.16)$$

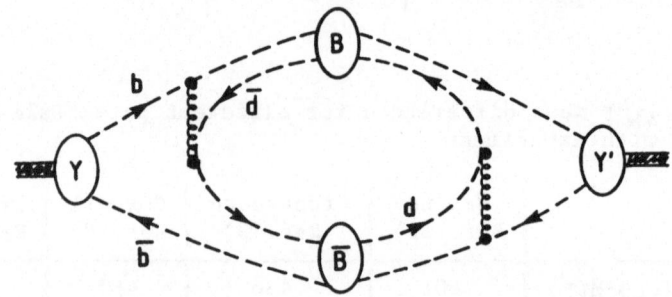

Fig. 7   Contribution to the Υ mass arising in second order perturbation theory due to light quarks.

Furthermore "coupled channel" corrections, first evaluated by the Cornell group[8], are important. They appear in second order perturbation theory as a consequence of heavy quark – light quark bound state intermediate states (cf. Fig. 7) and reflect vacuum polorization effects of light quarks. A recent calculation of Heikkilä, Törnqvist and Ono[20] yields

$$\Delta_{cc}\left(M_{\Upsilon'}-M_{\Upsilon}\right) \sim -20 \text{ MeV} \, . \quad (2.17)$$

In view of (2.16) and (2.17) the accuracy of potential models up to a few MeV appears almost suspicious! It may be due, however, to cancellations between various large contributions. The empirical fact of the flavour independence of the $(Q\bar{Q})$ potential, which is evident from Fig. 8, suggests that the total effect of relativistic and coupled channel corrections is small.

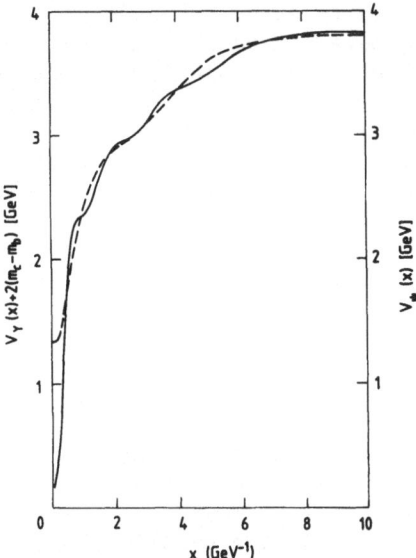

Fig. 8   Evidence for flavour independence of the (Q$\bar{\text{Q}}$) potential.
Comparison of the potentials V_ψ and V_Υ, constructed by
means of the inverse scattering method from ψ and Υ data,
from Ref. 21).

Even if the non-relativistic potential model is theoretically
self-consistent, one may still argue that the empirical (Q$\bar{\text{Q}}$)
potential is only an effective potential and has nothing to do
with the static energy of an infinitely heavy quark-antiquark pair
in QCD. This point of view has been advocated by the ITEP group
[c.f., e.g., Ref. 16)], based on the existence of the gluon
condensate,

$$\eta \equiv \langle 0 | : \frac{d_S}{\pi} G^a_{\mu\nu}(0) G^{a\,\mu\nu}(0) : | 0 \rangle \sim (330\ \text{MeV})^4 , \qquad (2.18)$$

which is said to be due to "vacuum fluctuations" of spacial and
temporal size

$$R_G \sim T_G \sim \eta^{-\frac{1}{4}} \sim 0.6\ \text{fm} . \qquad (2.19)$$

The validity of the potential picture requires that the
characteristic quark oscillation time,

$$T_Q \sim \frac{1}{M_{\psi'} - M_\psi} \sim \frac{1}{M_{\Upsilon'} - M_\Upsilon} \sim 0.3\ \text{fm} , \qquad (2.20)$$

is much larger than the gluon oscillation time, i.e., $T_Q \gg T_G$.
This is not the case for (2.19) and (2.20).

Obviously there is no exact two-body problem in any quantum field theory. Yet a pure dimensional analysis does not seem sufficient to argue away the possibility of the approximate validity of potential models, which may well be a consequence of some factors of 4π. Bell and Bertlmann[22] have studied the connection between potential models and the QCD sum rule approach but a simple link is still missing. It seems that we have to wait for the next test of flavour independence in $(t\bar{t})$ spectroscopy and for progress in lattice calculations for a final answer to the question of the theoretical significance of the empirical $(Q\bar{Q})$ potential.

3. ELECTROMAGNETIC AND HADRONIC TRANSITIONS

Until recently, it seemed that potential models had one serious problem: all models yield branching ratios for the electromagnetic transitions $\psi' \to \gamma\chi_J$ (cf. Table 4, Fig. 9) which are a factor 2-3 larger than the experimentally measured values[23]. Electromagnetic El-transitions between S- and P-states are given by

$$\Gamma_J = \frac{4}{27} e_Q^2 \alpha_{EM} k_\gamma^3 |\langle 1P|r|2S\rangle|^2 (2J+1) \quad , \qquad (3.1)$$

where e_Q, α_{EM} and k_γ are the quark charge, the electromagnetic fine structure constant and the photon momentum respectively. The partial width Γ_J is proportional to the square of an overlap integral involving the 1P and 2S wave functions. Since the early days of charmonium spectroscopy, it has been emphasized[24] that due to the sensitivity to details of the wave functions, the discrepancy between theory and experiment should not be taken too serious, and that even in atomic physics Hartree-Fock calculations of El rates are frequently off by 50%. Nevertheless, a discrepancy of almost 300% is rather disturbing and the puzzle has only recently been resolved by McClary and Byers[25]. They pointed out that, due to the node in the 2S wave function, the matrix element $\langle 1P|r|2S\rangle$ has a positive and a negative contribution which partly cancel each other. Thus a small change in the wave functions, due for instance to relativistic corrections, can have a large effect on the overlap integral. A quantitative analysis[25] which incorporates relativistic and coupled channel corrections removes indeed the discrepancy with experiment, as can be seen from Table 4. For comparison, we have also listed the remarkable results which Henriques, Kellett and Moorhouse[26] obtained already in 1976 in a relativistic treatment based on the Bethe-Salpeter equation.

Table 4 Branching ratios for the radiative decay $\psi' \to \chi_J$. For the first two columns, see Ref. 23).

State	Experiment	Theory, non-relativistic	Theory, relativistic McClary+Byers[25]	Theory, relativistic Henriques et al.[26]
J=2	7.7±1.7	18±3	10±2	6.5±1.2
J=1	8.8±1.9	23±4	11±2	9.8±1.8
J=0	9.7±2.2	27±5	7.5±1.5	10±2

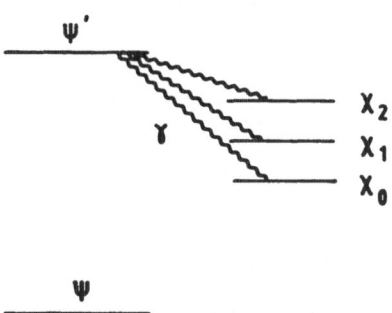

Fig. 9 Radiative transitions $\psi' \to \gamma \chi_J$ (J = 2,1,0).

In the T system relativistic corrections are expected to be much smaller than in the ψ system ($\langle v^2/c^2 \rangle_{J/\psi} \sim 0.23$, $\langle v^2/c^2 \rangle_T \sim 0.08$). Thus, it is comforting that recent measurements of the electromagnetic transitions $T' \to \gamma \chi^b_{1,2}$ are in agreement with potential model predictions as shown in Table 5.

Table 5 Branching ratios for the transition $T' \to \gamma \chi^b_J$. The dipole matrix element is taken from Ref. 13).

State	Experiment	Theory, non-relativistic
J=1	6.1±1.4	7.1±2.5
J=2	5.9±1.4	7.1±2.5

In addition to electromagnetic transitions hadronic transitions[27]-[31] between different quarkonium states, such as $\psi' \to \psi\pi\pi$, $\psi' \to \psi\eta$, $\Upsilon' \to \Upsilon\pi\pi$, etc. have been extensively studied experimentally as well as theoretically. Figure 10 shows a QCD graph which contributes to the transition $\Upsilon' \to \Upsilon\pi^+\pi^-$. Between the successive emission of the two gluons the $(Q\bar{Q})$ system propagates as a colour octet state. As the quark motion is slow, a multipole expansion[27] can be performed which leads to the scaling law[8]

$$\Gamma(\Upsilon' \to \Upsilon\pi\pi) = \left(\frac{\bar{r}_\Upsilon}{\bar{r}_\psi}\right)^4 \Gamma(\psi' \to \psi\pi\pi) \quad, \tag{3.2}$$

where $\bar{r}_{\Upsilon(\psi)}$ are the mean square radii of Υ and J/ψ. The relation (3.2) led to the prediction[1]

$$\Gamma(\Upsilon' \to \Upsilon\pi^+\pi^-)\Big|_{TH} = (5 \pm 1) \text{ keV} \quad,$$

in remarkable agreement with experiment[1],[2]

$$\Gamma(\Upsilon' \to \Upsilon\pi^+\pi^-)\Big|_{EXP} = (5.2 \pm 1.6) \text{ keV} \quad.$$

Predictions for η transitions[29] are also in agreement with experiment, whereas the theoretical expectation for the dipion mass spectrum is at variance with the experimentally measured distribution[1],[2].

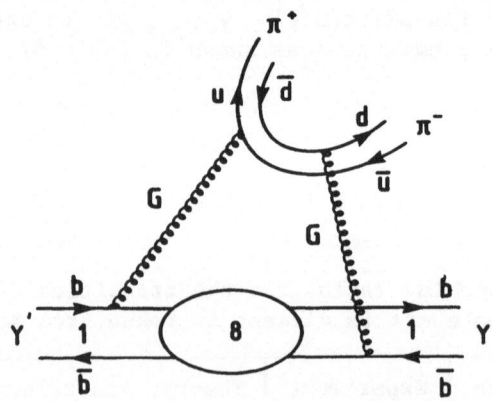

Fig. 10 Contribution to the hadronic decay $\Upsilon' \to \Upsilon\pi^+\pi^-$.

4.   SPIN-DEPENDENT FORCES

There exists an extensive literature[32] on spin-dependent
forces in heavy quark-antiquark systems. The most popular
approach has been to start from the Bethe-Salpeter equation and an
instantaneous kernel with some suitably chosen Lorentz structure,
and to deduce the corresponding Breit-Fermi Hamiltonian[33]. Since
the work of Henriques, Kellett and Moorhouse[26], it is known that
kernels which correspond to "scalar exchange" at large distances
and "vector exchange" at small distances describe well the fine
and hyperfine splittings of quarkonia. At a more fundamental
level, spin-dependent forces have been considered in the bag
model[34]-[36]. The results obtained are similar to those of the
"scalar+vector exchange" ansatz. Most closely related to QCD is
the approach of Eichten and Feinberg[37] and Gromes[38],[39]. These
authors start from the Wilson loop and relate the various pieces
of the spin Hamiltonian to correlation functions of electric and
magnetic field strengths; these correlation functions are then
either evaluated perturbatively or related to the static ($Q\bar{Q}$)
potential.

The basic structure of the spin Hamiltonian can also be
obtained in a very elementary way[40] by making use of the
intuitive flux tube picture of electric confinement and standard
techniques[41] used to derive the spin Hamiltonian for the hydrogen
atom or positronium from classical electrodynamics. For
simplicity, let us first consider a ($Q\bar{q}$) system with $m_Q \gg m_{\bar{q}} \gg \Lambda$
in the limit $m_Q \to \infty$. At small distances, $r \ll 1/\Lambda$, the antiquark
$\bar{q}$ moves in the static colour field generated by the quark Q. The
total energy of the system contains a piece ΔE which depends on
the spin $\vec{s}$ of $\bar{q}$[41],

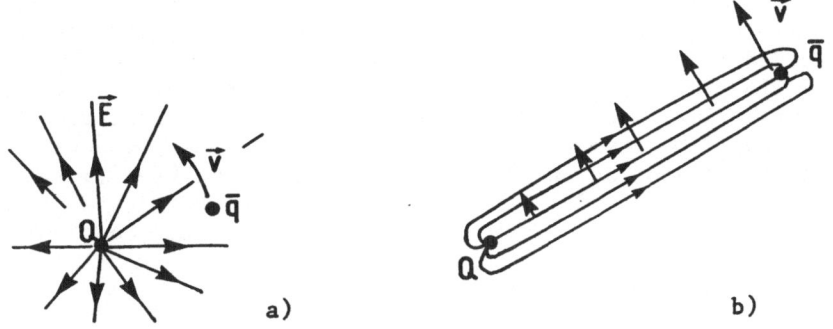

a)   b)

Fig. 11   a) Static colour electric field generated by the quark
             Q ($r \ll 1/\Lambda$);
          b) rotating electric flux tube ($r \gg 1/\Lambda$).

$$\Delta E = - \mu \vec{s} \cdot \vec{B}' + \vec{\omega}_T \cdot \vec{s} \quad , \tag{4.1}$$

where μ, $\vec{B}'$ and $\vec{\omega}_T$ are the magnetic moment, the magnetic field in the instantaneous co-moving rest frame of $\bar{q}$ and the Thomas frequency, which can be expressed in terms of the angular momentum $\vec{L}$ and the static $(Q\bar{q})$ energy $V_s(r)$,

$$\vec{\omega}_T = - \frac{1}{2m_{\bar{q}}^2} \vec{L} \frac{1}{r} \frac{dV_s}{dr} \quad . \tag{4.2}$$

In the rest frame of Q, one has a static electric field $\vec{E}$ and a vanishing magnetic field, $\vec{B} = 0$. In the co-moving frame of $\bar{q}$ this leads to

$$\vec{E}' = \vec{E} + O(v^2) \quad , \qquad \vec{B}' = - \vec{v} \times \vec{E} + O(v^2) \quad , \tag{4.3}$$

where $\vec{v}$ is the velocity of $\bar{q}$. Using $\mu = g/m_{\bar{q}}$ and $\vec{E} = -(1/g)\vec{\nabla}V_s$, one obtains from (4.1)–(4.3) the familiar spin-orbit interaction

$$\Delta E = \frac{1}{2m_{\bar{q}}^2} \vec{L} \cdot \vec{s} \frac{1}{r} \frac{dV_s}{dr} \quad . \tag{4.4}$$

In order to obtain this result, we have used the standard coupling of a spin to a magnetic field, i.e., Eq. (4.1), and our knowledge about the field configuration generated by a static quark. The spin dependent interaction energy for large Q-$\bar{q}$ separation can be obtained in a similar manner once the field configuration at large distances is known. This information is provided by the flux tube picture of electric confinement (cf. Fig. 11b), according to which a static tube contains only colour electric flux, i.e., $\vec{E} \neq 0$, $\vec{B} = 0$. If we assume that the fields in a flux tube segment moving with velocity $\vec{v}$ can be obtained by Lorentz transformation from the static case, we have in the rest frame of Q $\vec{B} = \vec{v} \times \vec{E} + O(v^2)$. Yet in the co-moving rest frame of $\bar{q}$, we obtain $\vec{B}' = 0$! From (4.1) and (4.2) we thus conclude

$$\Delta E = - \frac{1}{2m_{\bar{q}}^2} \vec{L} \cdot \vec{s} \frac{1}{r} \frac{dV_L}{dr} \quad , \tag{4.5}$$

where V_L is the static $(Q\bar{q})$ energy for distances $r \gg 1/\Lambda$. Equations (4.4) and (4.5) are identical except for the sign.

The generalization to a $(Q\bar{Q})$ system, i.e., $m_Q = m_{\bar{Q}} = m$, is straightforward and magnetostatic interactions are easily taken into account. Assuming, in the spirit of the "Coulomb+linear" potential model that the sum of the asymptotic expressions for $r \gg 1/\Lambda$ and $r \ll 1/\Lambda$ is a reasonable approximation also at intermediate distances, one obtains with $V_s = -(4/3)(\alpha_s/r)$, $V_L = kr$ [40]:

$$H_{SPIN} = - \frac{1}{2m^2} \vec{L} \cdot \vec{S} \frac{k}{r} + \frac{2}{m^2} \vec{L} \cdot \vec{S} \frac{\alpha_s}{r^3}$$

$$+ \frac{4}{3} \frac{\alpha_s}{m^2} \frac{3(\vec{s}_1 \cdot \hat{r})(\vec{s}_2 \cdot \hat{r}) - \vec{s}_1 \cdot \vec{s}_2}{r^3} \tag{4.6}$$

$$+ \frac{32}{9} \pi \frac{\alpha_s}{m^2} \vec{s}_1 \cdot \vec{s}_2 \delta^3(\vec{r}) \quad ,$$

$$\vec{S} = \vec{s}_1 + \vec{s}_2 \;,\; \vec{L} = \vec{r} \times \vec{p} \;,\; \hat{r} = \frac{\vec{r}}{r} \quad .$$

The Hamiltonian (4.6) is identical with the expression obtained from a "scalar+vector exchange" Bethe-Salpeter kernel [32]. Except for the sign of the first term, it also agrees with the result of Eichten and Feinberg [37]. According to a recent paper by Gromes [39], however, this sign difference results from an unjustified assumption about the correlation function between electric and magnetic fields and should be reversed. It thus appears that spin dependent forces of heavy quarks are essentially understood, and that the phenomenologically successful "scalar + vector exchange" ansatz, which corresponds to the naive flux tube picture of electric confinement, can be derived [37],[39] within QCD.

Table 6   ψ and Υ fine structure, total and relative P-state splittings [cf. Eq. (4.7)].   a) $\alpha_s(\psi) = 0.53$, $\alpha_s(\Upsilon) = 0.28$; b) $\alpha_s(\psi) = 0.341$, $\alpha_s(\Upsilon) = 0.227$.

	ΔM (MeV)	Flux tube model[a] Refs. 40),42)	Eichten + Feinberg[b] Ref. 37)	Experiment Refs. 23), 43)
(c c̄)	ΔM(1P)	134	131	140±5
	r(1P)	0.61	1.0	0.49±0.02
(b b̄)	ΔM(1P)	38	50	41±3
	r(1P)	0.73	1.0	0.93±0.3
	ΔM(2P)	31	36	33±5
	r(2P)	0.72	1.0	0.85±0.4

According to the Hamiltonian (4.6), spin-orbit forces are long ranged whereas tensor and spin-spin forces are short ranged. Contrary to the fine structure, the hyperfine splittings may therefore be suitable quantities for quantitative QCD tests. Predictions obtained from (4.6) for P-state splittings are compared with results of Eichten and Feinberg and experimental data in Table 6 where

$$\Delta M(nP) = M(n^3P_2) - M(n^3P_0) \quad,$$

$$r(nP) = \frac{M(n^3P_2) - M(n^3P_1)}{M(n^3P_1) - M(n^3P_0)} \quad, \tag{4.7}$$

denote the total mass difference and the ratio of upper to lower mass difference. The CUSB results[43] for the χ_b states favour the predictions of Eichten and Feinberg yet the errors are large and preliminary Crystal Ball data[44] read

$$\Delta M(\chi_b) \simeq 52 \pm 7 \text{ MeV} \quad, \quad r(\chi_b) = 0.58 \pm 0.3 \quad.$$

Beyond any doubt, fine structure splittings are a relativistic effect. From Table 6, one reads off

$$\frac{\Delta M(\chi_b)}{\Delta M(\chi_c)}\Big|_{EXP} \sim 0.29 \quad ,$$

whereas the velocities of b- and c- quarks, calculated in potential models, yield

$$\frac{\langle \frac{v^2}{c^2} \rangle_{\chi_b}}{\langle \frac{v^2}{c^2} \rangle_{\chi_c}}\Big|_{TH} \sim 0.28 \quad .$$

One important aspect of spin dependent forces remains to be tested experimentally. So far no singlet P-states ($^1P_1$) have been observed. The spin-spin forces given by the Hamiltonian (4.6) are short ranged, which one also expects from a comparison[45] of the $\psi'-\eta_c'$ and $\psi-\eta_c$ mass differences. The $^1P_1$ masses should be equal to the centre of gravity of the ^3P_J states, i.e.,

$$M(^1P_1) = \frac{5}{9} M(^3P_2) + \frac{3}{9} M(^3P_1) + \frac{1}{9} M(^3P_0) \quad . \quad (4.8)$$

A discovery of the ψ and Υ $^1P_1$-states at hadron colliders would complete our understanding of the spin forces of heavy quarks.

5. QUANTITATIVE QCD TESTS

In spite of considerable experimental and theoretical efforts, quantitative QCD tests[46],[47] have turned out to be very difficult. Scaling violations in deep inelastic scattering are obscured by "higher twist" contributions and α_s determinations from jet analyses seem to depend strongly on fragmentation models. It is conceivable that heavy quarkonia will lead to an accurate determination of the strong coupling α_s and the related QCD scale parameter $\Lambda_{\overline{MS}}$[46].

The domain of QCD perturbation theory are distances

$$r \ll \frac{1}{\Lambda} \sim 0.5 \text{ fm} \quad . \quad (5.1)$$

As we saw in Section 2, quarkonia populate the range of intermediate distances,

$$\langle r^2 \rangle^{\frac{1}{2}}_{J/\psi} \sim 0.4 \text{ fm}, \dots, \langle r^2 \rangle^{\frac{1}{2}}_{\psi'} \sim 0.9 \text{ fm},$$

(5.2)

$$\langle r^2 \rangle^{\frac{1}{2}}_{\Upsilon} \sim 0.2 \text{ fm}, \dots, \langle r^2 \rangle^{\frac{1}{2}}_{\Upsilon'''} \sim 0.9 \text{ fm}.$$

For some physical quantities, however, a factorization into a non-perturbative and a perturbative part can be achieved and reliable predictions are possible. The hadronic width of the η_b [48], for instance, is given by

$$\Gamma_{HAD}(\eta_b) \sim |\psi(0)|^2 \, |M(b\bar{b} \to 2G)|^2 ,$$

(5.3)

where $\phi(0)$ and $M(b\bar{b} \to 2G)$ are the $(b\bar{b})$ wave function at the origin and the on-shell scattering matrix element shown in Fig. 12. In general, hadronic and electromagnetic decay widths take the form (5.3). Their ratios are therefore determined by perturbation theory alone (at least up to a certain accuracy), whereas the absolute widths involve the non-perturbative quantity $\phi(0)$ which can be calculated in potential models. A complete discussion of hadronic and electromagnetic decays has been given by Remiddi [49].

At present the only determination of α_s from quarkonia is based on the hadronic width of the Υ resonance. The calculation of Mackenzie and Lepage [50] yields

$$\alpha_s^3 (0.48 \, M_\Upsilon) = \frac{\Gamma_{HAD}(\Upsilon)}{\Gamma_{\mu\mu}(\Upsilon)} \, \alpha_{EM}^2 \, \frac{81 \pi e_b^2}{10(\pi^2 - 9)} ,$$

(5.4)

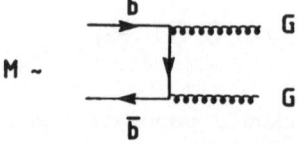

$$M \sim$$

Fig. 12 Perturbative amplitude $M(b\bar{b} \to 2G)$ for the decay $\eta_b \to$ hadrons.

from which one obtains[2]

$$\alpha_s(0.48\ M_\Upsilon) = 0.165 \pm 0.005 \ ,$$

$$\Lambda_{\overline{MS}} = 118\ {}^{+\ 16}_{-\ 15}\ MeV. \tag{5.5}$$

Additional, new information comes from the χ_b states where the branching ratios for the radiative transitions $1^3P_{2,1} \to \gamma\,1^3S_1$ have been measured. Together with the theoretical E1 widths (which are believed to be reliable, cf. Section 3), one obtains for the hadronic widths of the χ_b states[2]

$$\Gamma_{HAD}(1^3P_2) = 152\ {}^{+\ 63}_{-\ 50}\ keV \ ,$$

$$\Gamma_{HAD}(1^3P_1) = 41\ {}^{+\ 40}_{-\ 20}\ keV \ . \tag{5.6}$$

This yields the ratio

$$\frac{\Gamma_{HAD}(1^3P_1)}{\Gamma_{HAD}(1^3P_2)} \sim 0.27 \ , \tag{5.7}$$

which is compatible with the result of lowest order QCD,

$$\frac{\Gamma_{HAD}(1^3P_1)}{\Gamma_{HAD}(1^3P_2)} = \frac{5}{3\overline{u}}\ \alpha_s\ \ell u\,(m_b\,R_c) \sim 0.22 \ , \tag{5.8}$$

where we have used α_s of (5.5), $m_b = 4.9$ GeV, $R_c^{-1} = 0.4$ GeV.

More information on $\Lambda_{\overline{MS}}$ is expected from a measurement of the $\Upsilon - \eta_b$ hyperfine splitting[31]

$$\Delta M(\Upsilon - \eta_b) = \frac{8}{9} \frac{\Gamma_{ee}(\Upsilon)}{e_b^2 \, \alpha_{EM}^2} \, \alpha_{\overline{MS}}(m_b) \left[1 + 5.6 \frac{\alpha_{\overline{MS}}(m_b)}{\Pi} \right] \quad . \quad (5.9)$$

The pseudoscalar state η_b may be seen in the cascade shown in Fig. 13 for which Kuang and Yan have estimated a branching ratio of 0.5%[30]. A theoretically particularly clean quantity is the ratio of hadronic widths[52]

$$\frac{\Gamma_{HAD}(\chi^b_{J=0})}{\Gamma_{HAD}(\chi^b_{J=2})} = \frac{15}{4} \left(1 + 9.5 \frac{\alpha_s}{\Pi} \right) , \quad (5.10)$$

which, together with the $\Upsilon - \eta_b$ hyperfine splitting, is given in Table 7 for different values of $\Lambda_{\overline{MS}}$.

Table 7 $\Upsilon - \eta_b$ hyperfine splitting and χ_b hadronic widths ratio for different values of $\Lambda_{\overline{MS}}$.

$\Lambda_{\overline{MS}}$ (MeV)	100	200	300
$\Delta M(\Upsilon - \eta_b)$ (MeV)	27	35	42
$\Gamma_{HAD}(\chi^b_0)/\Gamma_{HAD}(\chi^b_2)$	5.5	5.8	6.1

A lower bound on the QCD scale parameter can been obtained from the empirical ($Q\bar{Q}$) potential. As down to 0.1 fm no sign of a Coulombic behaviour has become visible, $\Lambda_{\overline{MS}}$ cannot be arbitrarily small. In Fig. 14 the perturbative QCD potential for different values of $\Lambda_{\overline{MS}}$ is compared with the empirical ($Q\bar{Q}$) potential at small distances ($r < r_c$, $1/r_c^2\Lambda_{\overline{MS}}^2 = 100$). This leads to the estimate [13] $\Lambda_{\overline{MS}} > 100$ MeV.

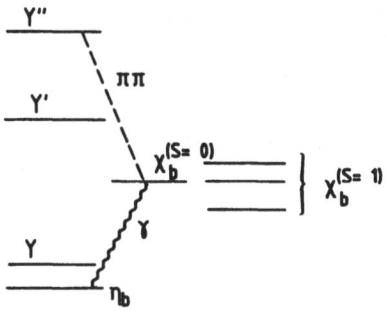

Fig. 13   Cascade decay $T" \to \chi_b^{S=0}\pi\pi \to \gamma\eta_b\pi\pi$.

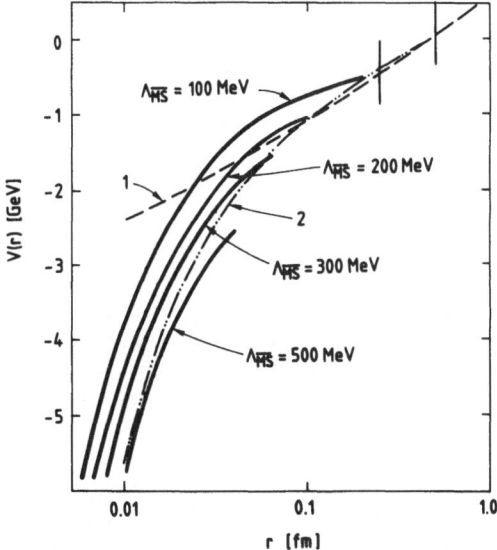

Fig. 14   Comparison of the perturbative QCD potential for different values of $\Lambda_{\overline{MS}}$ with the phenomenological ($Q\bar{Q}$) potential for distances $r < r_c = 1/10\Lambda_{\overline{MS}}$, from Ref. 13).

Heavy quarkonia are a promising testing ground for perturbative QCD. As they are non-relativistic systems, the bound state dynamics is very simple, and it may be possible to control and separate non-perturbative from genuine perturbative effects. Potential dangers for this programme are, however, already apparent; they include the dependence on the renormalization scheme[53),54)], relativistic[55)] and coupled channel[20)] corrections, and possible deviations from factorization. Nevertheless, agreement between various independent determinations of $\Lambda_{\overline{MS}}$ from heavy quarkonia would provide a non-trivial quantitative confirmation of quantum chromodynamics.

6. "EXTRA" STATES

Even more difficult than quantitative QCD tests appears the verification of a qualitative QCD prediction: the existence of states due to gluonic degrees of freedom which are not predicted by the quark model. So far, there exists no unequivocal evidence for glueballs[1].

Quarkonium families should also contain "extra" states. In the flux tube picture there are $(Q\bar{Q})$ states with an excited flux tube configuration. In the string model, which corresponds to the zero-width limit of the flux tube, such "vibrational" states (cf. Fig. 15) have first been discussed by Giles and Tye[56]. A specific prediction[57] for the first excited state in the Υ family with $J^{PC} = 1^{--}$,

$$M(\Upsilon_{VIB}) - M(\Upsilon) = 990 \pm 90 \text{ MeV},$$

$$\Gamma_{ee}(\Upsilon_{VIB}) = 0.20 \pm 0.15 \text{ keV}, \qquad (6.1)$$

turned out not to be correct. The experimental upper bound[3] for the leptonic width of a 1^{--} resonance between Υ'' and Υ''' reads $\Gamma_{ee} < 20$ eV. Thus the approximations made in Ref. 57) to estimate the properties of the "extra" states were not adequate.

Hasenfratz, Horgan, Kuti and Richard[34] have used the bag model to calculate properties of quark-antiquark-gluon bound states. They have predicted an axial vector ($J^{PC} = 1^{-+}$) with an excitation energy

Fig. 15 Ground state and excited state of a string (zero-width flux tube).

$$M(b\bar{b}G) - M(\Upsilon) = 1040 \pm 200 \text{ MeV} , \qquad (6.2)$$

which, unfortunately, is difficult to test experimentally. More recent calculations give somewhat larger excitation energies. Ono[58] expects $(Q\bar{Q}G)$ states with $\Delta M \sim 1.3$ GeV and a lattice calculation[59] also yields states with $\Delta M = (1.3 \pm 0.2)$ GeV for the Υ system.

The discovery of "extra" states in the ψ or Υ families would be of great importance. It could provide direct evidence for the gluonic degrees of freedom and more insight into the structure of bound states in QCD.

7. SUMMARY

We have discussed various aspects of quarkonium physics. The ψ and Υ spectroscopies are well described by potential models. The known part of the $(Q\bar{Q})$ potential between 0.1 fm and 1.0 fm is essentially logarithmic, yet the success of QCD-like potentials with respect to the Υ leptonic width and the centre of gravity of the recently discovered χ_b states suggests that the Coulombic region starts below 0.1 fm. The ζ-spectroscopy of the t-quark should therefore provide direct evidence for asymptotic freedom.

The long-standing puzzle of the charmonium E1 transitions has been resolved. The discrepancy between theory and experiment in the ψ family has been removed by incorporating relativistic corrections. For the Υ system, where the bound quarks move much slower than in the ψ system, non-relativistic models describe the data well.

What is the relation between the phenomenological $(Q\bar{Q})$ potential and the static energy of an infinitely heavy $(Q\bar{Q})$ pair in QCD? The phenomenological success of potential models, in particular the evidence for flavour independence, suggests that the difference between the two quantities is small, and that deviations due to relativistic and coupled channel corrections, as well as the gluon condensate are not very large. The final answer will be provided by lattice calculations and the $(\bar{t}t)$ spectroscopy.

Spin dependent forces appear to be essentially understood within QCD: starting from the Wilson loop, a Hamiltonian has been derived which agrees with the successful "scalar+vector exchange" ansatz and the intuitive flux tube picture. The measured ψ and Υ

fine structure splittings are in agreement with theoretical predictions.

There are reasons to believe that the Υ spectroscopy may lead to quantitative QCD tests. Valuable information will come from the $\Upsilon - \eta_b$ hyperfine splitting and the hadronic χ_b widths. Furthermore, quarkonium spectroscopies should contain "extra" states, due to gluonic degrees of freedom, which may be detectable at hadron colliders.

If our hopes with respect to quantitative QCD tests should be fulfilled, the Υ system would indeed be the "hydrogen atom" of QCD: so far the basic binding force and the spin forces have been studied; a test of radiative corrections would probe, like the Lamb shift, the underlying field theory at the quantum level.

ACKNOWLEDGEMENTS

I would like to thank J. Baacke, E. Eichten, D. Gromes, A. Martin and H.R. Rubinstein for helpful discussions and to express my gratitude to G. Grunberg, Y.J. Ng and S.-H.H. Tye for an enjoyable collaboration on some aspects of quarkonium physics.

REFERENCES

1) K. Gottfried, Proc. HEP-83, Brighton (UK), eds. J. Guy and C. Costain (1983).
2) P.M. Tuts, Proc. Lepton Photon Symp., Cornell University, eds. D.G. Cassel and D.L. Kreinick (1983)
3) See, for instance,
   K. Berkelmann, Phys. Rep. 98C:145 (1983);
   J. Lee Franzini and P. Franzini, Ann. Rev. Nucl. Part. Sci 33:1 (1983);
   E.D. Bloom and C.W. Peck, ibid. 33:143 (1983).
4) K.F. Einsweiler, Proc. HEP-83, Brighton (UK), eds. J. Guy and C. Costain, (1983).
5) E.D. Bloom and C.W. Peck, Ref. 3).
6) T. Barnes, these proceedings.
7) J. Kuti, Proc. Lepton Photon Symp., Cornell University, eds. D.G. Cassel and D.L. Kreinick (1983).
8) E. Eichten, K. Gottfried, T. Kinoshita, K.D. Lane and T.M. Yan, Phys. Rev. D17:3090 (1978); 21:203 (1980).
9) J.L. Richardson, Phys. Lett. 82B:272 (1979).
10) W. Buchmüller, G. Grunberg and S.-H.H. Tye, Phys. Rev. Lett. 45:103 (1980); 45:587(E) (1980).
11) C. Quigg and J.L. Rosner, Phys. Lett. 71B:153 (1977);
    M. Machacek and Y. Tomozawa, Prog. Theor. Phys. 58:1890 (1977).

12) A. Martin, Phys. Lett. 100B:511 (1981).

13) W. Buchmüller and S.-H.H. Tye, Phys. Rev. D24 (1981) 132.

14) G. Bhanot and S. Rudaz, Phys. Lett. 78B:119 (1978).

15) M.B. Voloshin, preprint ITEP-21 (1980).

16) M.A. Shifman, Proc. Lepton Photon Symp., Bonn, ed. W. Pfeil (1981).

17) M.A. Shifman, Ann. Rev. Nucl. Part. Sci. 33:199 (1983).

18) H.R. Rubinstein, private communication.

19) See, for instance,
E. Eichten, preprint Fermilab-Conf.-83/101 THY (1983);
Ref. 13), Appendix B.

20) K. Heikkilä, N.A. Törnqvist and S. Ono, Phys. Rev. D29:110 (1984).

21) C. Quigg and J.L. Rosner, Phys. Rev. D23:2625 (1981).

22) J.S. Bell and R.A. Bertlmann, CERN preprint TH.3769 (1983).

23) J.E. Gaiser, Proc. 2nd Moriond Workshop on New Flavours, Les Arcs, eds. J. Tran Thanh Van and L. Montanet, (1982).

24) K. Gottfried, Proc. Lepton Photon Symp., Hamburg, ed. F. Gutbrod (1977).

25) R. McClary and N. Byers, Phys. Rev. D28:1692 (1983).

26) A.B. Henriques, B.H. Kellett and R.G. Moorhouse, Phys. Lett. 64B:85 (1976).

27) K. Gottfried, Phys. Rev. Lett. 40:598 (1978).

28) A. Billoire, R. Lacaze, A. Morel and H. Navelet, Nucl. Phys. B155:493 (1979).

29) T.M. Yan, Phys. Rev. D22:1652 (1980).

30) Y.P. Kuang and T.M. Yan, Phys. Rev. D24:2874 (1981).

31) For a recent review, see
M.E. Peskin, preprint SLAC-PUB-3273 (1983).

32) For a recent review, see
J.L. Rosner, Univ. of Chicago preprint EFI 83/17 (1983).

33) For recent work, see Ref. 25);
P. Moxhay and J.L. Rosner, Phys. Rev. D28:1132 (1983);
H.J. Schnitzer, Phys. Lett. 134B:253 (1984).

34) P. Hasenfratz, R.R. Horgan, J. Kuti and J.-M. Richard, Phys. Lett. 95B:299 (1980).

35) K. Johnson, unpublished.

36) J. Baacke, Y. Igarashi and G. Kasperidus, Z. Phys. C13:131 (1982).

37) E. Eichten and F. Feinberg, Phys. Rev. D23:2724 (1981).

38) D. Gromes, Z. Phys. C22:265 (1984).

39) D. Gromes, Heidelberg preprint HD-THEP-84-5 (1984).

40) W. Buchmüller, Phys. Lett. 112B:479 (1982).

41) J.D. Jackson, Classical Electrodynamics, 2nd edition, J.Wiley and Sons, New York (1975) p. 541.

42) W. Buchmüller, Proc. 2nd Moriond Workshop on New Flavours, Les Arcs, eds. J. Tran Thanh Van and L. Montanet (1982).

43) C. Klopfenstein et al., CUSB collaboration, Phys. Rev. Lett. 51:160 (1983).

44) J.K. Bienlein, private communication.
45) A. Martin and J.-M. Richard, Phys. Lett. 115B:323 (1982).
46) A.J. Buras, Proc. Lepton Photon Symp., Bonn, ed. W. Pfeil (1981).
47) G.P. Lepage, Proc. Lepton Photon Symp., Cornell University, eds. D.G. Cassel and D.L. Kreinick (1983).
48) R. Barbieri, G. Curci, E. D'Emilio and E. Remiddi, Nucl. Phys. B154:535 (1979).
49) E. Remiddi, Proc. Workshop on Physics at Lear, Erice, eds. U. Gastaldi and R. Klapisch (1982).
50) P.B. Mackenzie and G.P. Lepage, Phys. Rev. Lett. 47:1244 (1981).
51) W. Buchmüller, Y.J. Ng and S.-H.H. Tye, Phys. Rev. D24:3003 (1981);
    R. Barbieri, R. Gatto and E. Remiddi, Phys. Lett. 106B:497 (1981);
    S.N. Gupta, S.F. Radford and W.W. Repko, Phys. Rev. D26:3305 (1982).
52) R. Barbieri, M. Caffo, R. Gatto and E. Remiddi, Phys. Lett. 95B:93 (1980).
53) S. Brodsky, G.P. Lepage and P.B. Mackenzie, Phys. Rev. D28:228 (1983).
54) G. Grunberg, Phys. Lett. 135B:455 (1984).
55) A.T. Aerts and L. Heller, Phys. Rev. D29:513 (1984).
56) R.C. Giles and S.-H.H. Tye, Phys. Rev. Lett. 37:1175 (1976).
57) W. Buchmüller and S.-H.H. Tye, Phys. Rev. Lett. 44:850 (1980).
58) S. Ono, Orsay preprint LPTHE 84/13 (1984).
59) N.A. Campbell, L.A. Griffiths, C. Michael and P.E.L. Rakow, Liverpool preprint LTH 113 (1984).

THE LEAR PHYSICS PROGRAMME

Pietro Dalpiaz

Istituto di Fisica, Università di Ferrara, Italy
Istituto Nazionale di Fisica Nucleare, Bologna, Italy

1. INTRODUCTION

The Low-Energy Antiproton Ring (LEAR) at CERN[1] is the low energy terminal of the CERN antiproton system, centered in the $\bar{p}$-source, consisting mainly of the old Proton Synchrotron (PS) of CERN and the Antiproton Accumulator (AA)[2].

LEAR is a stretched synchrotron with a circumference of 78 m and with four straight sections of 8 m each. It is installed in the South Hall of the CERN PS. It can be injected with bunches of $\bar{p}$, with 600 MeV/c of momentum, originated in the AA at 3.5 GeV/c and decelerated in the PS. Under the present conditions of the AA bunches of 4×10^9 $\bar{p}$ can be injected in LEAR continuously every 75 minutes. In the these conditions LEAR can produce an extracted beam of 10^6 $\bar{p}$/s in a spill of 1hr length with 100% duty cycle and with $\Delta p/p < 10^{-3}$ in the momentum range $100 < p_{\bar{p}} < 2000$ MeV/c. In the future, when the CERN ACOL[3] project will be ready, the intensity of LEAR $\bar{p}$'s could be substantially improved. The beam extracted from LEAR is split in to three other beams; each of them can be deviated into two experimental areas, where several experiments are installed in series. In this way 17 experiments are now installed in LEAR and a maximum of 3 can run simultaneously with antiprotons. Figure 1 shows the LEAR installation.

2. PHYSICS PROGRAM

The quality of $\bar{p}$ beams produced by LEAR gives the possibility of performing new experiments that can revive several physics topics and can also open up new subjects of physics.

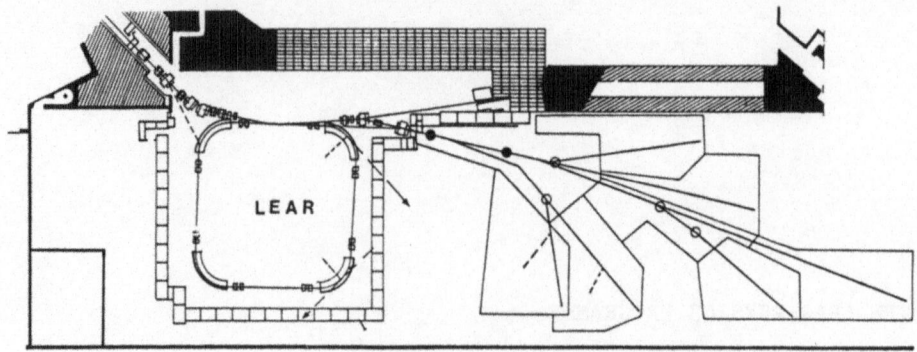

Fig. 1 -   LEAR and the experimental area

Table 1 shows a list of the 17 experiments installed in LEAR.

TABLE 1

Exp.	Title
PS170	Precision measurements of the proton electromagnetic form factors in the time-like region and vector spectroscopy
PS171	A study of $\bar{p}p$ interactions at rest in a H_2 gas target at LEAR
PS172	$\bar{p}p$ total cross-sections and spin effects in $\bar{p}p \to K^+K^-, \pi^+\pi^-, \bar{p}p$ above 200 MeV/c
PS173	Measurement of $\bar{p}p$ cross-sections at low $\bar{p}$ momenta
PS174	Precision survay of X-rays from $\bar{p}p(\bar{p}d)$ atoms using the initial LEAR beam
PS175	Measurement of the antiprotonic Lyman and Balmar X-rays of $\bar{p}H$ and $\bar{p}D$ atoms at very low target pressures
PS176	Study of X-ray and γ-ray spectra from antiprotonic atoms at the slowly extracted antiproton beam of LEAR
PS177	A search for heavy hypernuclei at LEAR
PS178	Study of antineutron production at LEAR
PS179	Study of the interaction of low-energy $\bar{p}$ and $\bar{n}$ with H, ^{2}H, ^{3}He, ^{4}He, Ne and ^{40}Ar nuclei using a streamer chamber in a magnetic field
PS182	Investigations on baryonium and other rare $\bar{p}p$ annihilation modes using high resolution π^o spectrometers
PS183	Search for bound $\bar{N}N$ states using a precision γ & charged pion spectrometer
PS184	Study of $\bar{p}$-nucleus interaction with a high resolution magnetic spectrometer
PS185	Study of threshold production of $\bar{Y}Y$ pairs in $\bar{p}p$ interactions at LEAR
PS186	Nuclear excitations by antiprotons and antiprotonic atoms
PS187	A good statistics study of antiproton interactions with nuclei
PS189	High precision mass measurements with a radiofrequency mass spectrometer - Application to the measurement of the $\bar{p}p$ mass difference

Several experiments concern precise measurements, others clarify old problems still open, and yet others are exploratory experiments.

We can divide the physics topics studied in LEAR into :

i)   – electromagnetic interaction.
ii) – strong interactions.
iii) – atomic physics.
iv) – nuclear physics.
v)   – flavour spectroscopy.

i) ELECTROMAGNETIC INTERACTIONS

PS170 (APPLE)[4] is a complex experiment installed in the
central branch of LEAR. It studies the production of e^+e^-
pairs from $\bar{p}p$ annihilation at low energy to :

- measure the differential cross-section for the process $\bar{p}p \rightarrow e^+e^-$
  to determine separately G_E and G_M, the electromagnetic
  form factors of the proton in the time-like region[5] in the
  range of $-6(GeV/c)^2 < q^2 < -4m^2$, in order to have the same
  uncertainties as the existing data in the space like region
  very well measured in the ep elastic scattering[6];
- search for vector mesons[7] in the range $m_\phi < m_{V^\circ} < m_{\rho^{\prime\prime}}$ with
  the reaction $\bar{p}p \rightarrow e^+e^- + \pi^0$.
  At rest the ratio e^+e^-/ total is of the order of 3×10^{-7},
  but at 2 GeV/c it is of the order of 10^{-9}.
The main characteristic of this experiment is the capability of
identifying electrons in a high hadronic background at high rates
with large solid angle acceptance. In fact it has a rejection power
of e/π of the order of 10^{-10} for an e^+e^- pair. The
experiment is running and produces first results[8].

ii) STRONG INTERACTIONS

Several experiments study strong interactions at LEAR. The main
subject is the spectroscopy of glue-balls and baryonium states.
Another subject is the study of threshold effects.

- PS171 (ASTERIX) is an experiment of large acceptance that studies
  the $\bar{p}p$ annihilation at rest in an axial magnetic field with the
  possibility of revealing the X-ray cascade, and therefore can
  measure events with several prongs, charged and neutral . In
  this experiment the initial state of annihilation is determined
  by the measurement of the X-ray cascade. It is a modern
  experiment to search in an exclusive way for glue-balls and
  baryonium states, and also to determine the quantum numbers of
  resonances already known. This experiment is running and
  producing the first results[9].
- PS172, more than a single experiment, is a complex of several
  experiments with the aim of searching for narrow and large width
  baryonium states. The research programme can be summarized as
  follows.

- To measure the total cross-section for $\bar{p}p$ in order to search bumps to study the possible existence of narrow width baryonium states. The LEAR beam gives the possibiliy of performing clean experiments in this subject. This measurement is running and producing the first results[10].
- To study the $\bar{p}$ scattering with the aim of producing a polarized beam of antiprotons.
- To study,with polarized target and $\bar{p}$ beam possibly polarized, the $\bar{p}p$ annihilation in $\pi^+\pi^-$ and K^+K^- to search large width baryonium states with the phase-shift analysis.

- PS173 is an experiment that measures $\bar{p}p$ total, elastic and charge exchange cross-section in order to search for narrow width baryonium states for $p_{\bar{p}}$<150 MeV/c. The experiment is running and producing the first results[11].
- PS180[12] is a one-arm electron pair spectrometer. The aim of the experiment is to study the inclusive spectrum of the γ-ray from $\bar{p}p$ annihilation at rest in order to search for narrow width baryonium states. The expected resolution for the γ-ray is 1 MeV. They also study the inclusive spectrum of charged π.
- PS182[13] studies the same process as experiment PS180. The γ-detector is in this case a BGO counter with a resolution of 2 MeV. This experiment is also equipped with a ring of glass Cerenkov counters to study the inclusive π^0 spectrum on the jacobian peak. The experiment is running and producing the first results.
- PS178 is an experiment[14] with a detector of antineutrons constructed with limited streamer tubes studying the $\bar{p}p$ charge exchange with the aim of constructing an antineutron beam useful for the study baryonium states and to measure with precision the antineutron mass.
- PS185 is an experiment[15] studying the variation of cross-section and the polarization of the process $\bar{p}p\rightarrow\bar{\Lambda}\Lambda$ at threshold. This measurement could give information on the polarization of the s-quarks via some models.

iii) ATOMIC PHYSICS

Three experiments are on the floor to study the X-ray cascade of a $\bar{p}p$ system stopping $\bar{p}$ in gas also at low pressure; these experiments study QED. The other two experiments study the X-ray cascade of a $\bar{p}$-nucleus system to study strong interactions effects on the cascade.

- PS171 (ASTERIX) have the large acceptance of the proportional chamber used to detect the X-ray cascade. They show the first result on the K and L lines[9].
- PS174[16] intends to detect K and L lines from $\bar{p}p$ and $\bar{p}d$ at low pressure with Si-detectors.
- PS175[17] is based on a cyclotron trap to stop $\bar{p}$ in H_2 and D_2 at the pressure of some Torr.

- PS176[18] and PS186[19] use the same appartus based on several Si-detectors to study the X-ray cascade the first of isotopically pure elements and the second of heavy elements.

iv) NUCLEAR PHYSICS

In order to study nuclear structure four experiments have been installed at LEAR.

- PS179[19] performs a systematic study at different $\bar{p}$ momenta of $\bar{p}$-N interaction with D_2, He, and Ne. The apparatus is based on a large streamer chamber in a magnetic field. It is very suitable for exploratory experiments due to the large acceptance. The experiment can also detect channels of astrophysical interest for the stellar evolution. The experiment is in running and proceduce the first results.
- PS184[20] with the strong focusing spectrometer SPASII, studies with a resolution of $\Delta p/p = 10^{-3}$ in an angular acceptance of 6° elastic and inelastic scattering of $\bar{p}$ on several nuclei. This experiment is running and producing the first results.
- PS187[21] systematically studies $\bar{p}$N interactions with a detector based on proportional chambers in a magnetic field. The experiment is running.
- PS177[22] studies with the recoil distance method the production of hypernuclei on heavy targets. The use of an $\bar{p}$ beam is based on the idea that in $\bar{p}$ annihilation a lot of kaons are produced in a small spot. The density of kaon is many times larger than can be obtained with kaon beams.

v) FLAVOURS SPECTROSCOPY

The new resonances i(1400), θ(1700) and ξ(2200) recently discovered at SPEAR with MARK III can be studied at LEAR, i and θ by PS171 (ASTERIX) and the ξ with PS170 (APPLE) and PS172. These resonances are candidates to be exotic states because they are in the decay channel of J/ψ. The determination of their quantum numbers is of great inportance[23].

To study the charmonium states not accessible to e^+e^- machines as is done at the ISR with the experiment with the jet-target[24] it is necessary to transform LEAR in to a $\bar{p}$p-minicollider[25]. The idea is interesting but it appears not to be realistic to transform LEAR into another machine with in view of the present pressure for the running time at LEAR.

More suitable seems the idea to construct another machine with 7 GeV/c maximum momentum. This machine, that we can call SUPERLEAR, could work as a stretcher to study charmonium states or as an $\bar{p}$p-minicollider to study the bottonium states[26-27].

The accurate measurements of the masses, splittings, and width
of the bottonium states is considered now the unique
quantitative test of QCD[28].

3. CONCLUSIONS

The present LEAR program is very heavy for the present
running time available. It is difficult to expect important
modifications of the apparatus installed in the present
situation but in any case important results are expected in
differet fields. The situation is promising for the end of 1987
when the ACOL project is terminated since by that time we expect
important experimental developments.

REFERENCES

In the following references "ERICE 82" means the book "Physics
at LEAR with low energy cooled antiprotons" Edited by U. Gastaldi
and R. Klapisch, PLENUM N.Y. and London 1984.

1 - P. Lefevre, ERICE 82, pg. 15
    D. Möhl, ERICE 82, pg. 27
    R. Cappi, R. Giannini and Hart, pg. 49
    D.J. Simon, ERICE 82, pg. 55
    J.L. Laclare, ERICE 82, pg. 69
    R. Giannini in this book.

2 - E. Jones ERICE 82 pg. 5 and references therein.

3 - E.J.N. Wilson editor CERN/83-10.

4 - G. Bardin et al., ERICE 82 pg. 347.

5 - P.F. Dalpiaz ERICE 82 pg. 329 and references therein.

6 - D.J. Drickey et al., P.R.L. 9, 521 (1962)
    D. Yount et al., P.R. 128, 1842 (1963)
    D. Frèrejacque et al.,P.R.L. 142, 922 (1966)
    T. Jenssens et al., P.R. 142, 922 (1966)
    K.W. Chen et al., P.R. 141, 1267 (1966)
    W. Bartel et al., P.R.L. 17, 608 (1966)
    W. Albrecht et al., P.R.L. 17, 1192 (1966)
    M. Goiten et al., P.R.L. 18, 1017 (1967)
    H. Behrend et al., N.C. 48A, 140 (1967)
    W. Albrecht et al., P.R.L. 18, 1014 (1967)
    D.H. Coward et al., P.R.L. 20, 292 (1968)

7 - J. Duclos, ERICE 82 pg. 339 and references therein.

8 - F. Petrucci et al., in this book.

9 - U. Gastaldi, ERICE 82, pg. 109.
   W. Dahme, ERICE 82, pg. 253.
   F. Feld, in this book.

10 - K. Bos, ERICE 82, pag. 427
   F. Bradamante, in this book.

11 - R. Rasome, ERICE 82, pg. 437
   T.A. Shibata et al., in this book.

12 - G.A. Smith et al., ERICE 82, pg. 289.

13 - K. Fransson et al., ERICE 82, pg. 281.

14 - C. Voci et al., ERICE 82, pg. 465.

15 - S. Limentani, CERN $\bar{p}$ LEAR-NOTE 49
   R.A. Einsenstein, ERICE 82 pg. 469 pg. 477
   H. Schmitt, ERICE 82 pg. 489.

16 - J.S. Davies et al., ERICE 82, pg. 143.

17 - L.M. Simon et al., ERICE 82, pg. 155.

18 - M. Poth et al., ERICE 82, pg. 567.

19 - G. Bendiscioli et al., ERICE 82, pg. 517.

20 - D. Garreta et al., ERICE 82, pg. 533.

21 - R.M. De Vries and H.J. Di Giacomo, ERICE 82, pg. 523.

22 - T. Johansson et al., ERICE 82, pg. 589.

23 - T. Barnes in this book.

24 - P. Dalpiaz K.f.K. 2836, 111 (1979)
   P. Dalpiaz et al., CERN/ISRC 79-23
   C. Baglin et al., CERN/ISRC/ 80-14.

25 - P. Dalpiaz K.f.K. 2836, 111 (1979).

26 - P. Dalpiaz, ERICE 82, pg. 725.

27 - U. Bizzari et al., ERICE 82, pg. 729.

28 - E. Remiddi, ERICE 82, pg. 711
   W. Büchmuller in this book.

ELECTRON-POSITRON PAIR PRODUCTION

IN pp̄ ANNIHILATION AT LEAR

Presented by F. Petrucci

G. Bardin[4], G. Burgun[5], R. Calabrese[1], G. Callegari[1]
G. Capon[2], R. Carlin[3], P. Dalpiaz[1], P. F. Dalpiaz[1]
J. P. de Brion[5], J. Derré[5], U. Dosselli[3], J. Duclos[4]
J. L. Faure[4], F. Gasparini[3], M. Huet[4], C. Kochowski[5]
D. Lafarge[5], S. Limentani[3], G. Marel[5], A. Meneguzzo[3]
E. Pauli[5], F. Petrucci[1], M. Posocco[3], M. Savriè[1]
A. Schuhl[4], G. Simone[6], L. Tecchio[6], and C. Voci[3]

[1]Istit. di Fisica dell'Univ. Ferrara and Istit. Nazionale
di Fisica Nucleare, Bologna; [2]CERN, Geneva; [3]Dipart. di
Fisica and Istit. Nazionale di Fisica Nucleare, Padova;
[4]Depart. de Physique Nucleaire des Hautes Energies, CEN
Saclay; [5]Depart. de Physique des Particules Elémentaires,
CEN Saclay; [6]Istit. di Fisica Superiore and Istit.
Nazionale di Fisica Nucleare, Torino.

We report here the first results of the PS170 (APPLE) experiment[1,2] installed on the LEAR beam, at CERN.

We study the reaction

$$p\bar{p} \rightarrow e^+e^-$$

for incident p̄ momenta ranging from 0 to 2 GeV/c in order to measure the electromagnetic form factors of the proton in the time-like region[3]. From the angular distributions we shall be able to determine G_E and G_M separately.

Besides, for vector meson spectroscopy[4], we detect also electron-positron pairs, in the following reaction:

$$p\bar{p} \rightarrow V^° + \pi^°$$
$$| \rightarrow e^+e^-$$

With p̄ at rest, the expected spectrum extends from ϕ to ρ''.

For these purposes the experimental set-up must satisfy the following requirements:

- momentum resolution better than 2%;
- large angular acceptance (~20%);
- rejection power of e/π around 10^{-8} per particle pair.

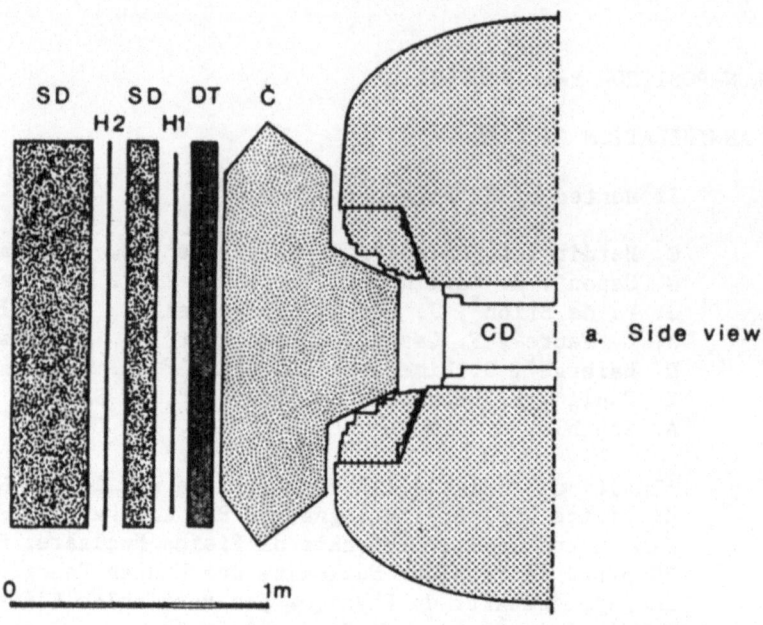

a. Side view

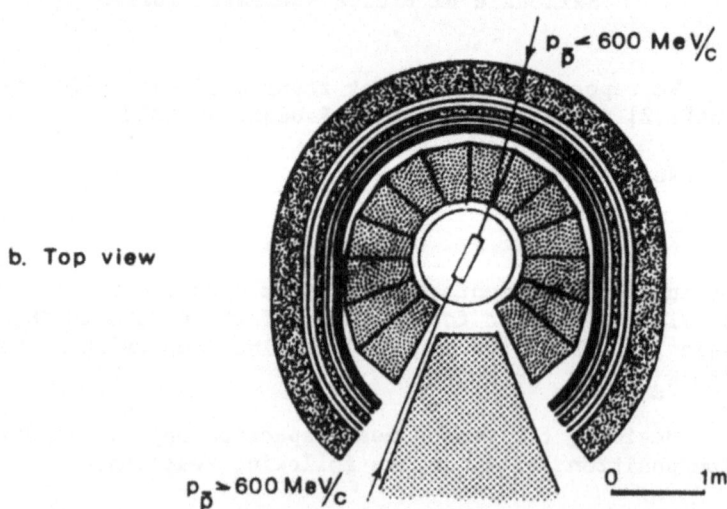

b. Top view

Fig. 1. Experimental set-up

Experimental Set-up

Figure 1a shows a sketch of our apparatus. It consists of a central detector (CD) in the uniform-field region of a 1.3 tesla magnet. Looking from the center, it is followed by:

- a 28-cell Cerenkov counter (Č);
- 100 drift tubes (DT);
- two hodoscopes (H1,H2) with 90 scintillators and 120 photomultipliers;
- 9 planes of an electromagnetic calorimeter (SD) equipped with limited streamer tubes, whose 8700 wires are read-out in 11000 digital channels.

The top view of the set-up, in Figure 1b, shows the possibility of getting the beam at both ends of the cylindrical target, depending on the antiproton momentum range. In fact, the whole apparatus can rotate around the vertical axis of the magnet and the vacuum line can be installed along the yoke of the magnet or through a slit in the calorimeter, on the opposite side.

Each detector surrounding the magnet is also split into two separate halves and may be opened for maintenance operations.

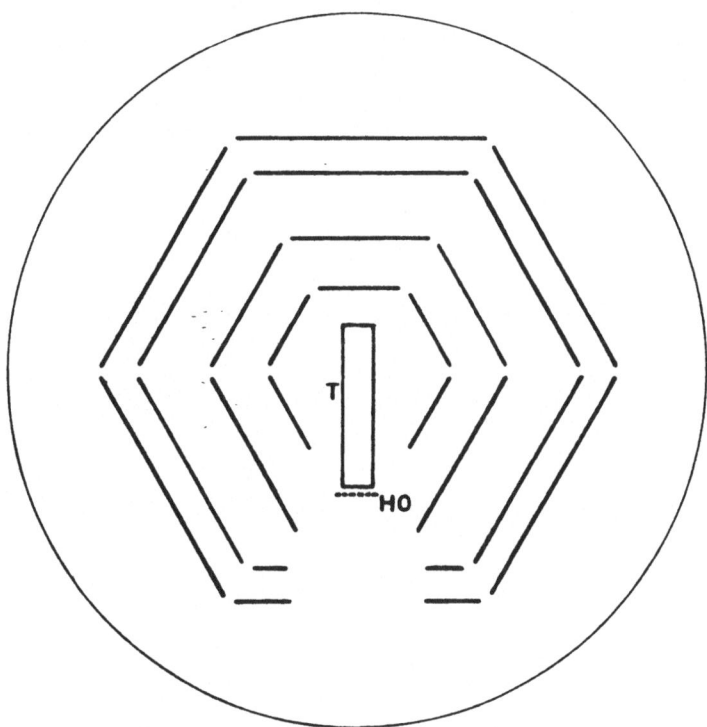

Fig. 2.   Central detector

In Figure 2 an enlarged view shows the liquid hydrogen target (T), 30 cm long, and the four planes of the central detector[5]. These are multi-wire proportional chambers with cathode read-out; the analog channels are about 3000 and the spatial resolution is ±0.3 mm.

The electron identification is done fully on-line by the Cerenkov counter and partially on-line and off-line with the electro-magnetic calorimeter. The final rejection factor against hadrons is at least 10^5 per particle. Taking account of both detectors, two body events are rejected at a level of 10^{10}.

A beam trigger is made by delayed coincidence of signals from two scintillators along the beam line (B_0, 30 meters upstream, and

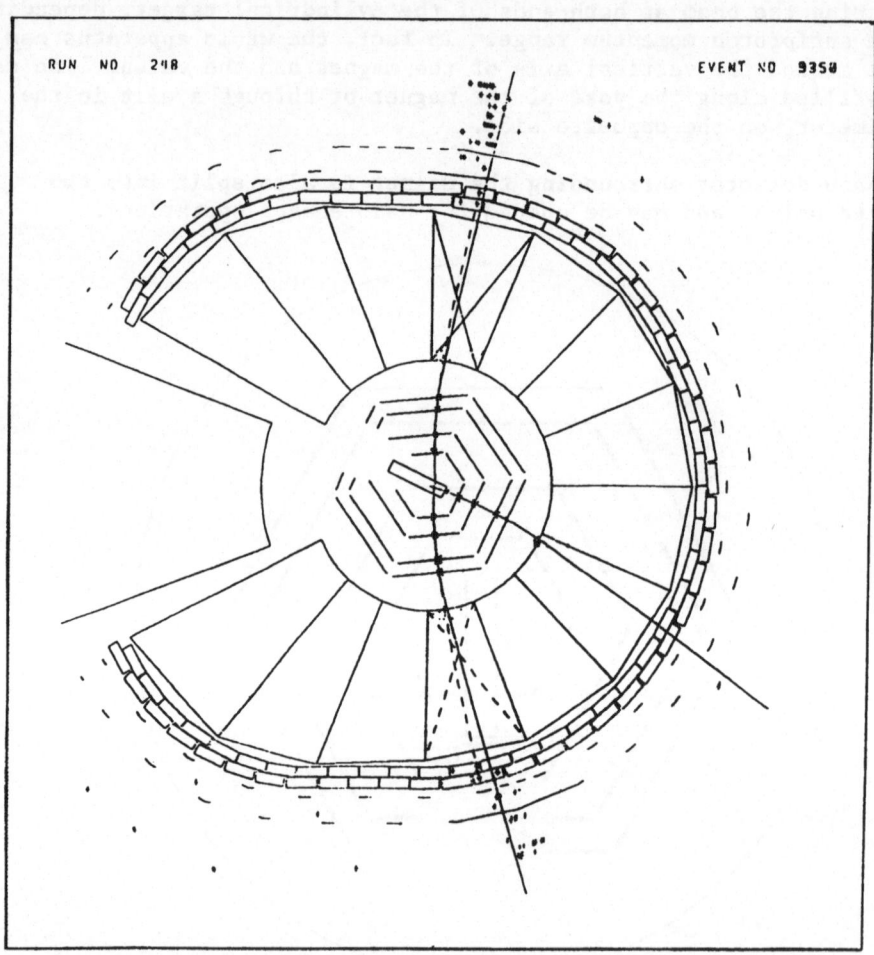

RUN NO 248   EVENT NO 9358

Fig. 3. Reconstruction of a $p\bar{p} \rightarrow e^+e^-$ event with $\bar{p}$ at rest

HO, just in front of the target) and by anticoincidence from counters (Ā) above, below and around the target:

Beam = $B_0 \cdot HO \cdot \bar{A}$

This signal enables a fast ECL trigger, formed with hodoscopes and Cerenkov output, in (H1·H2·C) triple coincidence for electron counts and (H1·H2) for hadron counts.

A typical display of an e^+e^- event is shown in Figure 3.

RESULTS

The first run at LEAR started in December 1983. We present preliminary results on a part of the run at 300 MeV/c. In this case

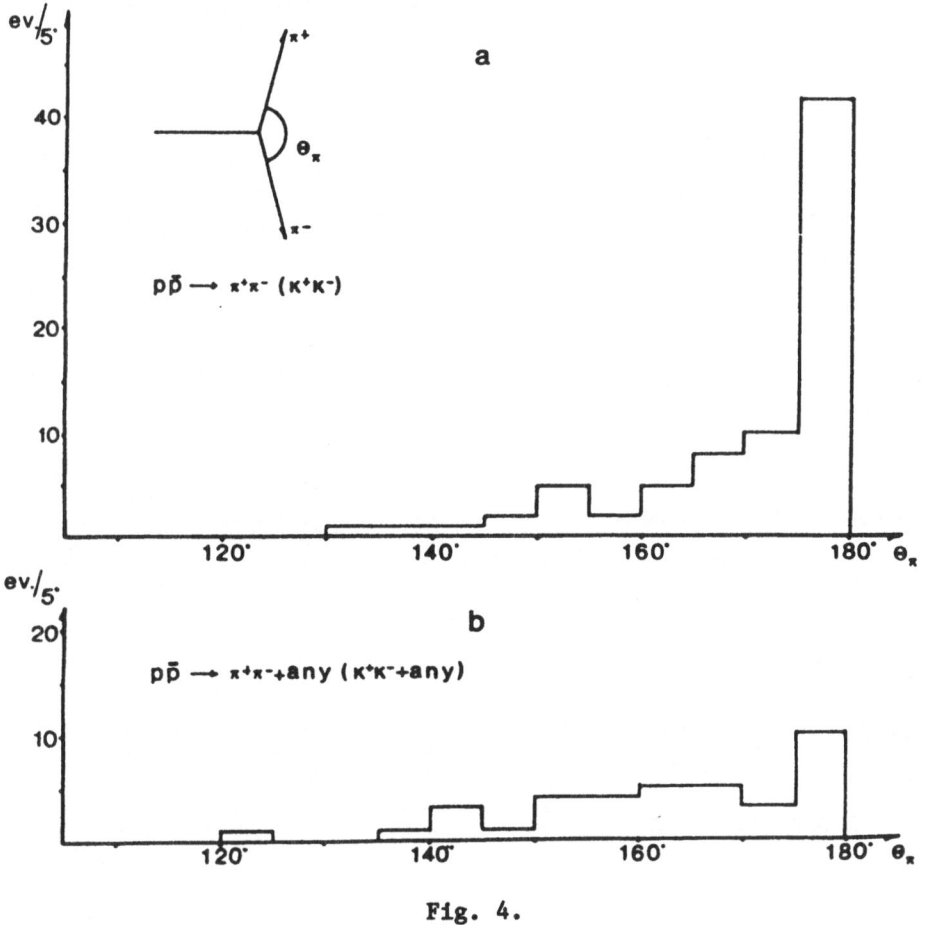

Fig. 4.

antiprotons stop nearly at the center of the target and we have events at rest and with $\bar{p}$ momentum ranging from 100 to 290 MeV/c.

Figure 4 shows the opening angle distribution of two hadrons in

$$p\bar{p} \to \pi^+\pi^- (K^+K^-)$$

reactions during a 1-hour spill. Coplanar ($\pm 10°$) events are shown in histogram 4a and 3-and-more-body events in Figure 4b.

Similar distributions for e^+e^- pairs collected during some spills are shown in Figure 5. About 109 events show only electron-positron pairs. Among these, about 80 pairs are collinear.

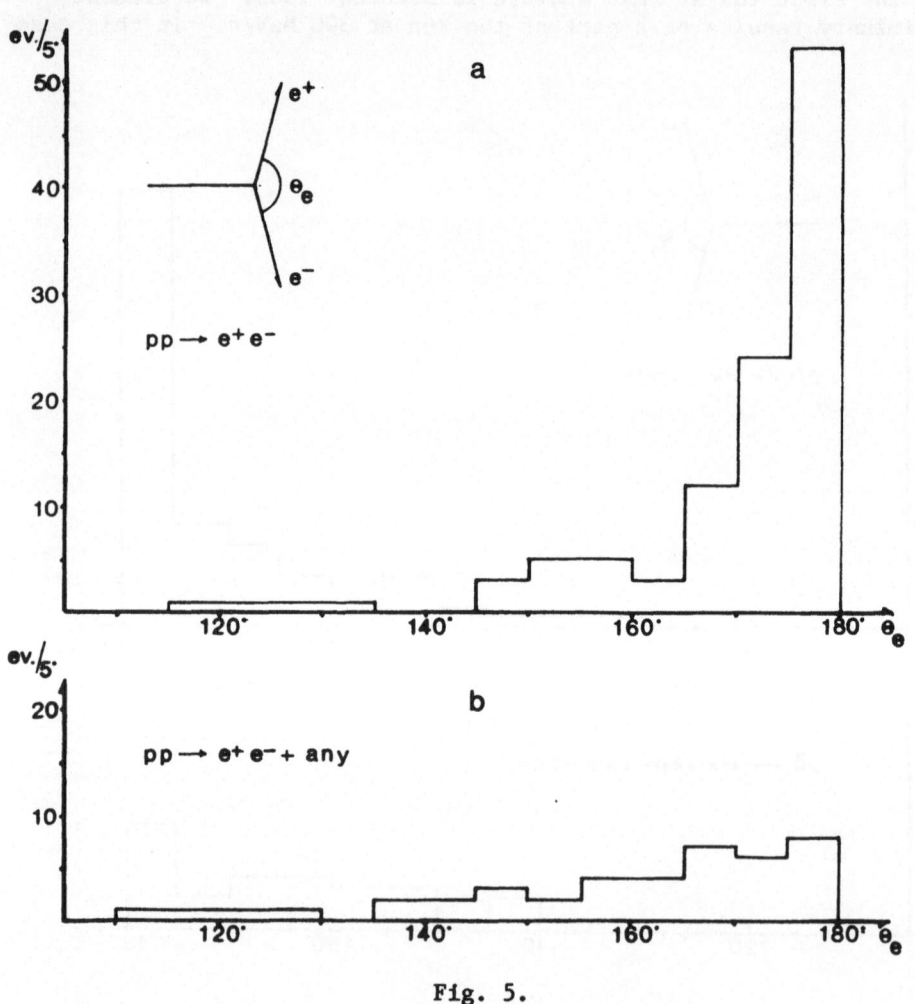

Fig. 5.

The acoplanarity background is deduced from counts between 120° and 140° in Figure 5a, and it is not larger than 10% of the total number of events.

Even if these are very preliminary results, the number of events collected in a few days of run is already at the level of the total statistics obtained until now at the e^+e^- storage rings. The future seems to be quite promising for a precise measurement of the nucleon form factors in the LEAR energy range.

REFERENCES

1. P. Dalpiaz, CERN/PSCC/79-56/PSSC/17, 10 Dec. 1979. J. P. de Brion, P. Dalpiaz, P. F. Dalpiaz, J. Derré, J. Duclos, J. L. Faure, F. Gasparini, S. Limentani, A. Magnon, A. Meneguzzo, M. Nigro, E. Pauli, C. Peroni, C. Pigot, M. Posocco, M. A. Schneegans, C. Schuhl, L. Tecchio, and C. Voci, Proposal CERN/PSCC/80-95/PSSC/P25, 29 Aug. 1980.

2. G. Bardin, G. Burgun, P. Dalpiaz, P. F. Dalpiaz, J. P. de Brion, M. De Giorgi, J. Derré, J. Duclos, J. L. Faure, F. Gasparini, M. Huet, S. Limentani, A. Meneguzzo, E. Pauli, M. Posocco, L. Tecchio, and C. Voci, in: "Physics at LEAR with Low-Energy Cooled Antiprotons," U. Gastaldi and R. Klapisch, eds., p.347, Plenum Press (1984).

3. P. F. Dalpiaz, in: "Physics at LEAR with Low-Energy Cooled Antiprotons," U. Gastaldi and R. Klapisch, eds., p.329, Plenum Press (1984).

4. J. Duclos, in: "Physics at LEAR with Low-Energy Cooled Antiprotons," U. Gastaldi and R. Klapisch, eds., p.339, Plenum Press (1984).

5. F. Gasparini, in: "Physics at LEAR with Low-Energy Cooled Antiprotons," U. Gastaldi and R. Klapisch, eds., p.353, Plenum Press (1984).

pp̄ ANNIHILATIONS AT REST IN HYDROGEN GAS:

REPORT ON PRELIMINARY RESULTS OF THE ASTERIX EXPERIMENT AT LEAR

The ASTERIX* Collaboration

S. Ahmad,[4] C. Amsler,[6] R. Armenteros,[4] E.G. Auld,[5] D. Axen,[5]
D. Bailey,[1] S. Barlag,[1] G. Beer,[5] J.C. Bizot,[4] M. Caria,[6]
M. Comyn,[5] W. Dahme,[3] B. Delcourt,[4] M. Doser,[6] K.D. Duch,[2]
K. Erdmann,[5] F. Feld,[3] U. Gastaldi,[1] M. Heel,[2] B. Howard,[5]
R. Howard,[5] J. Jeanjean,[4] H. Kalinowsky,[2] F. Kayser,[2]
E. Klempt,[2] R. Landua,[1] G. Marshall,[1] H. Nguyen,[4] N. Prevot,[4]
L. Robertson,[5] C. Sabev, U. Schaefer,[3] R. Schneider,[2]
O. Schreiber,[2] U. Straumann,[2] P. Truoel,[6] B.L. White,[5]
W.R. Wodrich,[3] and M. Ziegler.[2]

(Presented by F. Feld)

INTRODUCTION AND CONCLUSIONS

The ASTERIX experiment has been set up to study pp̄-annihilations at rest into qq̄-mesons and other boson resonances like glueballs (gg, ggg), hybrids (qq̄g), baryonia (qqq̄q̄) and NN̄ bound states from S and P states of the pp̄ atom. A detailed description of the apparatus can be found in the proposal[1] and in the contributions to the LEAR Erice Workshop 1982[2,3]

The installation of the experiment started in 1982; the first antiprotons for tests arrived in July 1983 and during December 1983 we had our first production runs with an antiproton beam of 308 MeV/c before moderation. In about 50 spills of 1 hour we recorded a total of about 4×10^6 events on tape. For part of the data the trigger required an antiproton stopping in the target, while for the remainder in additon an X-ray candidate was requested. During January 1984 the events were reconstructed and

*) Antiproton STop Experiment with tRigger on Initial X-rays.
   CERN,[1] Mainz,[2] Munich,[3] Orsay[4] (LAL), TRIUMF-Vancouver-Victoria,[5]
   Zurich[6]

about 8×10^5 events were found to have a reconstructed annihilation vertex inside the Hydrogen target. The analysis of these data, which is still in progress, has so far led to the following preliminary results:

The 2P level of the $p\bar{p}$ atom is reached in the atomic cascade in H_2 gas at NTP with a probability of about 10% with emission of an X-ray of the L series ($L_\alpha = 1.7$ keV, $L_\beta = 2.4$ keV, $L_\infty = 3.1$ keV)

In fully reconstructed annihilation channels involving besides charged particles 0 or 1 neutral particle, we observe well known mesons (π, K, η, ρ, ω, f, A_2) at their nominal masses.

When comparing relative branching ratios (e.g. $p\bar{p} \to \rho^0 \pi^0/$ $p\bar{p} \to f^0 \pi^0$) obtained in bubble chamber experiments[4] (assumed to be S-wave annihilation, because of strong Stark mixing in higher atomic levels leading to annihilation from nS levels) with our data once requiring a stopped $\bar{p}$ only (mixture of S- and P-wave annihilation) and secondly asking also for the L X-ray to be detected in coincidence (P-wave annihilation), we find a continuous change. Preliminary we conclude that one has $\sim$ 10-40% S-wave and $\sim$ 60-90% P-wave annihilation in atmospheric H_2 gas.

In the following sections we recall briefly the main features of the experiment and then present some of our observations. We stress the preliminary nature of the spectra that we show.

For the purpose of this report all observed charged tracks are called Pions. No corrections have been made for the small fraction of Kaon and electron (Dalitz-pairs) tracks contained in the spectra.

EXPERIMENTAL SET-UP

A schematic view of the detector is shown in Fig. 1. Antiprotons enter a solenoidal magnet along its axis and are brought to rest in a cylindrical hydrogen gas target. This target is surrounded by a cylindrical projection chamber (XDC for X-ray Drift Chamber) which measures energy and conversion point of the X-rays emitted during the de-excitation of the $p\bar{p}$ atom. This chamber also gives the initial track element for charged particles emerging from the annihilation. The continuation of these tracks is identified by seven cylindrical multi-wire proportional chambers (C1, C2, Q1, Q2, P1, Q3, P2). Gammas produced in the annihilation are partially converted in lead sheets (0.9 radiation length) placed at the two endcaps and immediately in front of the chamber Q3. The hexagonal lead sheets at the endcaps are preceeded by a MWPC to veto against charged particles and followed by two MWPC's to identify the electromagnetic shower. From the gamma conversion point, determined in this way, and the annihilation vertex reconstructed

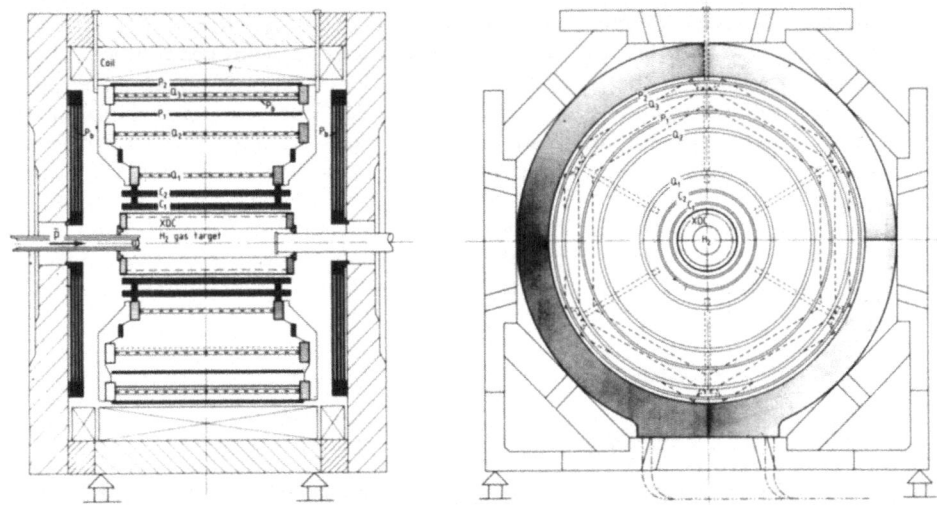

Fig. 1. Schematic side and front view of the ASTERIX detector:
XDC: X-ray Drift Chamber (three-dimensional projection
chamber); C1, C2, Q1, Q2, Q3 and endcaps: multiwire proportional chambers with anode and cathode strip readout;
P1, P2: multiwire proportional chambers with anode readout

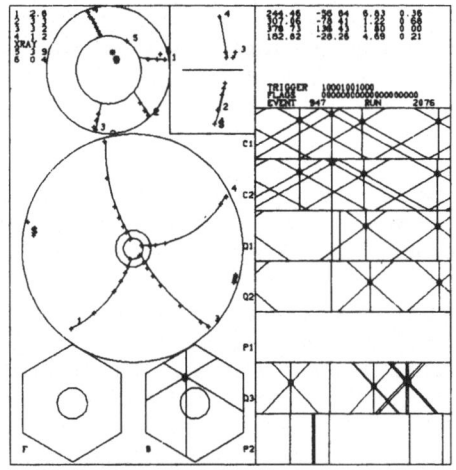

Fig. 2. Reconstructed event:
top left: magnified view of the charge observed in the
XDC. Four tracks and an X-ray converted near the mylar
foil are clearly visible; centre left: the four tracks
reconstructed using the hit wires pattern of the cyl.
MWPC's (right) and a reconstructed γ conversion point
(indicated by $); a second γ has converted in the backward
endcap detector (bottom left).

from the tracks of the charged annihilation products, we obtain the direction of flight of the gammas. For illustration we show an event as seen by the computer in Fig. 2. The distribution of reconstructed vertices for all events of one run is displayed in Fig. 3. Antiprotons stopped in the H_2 gas volume are clearly distinguished from those stopping in the entrance and exit scintillators as well as in the XDC counter gas (argon) and the Mylar foil separating hydrogen and argon. The energy of the X-rays is determined with a resolution of $\pm$ 11% at 5.5 KeV, and the momenta of the charged particles are measured with a resoluton of $\pm$ 4% at 500 MeV/c.

PRELIMINARY RESULTS

Protonium Spectroscopy

Figure 4a shows the X-ray spectrum obtained during $\bar{p}_{STOP}$ trigger runs. The strong signal in the energy region between 1 and 4 keV corresponds to L transitions populating the atomic 2P level. The L_α line at 1.7 keV is clarly distinguished from the L_β.... L_∞ lines. The radial distribution of the associated absorption points (Fig. 4b) confirms that these X-rays are produced inside the H_2 target.

The fact that the spectrum contains almost exclusively L-line events reconfirms the earlier observation by E. Auld et al.[5] that the 2P level (which has a radiative width of 10^{-11} sec and therefore does not live long enough to mix with the 2S level by collisional Stark mixing in 1 atm gas) decays dominantly via annihilation.

The events below the L_α line are most likely M X-rays emitted from high levels that are still able to traverse the 6μ Mylar foil separating the counter gas from the target gas.

X-rays emitted from transitions to the atomic 1S level (K-lines) are expected at energies of about 9 keV. Only few events are observed in this energy region. It should be mentioned that the X-ray detector was operated initially at a high gas gain to ensure a proper identification of the L-transitions to study annihilations from the atomic 2P level. This operating condition led to a deteriorated energy scale for energies larger than 6 keV. In subsequent runs the gas gain was lowered, but the analysis of this data is still in progress. However, a weak population of the atomic 1S-level can already be concluded from the scarcity of events observed above 6 keV.

In Fig. 4a we also compare the events with an associated reconstructed vertex in H_2 with those events where the vertex was found in the argon gas of the XDC. For these events the 3 KeV

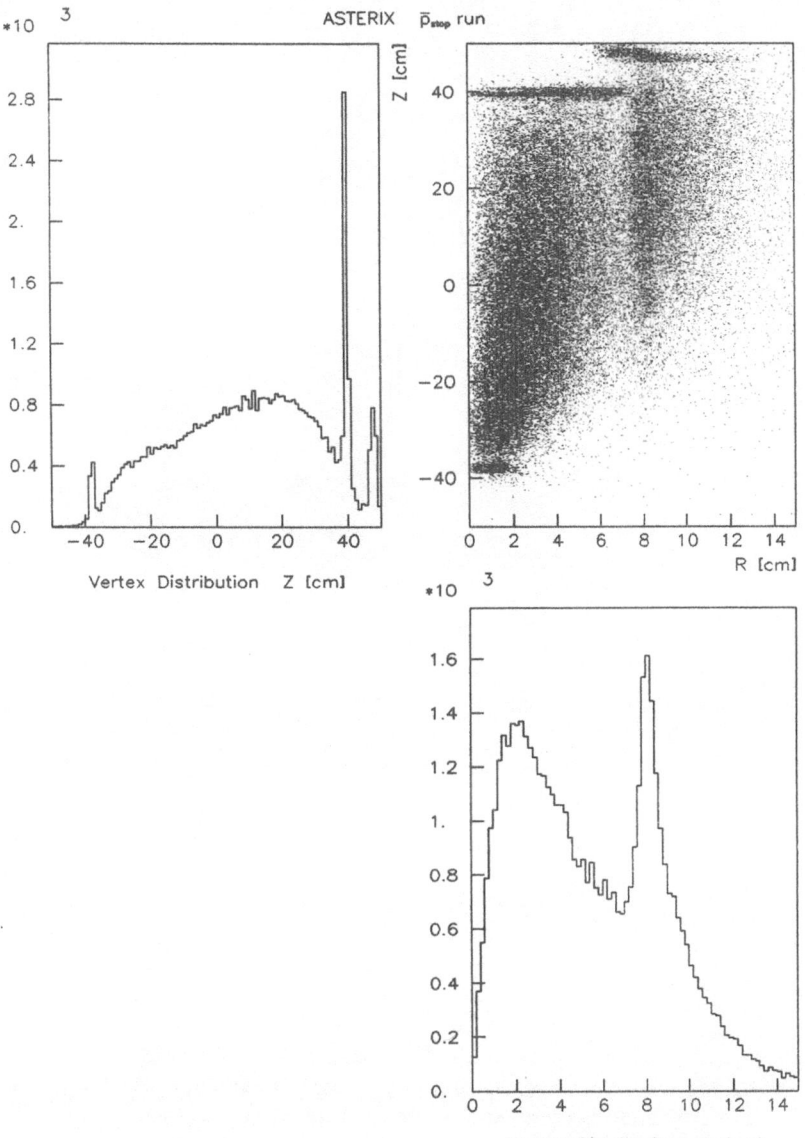

Fig. 3. Distribution of reconstructed annihilation vertices for one run. The three sharp peaks in the z-distribution (top left) represent annihilations occurring in the entrance (T2) and exit (T4) scintillators and the aluminium frame of the XDC, respectively. Annihilations in the thin (6μ) Mylar foil separating the H_2 target from the XCD counter gas lead to the sharp peak in the r-distribution.

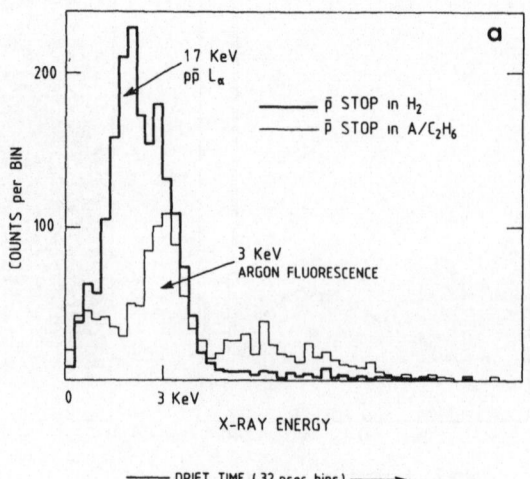

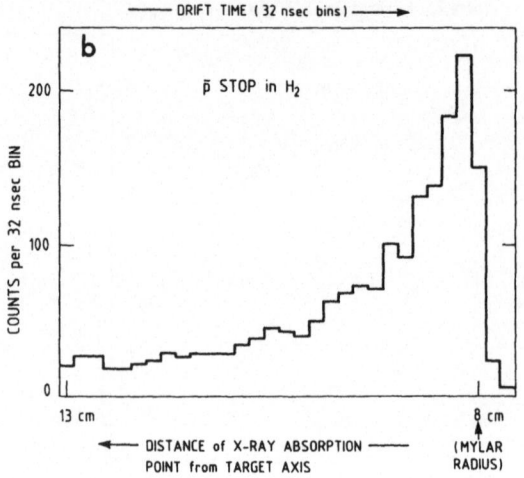

Fig. 4. X-ray energy spectrum (a) and distribution of associated
absorption points (b) for p$\bar{\text{p}}$ annihilations with identified
annihilation vertex inside a fiducial volume in the H_2 gas
target. Nearly background free, the spectrum is dominated
by the transitions feeding the atomic 2P-level. For comparison the spectrum obtained for events with identified
vertex inside the XDC counter gas volume is shown. The
3 keV argon fluorescence line serves for on-line energy
monitoring.

284

argon fluorescence line due to scattered antiprotons is apparent. This line serves as an online in beam energy calibration monitor.

Two Collinear Tracks

In Fig. 5 we show the momentum spectrum for events with two collinear tracks of total charge 0. Two peaks expected for $p\bar{p} \rightarrow \pi^+\pi^-$ and $p\bar{p} \rightarrow K^+K^-$ events are well separated. We find for the ratio $p\bar{p} \rightarrow K^+K^-$ to $p\bar{p} \rightarrow \pi^+\pi^-$ $\sim$ 15% compared to $\sim$ 30% found in bubble chamber experiments[4].

The $\pi^+\pi^-$ X-channel

Figure 6 displays a scatterplot (including projections) of the missing energy and the missing mass squared recoiling against two particles of opposite charge. The peaks in the missing mass spectrum correspond to the neutrals π^0, η^0, ρ^0/ω^0 respectively. In the scatterplot the energies of the recoiling neutrals are indicated for the case in which $\pi^+\pi^-$ were the decay products of ρ^0 and f^0. The dashed line represents the fraction of events which passed the kinematics fit to the $p\bar{p} \rightarrow \pi^+\pi^-\pi^0$ hypothesis. The resulting three-body Dalitz plot (Fig. 7) shows that this reaction is dominated by ρ resonance production. The limited solid angle of our detector imposes a smaller probability for detecting ρ^0 as compared to $\rho^\pm$, thus explaining the apparent smaller fraction of $\rho^0\pi^0$ events. This limit applies only up to invariant masses of the $\pi^+\pi^-$ pair of $\sim$ 1.0 GeV. Above the ρ^0 a strong peak is seen, which was not seen previously in bubble chamber experiments[4], corresponding in width and mass to the f^0. When imposing the presence of an L X-ray, which feeds the atomic 2P state, in coincidence the fraction of $f^0\pi^0$ events increases compared to the number of $\rho^0\pi^0$ events. A detailed fit of the Dalitz plot in view of the different contributing amplitudes is currently in progress.

The $2\pi^+2\pi^-$ X-channel

For the events with four observed tracks of total charge 0 we show the scatterplot of missing mass squared versus missing energy recoiling against the 4 pions (Fig. 8). We observe mainly 4-pion ($2\pi^+2\pi^-$) and 5-pion ($2\pi^+2\pi^-\pi^0$) events. These two channels can be identified nearly background free by kinematical fitting. In this first run we obtained $\sim$ 3500 $p\bar{p} \rightarrow 2\pi^+2\pi^-\pi^0$ events for which we show in Fig. 9 the invariant mass combinations of the 3-pion versus the 2-pion subsystems. In the projections the η^0, ω^0 and ρ^0 can be distinguished clearly. The scatterplot also shows a strong bump of $p\bar{p} \rightarrow \omega^0\rho^0$ production.

Finally we should like to mention that selecting annihilations with Kaons present in the final state is in progress. We have so far identified about 10^3 $p\bar{p} \rightarrow \pi^+\pi^-K^+K^-$ events and a few $p\bar{p} \rightarrow K^0\bar{K}^0$ events.

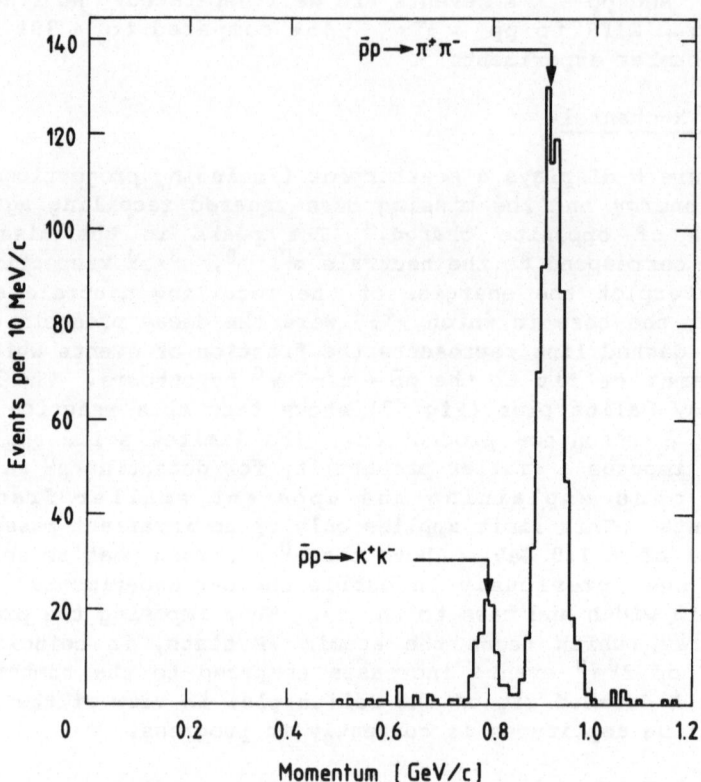

Fig. 5. Momentum spectrum of two collinear particles of opposite charge. Two peaks corresponding to annihilatons into $\pi^+\pi^-$ and K^+K^- respectively, are clearly visible at their expected positions.

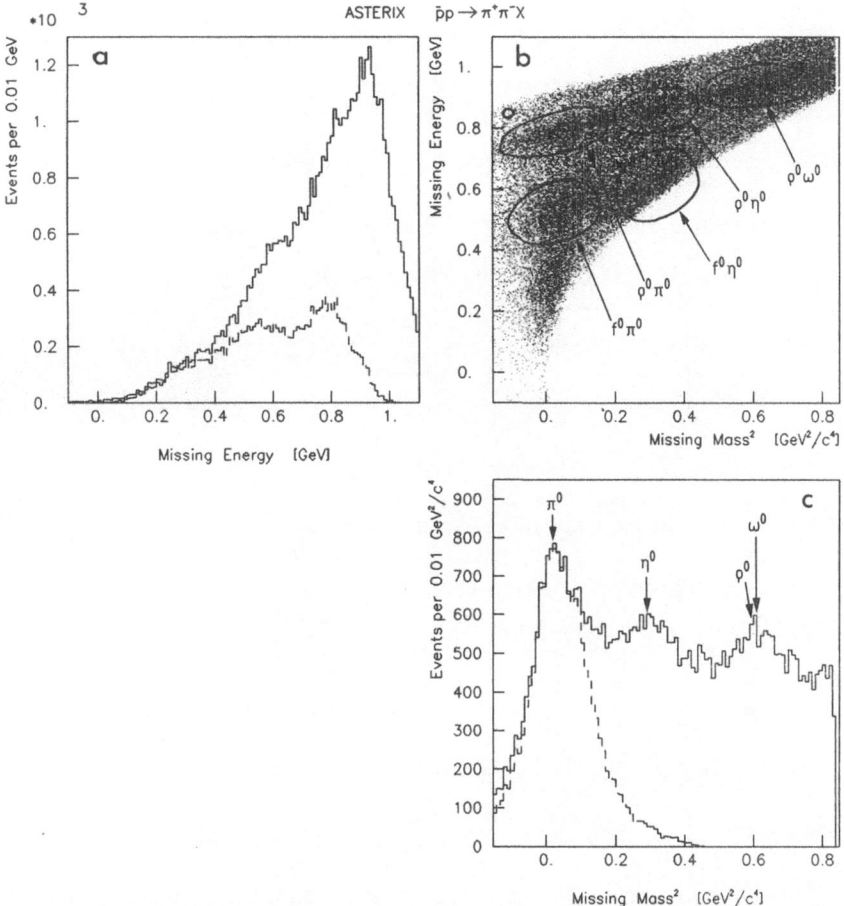

Fig. 6. Scatterplot (b) and projections of the missing mass squared (c) and the missing energy (a) recoiling against two particles of opposite charge. Energies and masses of possible two-body annihilations are indicated. The dashed line shows the fraction of events that passed the kinematics fit to the $p\bar{p} \rightarrow \pi^+\pi^-\pi^0$ hypothesis.

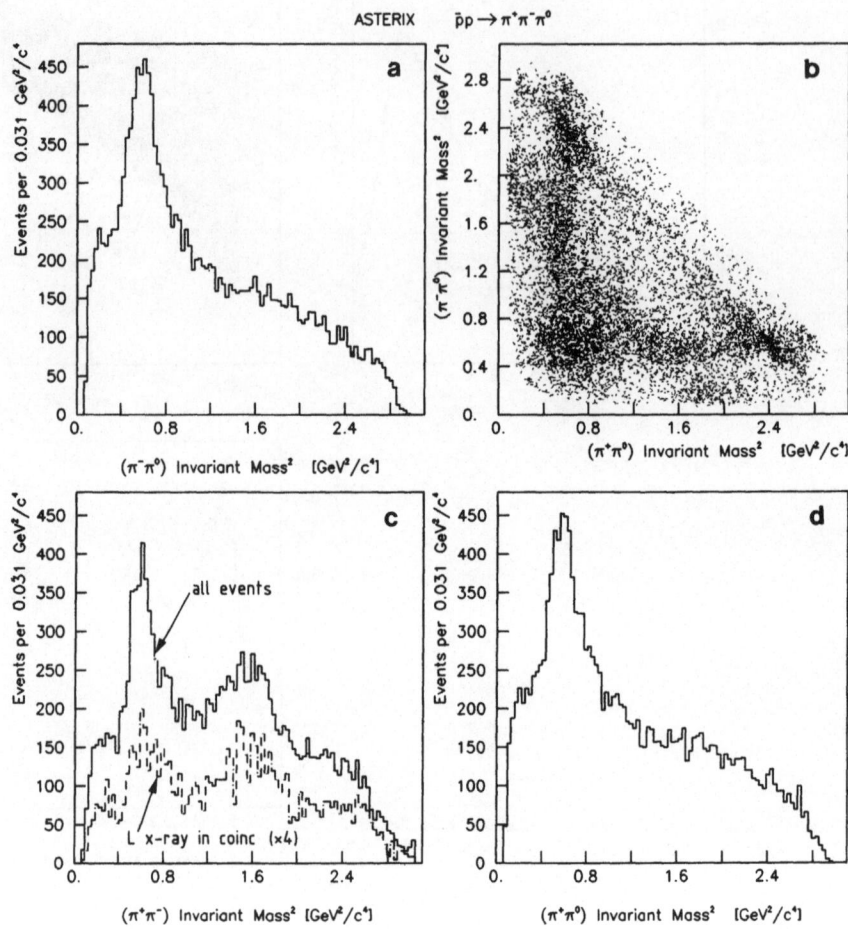

Fig. 7. Dalitz plot and invariant mass squared projections of pion pairs for the events that passed the kinematics fit to the $p\bar{p} \rightarrow \pi^+\pi^-\pi^0$ hypothesis. Large $\rho^{0\pm}$ and f^0 production can be seen. The dashed dotted line in Fig. 7c shows the spectrum obtained with an L X-ray detected in coincidence.

288

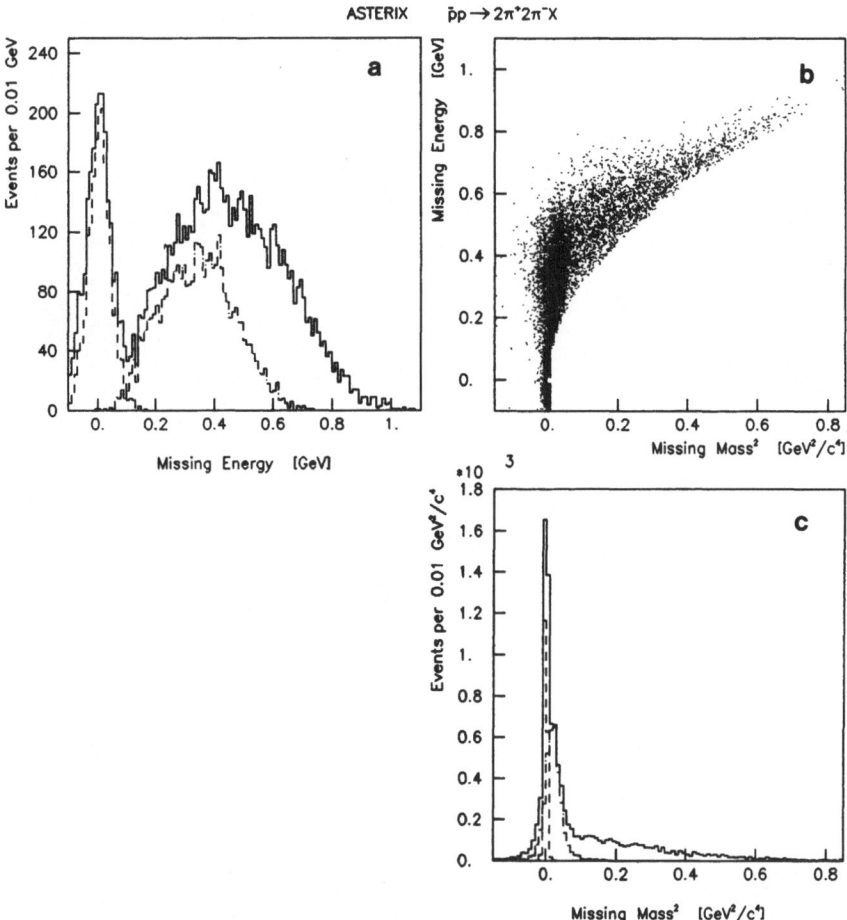

Fig. 8. Scatterplot (b) and projection of the missing mass squared
(c) and the missing energy (a) recoiling against two pion
pairs of opposite charge. The dashed and dashed-dotted
lines indicate the fractions of events that passed the
kinematics fit to the $p\bar{p} \to 2\pi^{+}2\pi^{-}$ and the $p\bar{p} \to 2\pi^{+}2\pi^{-}\pi^{0}$
hypothesis, respectively. Good separation is obtained.

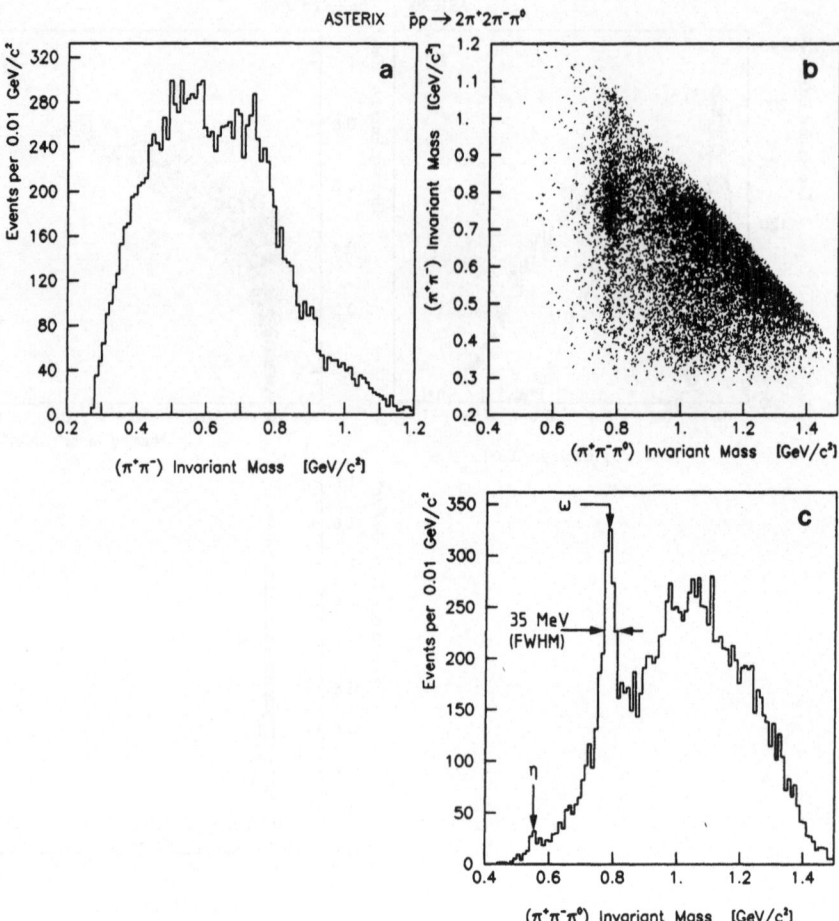

Fig. 9.   Scatterplot (incl. projections) of the invariant mass of three pion combinations versus two pion combinations for events which passed the kinematics fit to the $p\bar{p} \rightarrow 2\pi^+\pi^-\pi^0$ hypothesis.   Large ω production is observed.   The ω recoils predominantly against a ρ.

REFERENCES

1. R. Armenteros et al., A study of p$\bar{\text{p}}$ interactions at rest in a
    H$_2$ gas target at LEAR, CERN/PSCC 80-101 (1980).

2. S. Ahmad et al., Protonium Spectroscpy and Identification of
    P-Wave and S-Wave Initial States of p$\bar{\text{p}}$-Annihilations at
    rest with the ASTERIX Experiment at LEAR, in: "Physics at
    LEAR with Cooled Low Energy antiprotons", U. Gastaldi and
    R. Klapisch, ed., Plenum Press, New York (1984) 109.

3. S. Ahmad et al., (q$\bar{\text{q}}$) Spectroscopy and Search for Glueballs,
    Baryonia and other Boson Resonances in p$\bar{\text{p}}$-Annhilations at
    Rest with the ASTERIX Experiment at LEAR, in:"Physics at
    LEAR with Cooled Low Energy antiprotons", U. Gastaldi and
    R. Klapisch, ed., Plenum Press, New York (1984) 253.

4. R. Armenteros and B. French, N$\bar{\text{N}}$ Interactions, in: "High Energy
    Physics", E.H.S. Burhop, ed., Academic Press Inc., New York
    (1969).

5. E. Auld et al., Phys. Lett. 77B:454 (1978).

FIRST PHYSICS RESULTS FROM EXPERIMENT PS 172 AT LEAR

Franco Bradamante

Institute of Physics
University of Trieste, Italy

INTRODUCTION

This report is meant to illustrate the first physics results obtained by Experiment PS 172 which started running with $\bar{p}$'s in August 1983, when LEAR first ejected a 300 MeV/c cooled antiproton beam into the South Hall at CERN.

The experiment is the joint effort of four teams, from Nikhef, Amsterdam, Geneva University, Trieste University and INFN, Queen Mary College, London, and University of Surrey, who like to name themselves SING collaboration (from Switzerland-Italy-Netherland and Great Britain). The full list of the participants is given in Ref. 1, all of whom I want to thank for permission of showing our data before them being published.

The aims of the experiment are manyfold:

1.- Measurement of δ_{tot} $(\bar{p}p)$ at small momentum steps (~ 5 MeV/c) in the momentum range from 250 to 800 MeV/c, to look for narrow resonances;

2.- measurement of the ratio of the real to imaginary part of the forward elastic $\bar{p}p$ scattering amplitude at a few momenta in the range from 200 to 600 MeV/c;

3.- measurement, at a few $\bar{p}$ energies, of the polarization parameter in elastic $\bar{p}p$ and $\bar{p}C$ scattering at small angles, and of the analysing power of Carbon;

4.- measurement, at 50 MeV/c steps in the range 500 to 1000 MeV/c, of the differential cross section and of the polarization parameter, of the $\bar{p}p$ annihilation into $\pi^+\pi^-$ and K^+K^-;

5.– measurement, at 50 MeV/c steps, and possibly in all the momentum range spanned by LEAR, of the differential cross-section, of the polarization parameter and of the polarization transfer parameters in elastic $\bar{p}p$ scattering.

To fulfill these aims we have assembled two separate pieces of equipment, installed in the C2 beam-line at LEAR, the first centered around a liquid hydrogen target (measurements 1, 2 and 3), the second one around a polarized proton target. The detectors consist of scintillator and Cerenkov counters and of multi-wire-proportional Chambers, part of which with both anode and cathode read-out. A throughout description of the techniques and of the physics goals can be found in the Proceedings of the Erice Workshop held in May 1982 [2], and will not be given here.

THE MEASUREMENTS

Up to now we have had three short physics runs, first 3 days in October 1983, when LEAR was operated at 600 MeV/c, then one

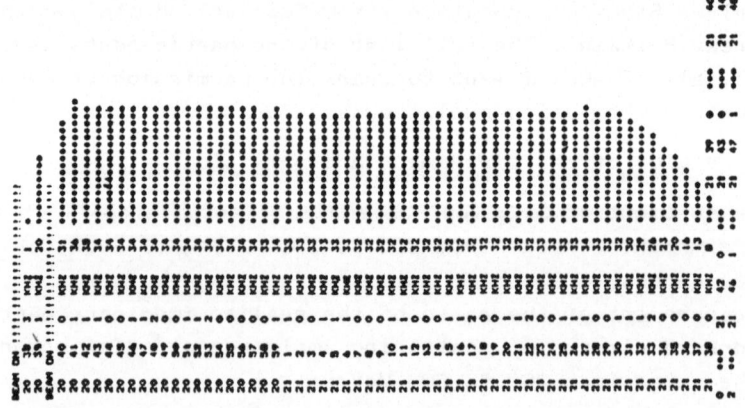

Fig. 1. Monitor of the $\bar{p}$ intensity in the C2 beam-line during a "typical" one-hour spill in March 1984

day in December, at 300 MeV/c, and finally one more day in March 1984, again at 300 MeV/c. In spite of the usual problems occurring when a new machine is being operated, the runs were successful, due mainly to the good quality of the extracted $\bar{p}$ beam. Fig. 1 shows an on-line record over one spill of the intensity of our beam, the C2 line, in the March 1984 run. The intensity is essentially constant over the whole extraction, which now lasts one hour, and it averages 10^5 $\bar{p}$/sec in good operating conditions.

In October we made a first scan of σ_{tot} in the S-meson region, and definitive results were produced in January 1984. In December and March we measured the elastic $\bar{p}p$ scattering at small angles, in the Coulomb-Nuclear interference region, at 272, 200 and 233 MeV/c respectively.

$\bar{p}p$ TOTAL CROSS-SECTION

Excellent reviews of both the experimental and the theoretical situation for baryonium physics exist in the literature (see, for instance, Ref. 3), to which I refer the interested reader. Here let me just remind that the first baryonium candidate, the S-Meson, was originally seen as a bump in the $\bar{p}p$ total cross-section.[4]

In a "conventional" transmission experiment, using an 83 mm long liquid hydrogent target and an array of 4 thin (1.5 mm) scintillator counters we have measured the $\bar{p}p$ total cross-section in the momentum range from 388 to 599 MeV/c. The beam was extracted from LEAR at 610.8 MeV/c, and lower momenta were obtained using a carbon degrader of variable thickness at an intermediate focus of the C2 line.

A full description of the experimental technique, of the apparatus, of the data reduction and analysis and of the results is given in Ref. 1) and will not be repeated in this written version of my talk. For self-consistency I will just show in Fig. 2 a plan view of the apparatus and in Fig. 3 the results.

No narrow bump is visible in Fig. 3, in particular in the S-region (p~ 500 MeV/c). The data show a smooth behaviour, well reproduced by the function a+b/p, in agreement with the recent results at KEK [5,6]. From our data we can place a limit of 2 mb MeV with 90% confidence on the strength of a resonance narrower than the resolution of the beam and hydrogen target (δ =1.4 MeV in mass). Our data are also incompatible with the broad (~ 22 MeV) enhancement seen by Hamilton et al.[7].

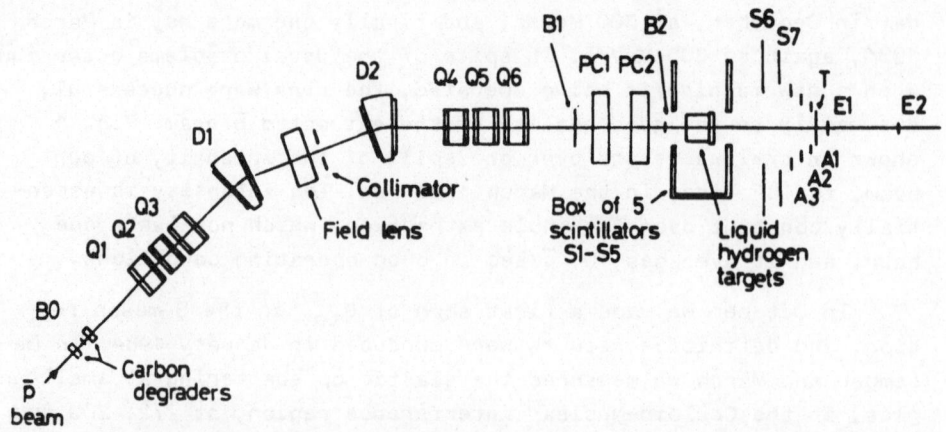

Fig. 2. Schematic diagram of the experimental lay-out (not to scale). Q's are quadrupoles, D's are bending magnets. S's are
scintillators used as vetoes, T and A's are the transmission counters.

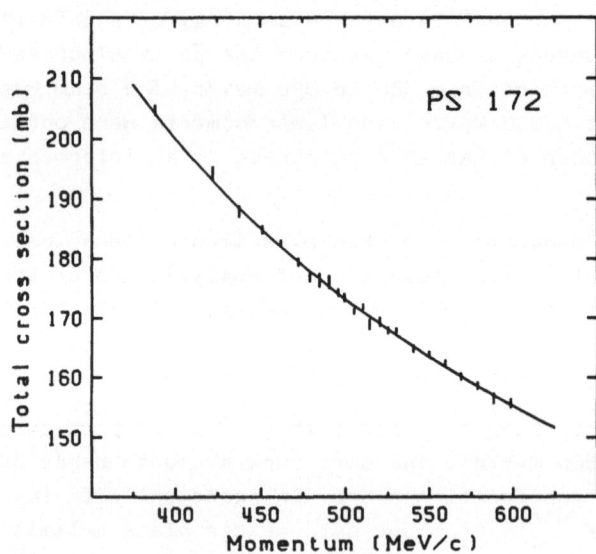

Fig. 3. Total cross-section as a function of laboratory momentum
fitted by a function (a + b/p) where a = 65.78 ($\pm$ 1.71),
b = 53759 ($\pm$ 845). χ^2 = 40.9

Some years ago it has been suggested that the real part of the elastic p̄p scattering amplitude Fp̄p has a zero at p ~ 300-500 MeV/c and that $\rho = \text{Re } F_{\bar{p}p}/\text{Im } F_{\bar{p}p}$ becomes large and negative for momenta smaller than 300.

In particular, it was shown by W. Grein [8] that the behaviour of ρ at small momenta should be dominated by the possible existence of pole terms below the p̄p threshold, so that, conversely, measurements of ρ would provide important information on the existence of such pole terms. Fig. 4, from Ref. [9], shows an updated version of these predictions, based on dispersion-relation calculations, together with existing measurements of ρ.

To measure we have followed the usual approach, namely a precision measurement of the differential cross section of the

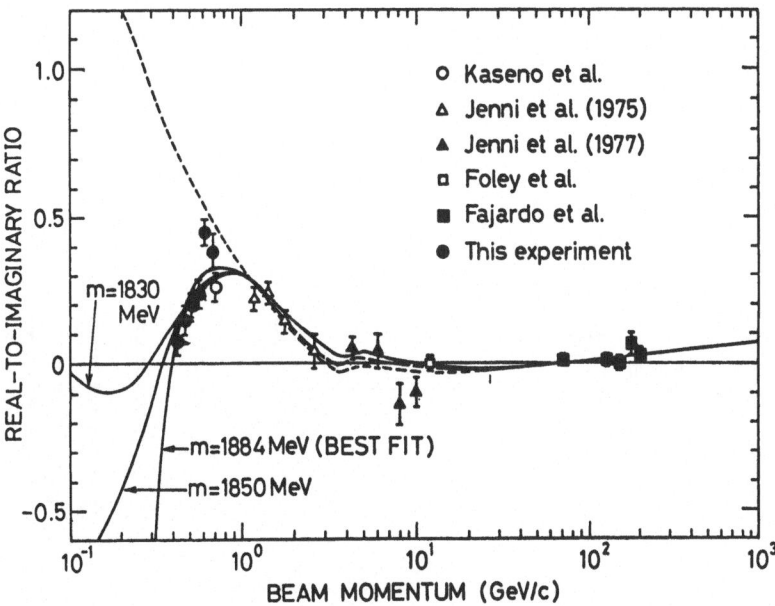

Fig. 4. Compilation of measurements of the real-to-imaginary ratio of the p̄p forward elastic amplitude. For references as well as explanation of the different calculations see ref.9.

elastic p̄p scattering at small angles, where the nuclear scattering
is comparable in magnitude to the Coulomb scattering (mostly real),
so that the exact shape of the angular distribution depends strongly on the relative phase between the two amplitudes [10]. The phase
of the Coulomb amplitude being known, the measurement of the interference allows the determination of the (unknown) phase of the nuclear amplitude in the forward region (t, the four-momentum transfer, in the interference region ranges from 10^{-3} to 4.10^{-3} GeV^2).

We have performed the measurement using a small liquid hydrogen target (11.7 mm long) and the simple MWPC's and scintillator
counters arrangement shown in Fig. 5. Two MWPC telescopes give the
measurement of the scattering angle, while TOF and pulse amplitude
in the forward counter R allow selection of p̄'s after the scattering (further reduction of the contribution from p̄p annihilation is
provided by the information of the Veto Box counters around the
target). A <u>preliminary</u> angular distribution obtained at 272 MeV/c

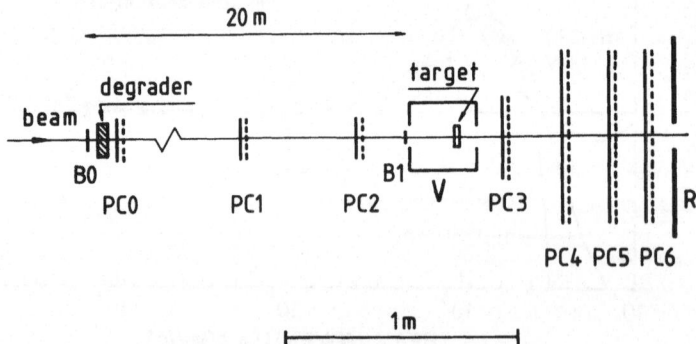

Fig. 5. Schematic diagram of the experimental lay-out for the measurement of small-angle p̄p scattering. PC1....PC6 are
MWPC's, B_0 B_2 and R are scintillator counters, V is the
veto box surrounding the liquid hydrogen target.

is shown in Fig. 6 just to give an idea of the quality of the measurement. The data are not yet corrected, in particular the empty target data are subtracted from the full target data without any correction for the different multiple scattering, nor there is any correction for the geometrical acceptance, so it is premature to

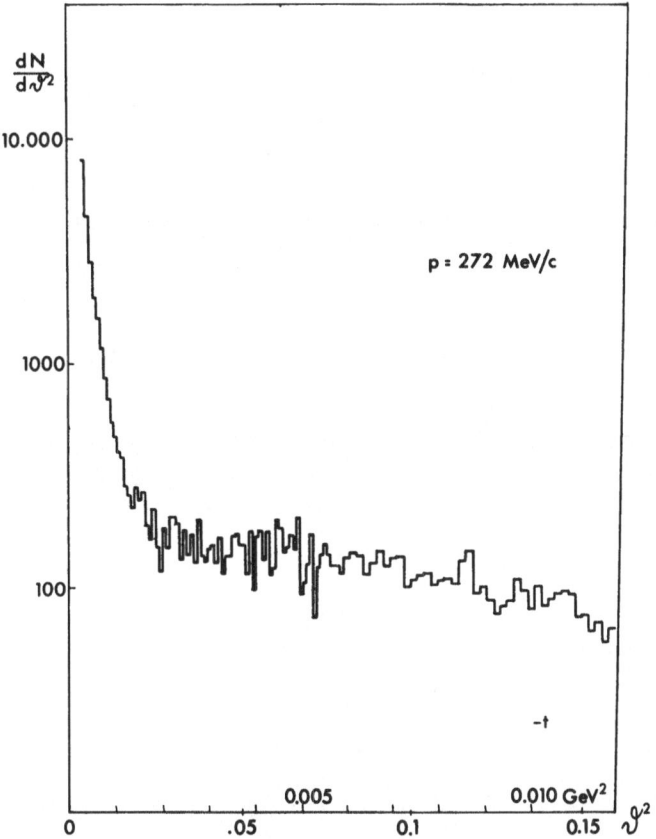

Fig. 6. Preliminary angular distribution for $\bar{p}p$ elastic scattering at 272 MeV/c.

extract a value for ρ. Very crudely, from the shape of the angular distribution one can infer that ρ is either zero, or slightly negative, but clearly more analysis is needed and it is indeed in progress.

REFERENCES

1) A.S. Clough, C.I. Beard, D.V. Bugg, J.A. Edgington, J. Hall; K. Bos, J.C. Kluyver, R.A. Kunne, L. Linssen; R. Birsa, F. Brada-mante, S. Dalla Torre-Colautti, M. Giorgi, A. Martin, A. Penzo, P. Schiavon, A. Villari; S. Degli Agosti, E. Heer, R. Hess, C. Lechanoine-Leluc, Y. Onel, D. Rapin, (SING Collaboration), "E-vidence against the S-Meson", submitted to Physics Letters, March 1984.

2) J. Bailey et al. "A polarized antiproton beam at LEAR", Physics at LEAR with Low-Energy Cooled Antiproton, PLENUM Press, New York (1984), Proceedings of the LEAR-Erice Workshop (9-16 May 1982), p. 455.

   J. Bailey et al. "Status Report on Experiment PS 172", Proceed-ings of the LEAR-Erice Workshop (9-16 May 1982), p. 427

   D. Rapin "Spin-Effects in $\bar{p}$-p Scattering", Proceedings of the LEAR-Erice Workshop (9-16 May 1982), p. 447

3) L. Montanet, G.C. Rossi and G. Veneziano, Phys. Rep. C63 (1980) 149

4) A.S. Carrol et al. Phys. Rev. Lett. 32 (1974) 247. Actually, the S-bump had already been seen at (1929 $\pm$ 4) MeV in the for-mation experiment of G. Chikovani et al., Phys. Lett. 22 (1966) 233

5) T. Kamae et al. Phys. Rev. Lett. 44 (1980) 1439

6) K. Nakamura et al. Phys. Rev. D, 29 (1984) 349

7) R.P. Hamilton et al. Phys. Rev. Lett. 44 (1980) 1182

8) W. Grein, Nucl. Phys. B131 (1977) 255

9) H. Iwasaki et al., Phys. Lett. B103 (1981) 247

10) See, for instance, P. Jenni et al., Nucl. Phys. B 94 (1975) 1.

ANTIPROTON-PROTON REACTIONS IN THE MOMENTUM RANGE

FROM 250 TO 600 MeV/c

W. Brückner, H. Döbbeling, F. Güttner, D. von Harrach,
H. Kneis, S. Majewski, M. Nomachi, S. Paul, B. Povh,
R.D. Ransome, T.-A. Shibata, M. Treichel and Th. Walcher

Max-Planck-Institut für Kernphysik, Heidelberg and
Universität Heidelberg, Fed. Rep. Germany

1. INTRODUCTION

The Low Energy Antiproton Ring (LEAR) opened up a new possibility
to study antiproton-proton interactions with good precision. The
first experimental results of the Heidelberg group at LEAR (PS173)
are reported.

There are three branches in low-energy $\bar{p}p$ reactions: elastic
scattering ($\bar{p}p \rightarrow \bar{p}p$, 35%), charge exchange ($\bar{p}p \rightarrow \bar{n}n$, 12%) and annihilation ($\bar{p}p \rightarrow$ mesons, 53% at 500 MeV/c). The Heidelberg group has
constructed a detector complex to measure these three branches exclusively[1]. This is based on the experience of two previous $\bar{p}p$ experiments at CERN Proton Synchrotron secondary beam lines[2].

The $\bar{p}p$ reactions are expected to supply us with fundamental
knowledge on strong interactions between baryons. The NN interactions
are described by an attractive long-range force and a repulsive short-
range force. There are two different pictures. In the one-boson-
exchange model the force is mediated by π, ρ, and ω exchanges. The
NN potential is then related to the $N\bar{N}$ potential by G-parity transformation[3,4]. The ω-exchange potential, which is the source of the
hard core in NN interactions, becomes attractive in $N\bar{N}$ interactions.

Recently descriptions of NN interactions have been developed in terms of quarks and gluons[5]. Although these descriptions are not yet very successful the N̄N cross-sections offer the hope of a better understanding of the mechanism of quark interactions at low energy. In particular, the annihilation radius which can be extracted from the data at low energy should be closely related to the overlap of the confinement region or the bag radius. Such information is not unambiguously accessible from NN data since the entrance and the exit channel always contain the same particles.

From the charge-exchange process ($\bar{p}p \rightarrow \bar{n}n$) we will learn isospin dependence of $\bar{p}p$ reactions since the isospin amplitude is $T_3 + T_1$ for elastic scattering while it is $T_3 - T_1$ for charge-exchange reactions, where T_3 and T_1 are isospin triplet and singlet states, respectively.

2. EXPERIMENTAL METHODS

Antiprotons were extracted from LEAR at 609 or 309 MeV/c and were degraded to the desired momenta by a carbon degrader in the beam line. The beam particles were defined by the coincidence of two plastic scintillation counters at 20 m and 0.8 m upstream from the target. The diameter of the second counter was 17 mm, and by its use, beam particles with a divergence of less than half a degree were selected. The beam intensity was typically 10^4/s on the target. The liquid-hydrogen targets were of 20 or 6 mm thickness.

Cross-sections of the detector are shown in Fig. 1. Angular distributions of elastic scattering were measured with a cylindrical multiwire proportional chamber (MWPC) and forward hodoscopes (FHDs). Angular distributions of antineutrons were measured with antineutron calorimeters (ANCs). The charged annihilation cross-section was measured with the FHDs, backward hodoscopes (BHDs) and upper and lower hodoscopes (UHDs and LHDs). Gammas from π^0 decay were detected by lead-glass counters (PbGs).

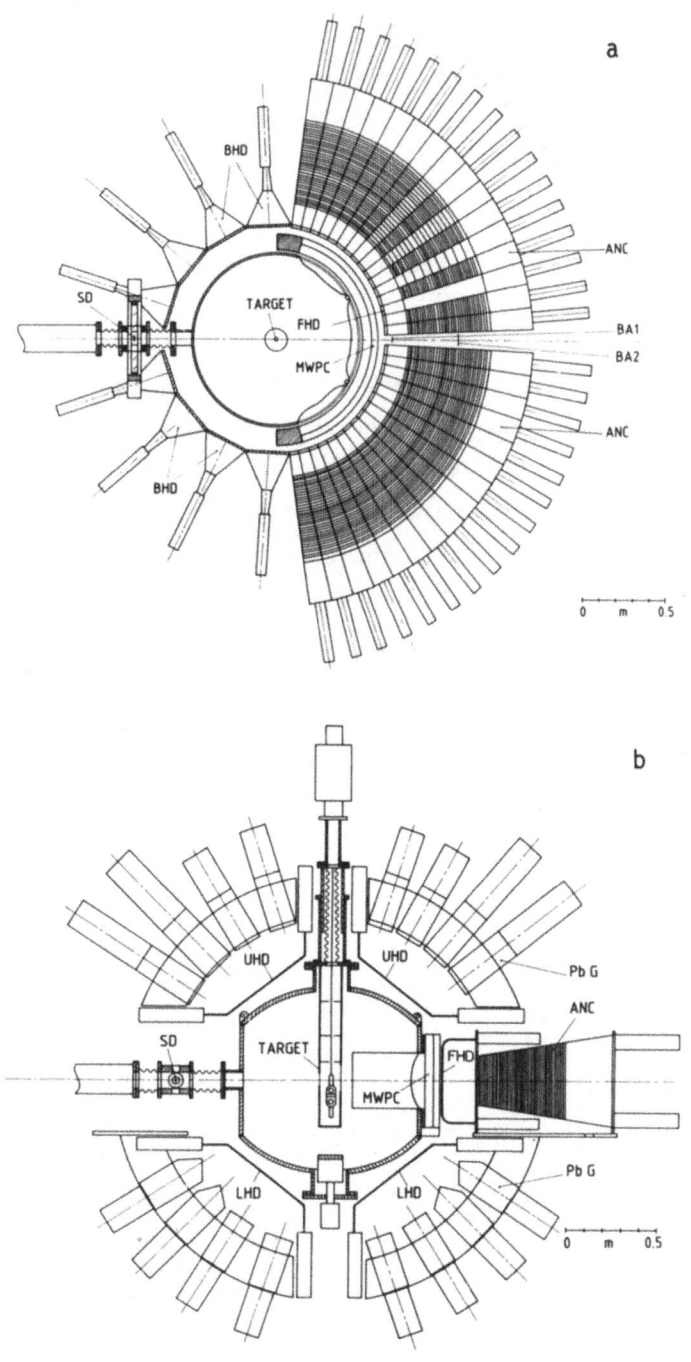

Fig. 1. (a) Horizontal and (b) vertical cross-sections of the detector.

The cylindrical MWPC, the FHDs, and the ANCs cover the angular range from 2.5° to 87.5° in the lab. system. Each ANC consist of a sandwich of 50 layers of 6 mm plastic scintillator and 4 mm iron plates viewed by two wavelength shifters.

The timing and pulse height of these detectors were calibrated by fast laser pulses distributed by fibre-glass cables.

3. EXPERIMENTAL RESULTS

Angular distributions of elastic scattering have been measured in the momentum range from 254 MeV/c (E_L = 34 MeV) to 606 MeV/c (E_L = 179 MeV). An example at 300 MeV/c is shown in Fig. 2. Charged-annihilation cross-sections were measured in the same momentum range. Since the detector covers 73% of the total solid angle the detection efficiency for charged annihilation was typically 95% according to Monte Carlo simulation. Charged annihilation cross-sections in the low-momentum region corrected for the finite acceptance are shown later in Fig. 5.

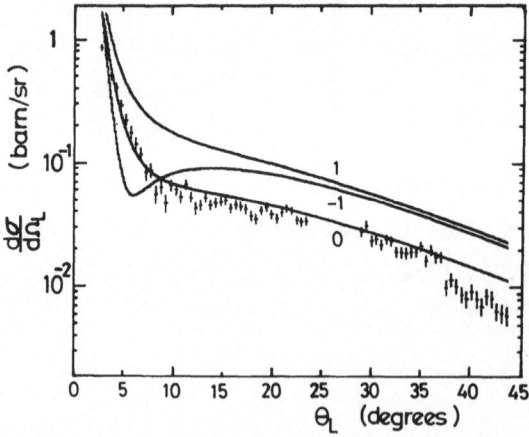

Fig. 2. The angular distributions of $\bar{p}$p elastic scattering at 300 MeV/c. The curves show the estimate with ρ = 1, 0, and -1.

4. DISCUSSION OF THE RESULTS

4.1. Angular Distributions of Elastic Scattering

In Fig. 2 one can see a sharp rise of the elastic angular distribution at small angles $\left[\theta_L < 8° \text{ or } t < 2 \times 10^{-3} \text{ (GeV/c)}^2\right]$ due to Coulomb scattering and another component with a less steep slope at large angles due to the scattering by nuclear force. In Fig. 3 angular distributions are plotted at 300 and 500 MeV/c as a function of four-momentum transfer $t = 2p_{cm}^2 (1 - \cos \Theta_{cm})$. The slope of nuclear scattering becomes steeper at lower momentum.

Forward angular distribution can be parametrized as follows:

$$\frac{d\sigma}{dt} = \frac{d\sigma_C}{dt} + \frac{d\sigma_I}{dt} + \frac{d\sigma_N}{dt}$$

$$= 4\pi \left(\frac{\alpha\hbar c}{\beta t}\right)^2 F(t)^2$$

$$+ \left(\frac{\alpha\sigma_{tot}}{\beta t}\right) F(t) \exp(-\tfrac{1}{2}bt)(\rho \cos \delta - \sin \delta)$$

$$+ \left(\frac{\sigma_{tot}}{4\hbar c\sqrt{\pi}}\right)^2 (1 + \rho^2) \exp(-bt) , \qquad (1)$$

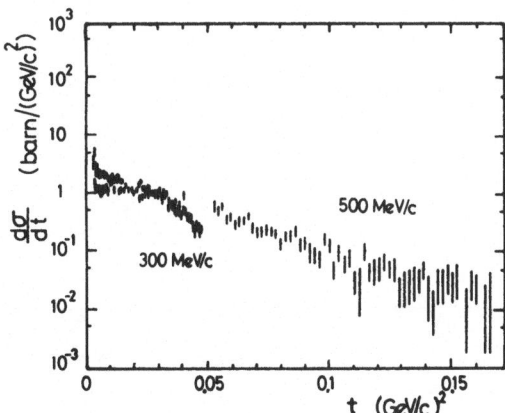

Fig. 3. The elastic scatterings at 300 and 500 MeV/c plotted as a function of t.

where $d\sigma_C/dt$ is Coulomb scattering, $d\sigma_I/dt$ Coulomb-nuclear interference, $d\sigma_N/dt$ the nuclear scattering, α the hyperfine coupling constant, β the laboratory velocity of $\bar{p}$, $F(t)$ the Coulomb form factor
of the proton, b the slope parameter, ρ the real-to-imaginary ratio
of the forward nuclear scattering amplitude, and δ the Coulomb phase.
The spin dependence was neglected[6]. At large angles only the third
term remains non zero:

$$\frac{d\sigma}{dt} = \frac{d\sigma_N}{dt} = \left(\frac{\sigma_{tot}}{4\hbar c \sqrt{\pi}}\right)^2 (1 + \rho^2) \exp(-bt) . \qquad (2)$$

In Fig. 4 are plotted $d\sigma/dt$ $(t = 0) = |f_n|^2$ obtained by fitting
data with Eq. (2). Real and imaginary parts of $f_n(0)$ can be determined separately by Eq. (1) with three parameters: σ_{tot}, ρ, and b.
The curves in Fig. 2 show examples at 300 MeV/c with $\rho = -1$, 0, and 1.
Here σ_{tot} and b are set at 240 mb and 36 $(GeV/c)^{-2}$, respectively.
The ρ parameter at 300 MeV/c is found to be close to zero. The ρ
parameter at low momenta is of particular interest since it is sensitive to possible poles near $\bar{p}p$ threshold, as predicted by the dispersion-relation calculation[7].

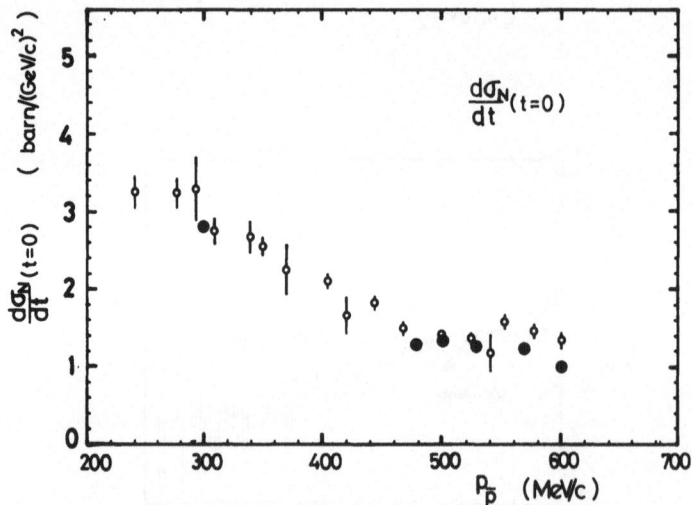

Fig. 4. The nuclear scattering amplitudes at t = 0 obtained by fitting the data with Eq. (2). Closed circles are present data and
open circles are taken from Ref. 9.

4.2 Charged-Annihilation Cross-Sections

Several authors[3,4] have estimated the features of $\bar{p}p$ reactions
by using the one-boson-exchange model. Boson-exchange potentials
were fixed by NN data and were converted to the $N\bar{N}$ case by G-parity
transformation. An imaginary potential i W(r) was introduced for
annihilation.

Curves in Fig. 5 are the predictions of total, annihilation, and
charged-annihilation cross-sections[3]. The annihilation cross-section
increases faster than total cross-section as beam momentum decreases,
suggesting that the annihilation branch becomes dominant in low
partial waves.

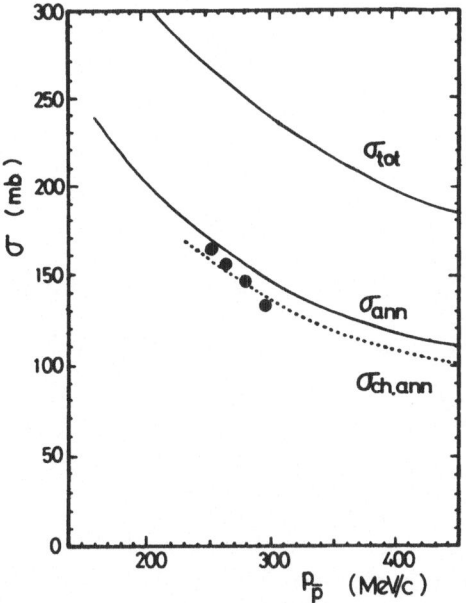

Fig. 5. The charged-annihilation cross-sections. The estimates for
the total, annihilation, and charged-annihilation cross-sections
given in Ref. 3 are also shown.

The quark-rearrangement model[8] worked well in the stopped $\bar{p}p$ annihilation where the initial state is an S wave. Efforts are now being made to extend it to annihilation in flight. Both stopped and in-flight annihilations are expected to be understood in the same model and can be compared with our measurements.

5. SUMMARY

Systematic measurements of $\bar{p}p$ reactions down to 250 MeV/c were done for the first time. In the elastic scattering the Coulomb nuclear interference method turned out to work well in determining the real-to-imaginary ratios of nuclear-scattering amplitude. The charged-annihilation cross-sections give a test of models for the annihilation mechanism at low energy. We expect to study basic $N\bar{N}$ interactions in more detail as the analyses and new data taking proceed in 1984.

REFERENCES

1. W. Brückner et al., in 'Proc. Workshop on Physics at LEAR with Low Energy Cooled Antiprotons', Erice, 1982 (Plenum, New York, 1984) (eds. U. Gastaldi and R. Klapisch), p. 437.
2. W. Brückner et al., Phys. Lett. 67B:222 (1977).
   C. Amsler et al., in 'Proc. Workshop on Physics at LEAR with Low Energy Cooled Antiprotons, Erice, 1982 (Plenum, New York, 1984) (eds. U. Gastaldi and R. Klapisch), p. 375.
3. R.A. Bryan and R.J.N. Phillips, Nucl. Phys. B5:201 (1968), and B7:481 (E) (1968).
4. P.H. Timmers et al., THEF-NYM-83.06, 1983, University of Nijmegen, Holland.
5. M. Oka and K. Yazaki, Nucl. Phys. A402:477 (1983).
   A. Faessler and F. Fernandez, Phys. Lett. 124B:145 (1983).
   G. Elster and K. Holinde, Phys. Lett. 136B:135 (1984).
6. M. Lacombe et al., Phys. Lett 124B:443 (1983).
7. W. Grein, Nucl. Phys. B131:255 (1977).
8. M. Maruyama and T. Ueda, Phys. Lett. 124B:121 (1983).
   A.M. Green and J.A. Niskanen, Nucl. Phys. A412:448 (1984).
   S. Furui and A. Faessler, contributed paper to 10th Int. Conf. on Few Body Problems in Physics, Karlsruhe, 1983.
9. J.E. Enstrom et al., Particle Data Group, $N\bar{N}$ and $\bar{N}D$ Interactions - A Compilation, LBL-58, 1972.

X-RAYS FROM PROTONIUM - PS174 PROGRESS REPORT

Tim Gorringe

University of Birmingham, NIKHEF, Rutherford
Laboratory, William and Mary College
USA Collaboration

INTRODUCTION

The purpose of this experiment is to measure the strong interaction shift and width of the atomic 1S groundstate of antiprotonic-
hydrogen by observing its K series X-rays. This gives directly the
$\bar{p}$-p complex scattering length. Also from a measurement of the L
X-rays we can determine the 2P annihilation width and investigate the
atomic cascade. $\bar{P}$-helium runs provide us with a means of optimizing
and periodically checking the working of our equipment in addition to
their physics interest (strong $\bar{p}$-He interaction and atomic cascade).
To date we have stopped antiprotons in gaseous hydrogen (30K and
300K), deuterium (30K) and helium (30K, 100K and 300K).

The $\bar{p}$-p atom has atomic level energies and sizes scaled by the
ratio of ($\bar{p}$ reduced mass/electron mass) relative to ordinary hydrogen. This results in K X-ray energies of 9.4-12.5 KeV and L X-rays
of 1.7-3.0 KeV with an atomic groundstate radius of about 50 fm.

In addition to the electromagnetic interaction between proton
and antiproton there is also the strong interaction. This shifts
and broadens the pure electromagnetic levels - the magnitude being
roughly proportional to the wavefunction overlap since the strong
interaction is short ranged compared to the atomic size. A popular
method of estimating absolute magnitudes is via an optical model
calculation[1]. These predict for the 1S state shifts and widths
of typically 0.5-1.0 KeV with a singlet-triplet splitting of about
250 eV. An optical model approach may however not be appropriate due
to its use of potentials to describe short range behavior. Also
these quantities would be very sensitive to the existence of possible
baryonium states near threshold.

EXPERIMENTAL DETAILS

Antiprotons from LEAR at 308 MeV/c are slowed down by a double wedge shaped aluminium degrader before passing through windows in the vacuum and target vessels, to stop in the target gas (see Figure 1). A coincidence between two scintillators with a third acting in veto mode as a collimator tells us an antiproton has arrived. The target

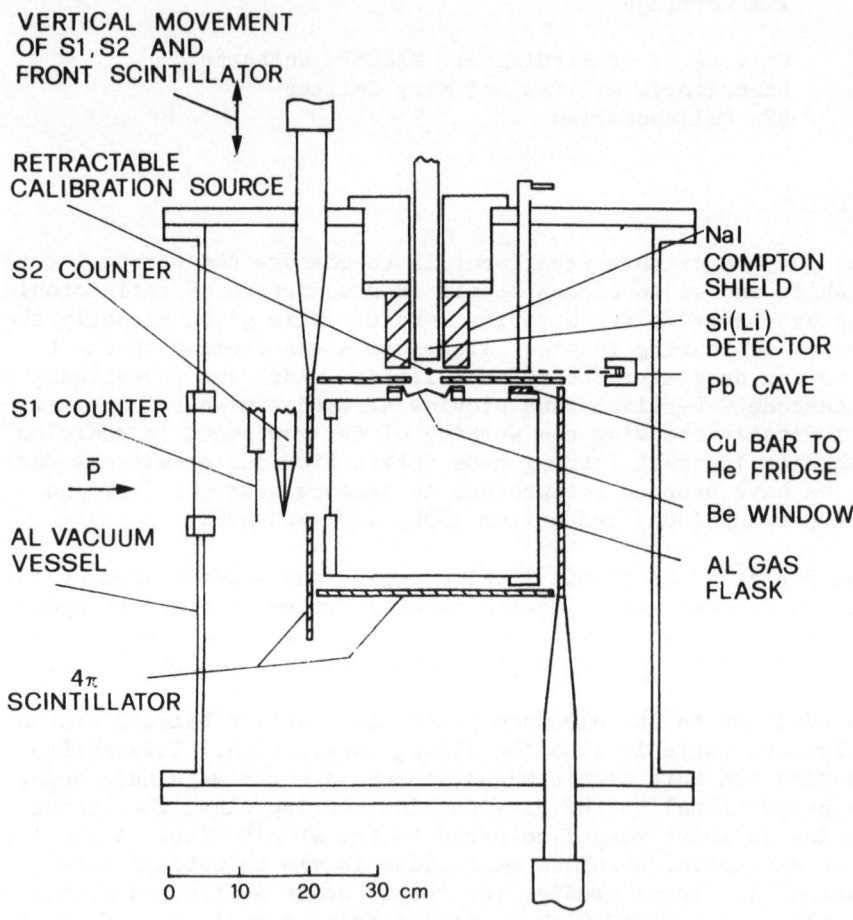

Fig. 1. P̄P target and counters

vessel is a cylinder of 25 cm diameter and 25 cm length containing
gas at atmospheric pressure and any temperature between 30K and 300K.
A remote helium refrigerator maintains the desired temperature. A
lithium drifted silicon detector of in-beam resolution 300 eV at
6.4 KeV and area 300 sq.mm looks down at the target through an X-ray
window. The window consists of 6 micron mylar supported on a 1 mm
thick pure aluminium grid. Surrounding the Si(Li) detector is a
segmented NaI annulus 12 cm in height and 2.5 cm thick. This can be
used in a veto mode to suppress the continuum background due princi-
pally to Compton scattering of gamma rays in the Si(Li) detector.
Three counter telescopes look at the front, center, and rear of the
target volume to enable measurement of the stopping distribution by
detecting charged particle annihilation products. In-beam calibration can be obtained by means of a ^{57}Co source (energies 6.4, 7.1 and
14.4 KeV) which can be remotely moved into position just below the
Si(Li) detector.

Several experiments have looked for $\bar{p}$-p X-rays, but only the
L series have ever been seen. There are many experimental difficulties and in this light a discussion of these is quite appropriate.

For S.T.P. hydrogen gas total K X-ray yields per protonium atom
from cascade calculations are expected to be between 0.5 and 5 percent, and a factor of about 2 less at 30K[2]. The low yield is a
consequence of Stark mixing between l substates populating the S and
P states from where annihilation is strong. Protonium is small
(compared to ordinary hydrogen) and neutral – hence it is easy for it
to pass through the strong electric fields close to nuclei of neigh-
boring molecules. These fields cause Stark transitions to occur
between l sublevels hence continually populating the S and P states
leading to $\bar{p}$-p annihilation in high n states before X-ray emission.
For $\bar{p}$-He this effect is not nearly so strong since it moves around as
a positive ion and Coulomb repulsion prevents such close approach to
nuclei.

Previous experiments used secondary beams with low $\bar{p}$ fluxes
(1000 pbars/sec), poor momentum resolution (several %), and heavy
pion contamination. The new LEAR machine provides us with about 100
times the $\bar{p}$ flux, no pion contamination, and a much improved momentum
resolution (about 0.2%) which allows us to stop 80% of the beam in
gas at 30K, a factor of 10 higher than before. At 200 MeV/c we could
stop 80% of the beam in 300K gas and thereby gain from the higher
X-ray yield due to the density dependence of stark mixing.

Increasing the X-ray rate alone does not solve our problems
since we also increase the continuum background by the same factor.
This background has two origins, below several KeV electronic noise
is dominant, whilst elsewhere Compton scattering of gamma rays in the
Si(Li) detector.

To reduce the Compton background which is due to prompt gammas in the energy region 50 to 500 KeV coming from the showering of much higher energy gammas (produced in the annihilation) in the target, the Si(Li) is surrounded by annulus of NaI to detect the Compton scattered photons. This can act as a veto on Si(Li) detector signals and offers a background reduction factor of up to 6 depending on the energy region. As well as background reduction genuine X-ray events are also vetoed by a gamma ray or charged particle entering the NaI in coincidence with them. About 30% of X-rays are lost in this way.

Signals from the detector/FET/pre-amp assembly are amplified and shaped by a spectroscopic amplifier (LINK 2010 Pulse Processor). Since this employs time variant filtering it has a low level discriminator on the input stage which initiates the processing of a detector signal if it fires. The level at which this is set is a compromise between wishing to be as sensitive as possible to X-rays in the 2 KeV region and not processing vast quantities of noise which will increase our deadtimes and wash out any X-ray structure at low energies. Some time during our preliminary running was used to optimize the setting of this discriminator.

On average 3 charged pions are produced per $\bar{p}$-p annihilation. Operating the Si(Li) detector and associated electronics in the environment of such a high particle flux is a major problem. The detector used a pulsed optical feedback method of charge neutralization on the FET. This is initiated when an equivalent of 0.5 MeV is deposited in the silicon which is less than the energy lost by a minimally ionizing pion passing through the crystal. Each incident pion therefore causes an optical reset and all the associated deadtimes required to allow the detector and electronics to settle. The detector system therefore had to be designed to handle these fluxes.

PRELIMINARY RESULTS

The p helium spectra of Figure 2 were collected in one spill (approximately 1 hour of beam time) and show clearly the N, M and L lines. The 5-4 line tells us we are sensitive down to 1.7 KeV. Preliminary results suggest yields at various temperatures in good agreement with the cascade calculations of Landua[3] making us confident we know our detector acceptance well. As well as several short helium runs a long (8 spills) run was made to determine the $\bar{p}$-He strong interaction 2P shift and width, which were found to be 7.4±4.9 eV and 35±15 eV respectively.

Table 1 gives a summary of hydrogen and deuterium runs in chronological order, it includes preliminary estimates of K and L X-ray yields.

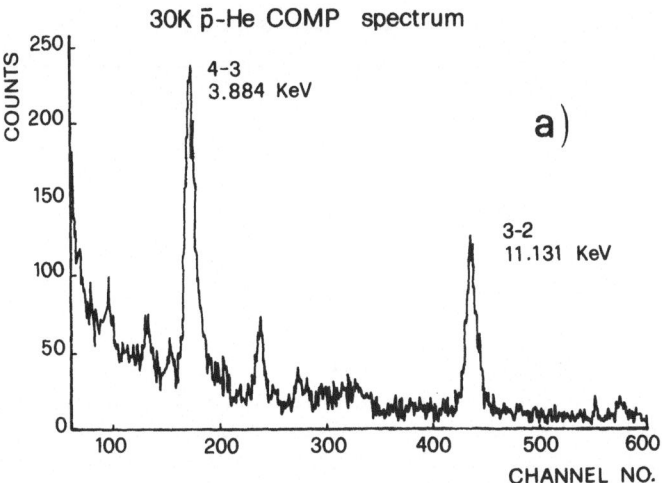

30K p̄-He COMP spectrum

a)

4-3
3.884 KeV

3-2
11.131 KeV

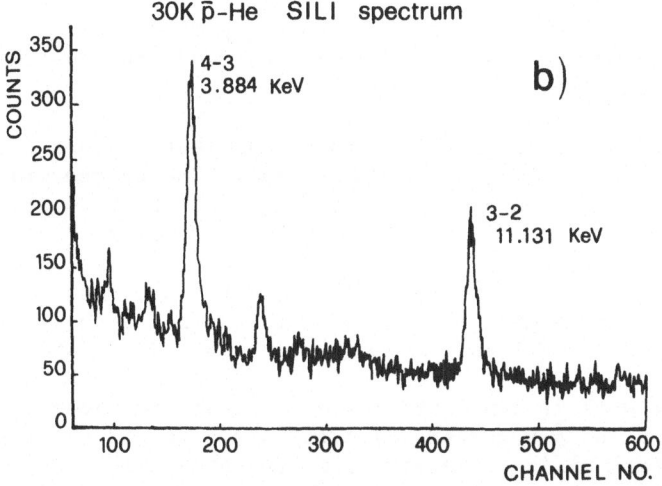

30K p̄-He SILI spectrum

b)

4-3
3.884 KeV

3-2
11.131 KeV

Fig. 2. Compton background suppressed (top) and unsuppressed
(bottom) p-He spectra showing the N, M, and L lines

Figure 3 contains the compton suppressed spectra (with a 500 ns
prompt time window) for runs H.2 and D.1, 30K hydrogen and deuterium
respectively. Comparison of the two spectra show a clear excess of
counts in the hydrogen spectrum above background in the region
10-12 KeV with a yield of about 0.1% per p̄-H atom. The presence of

Table 1. Preliminary Yields for $\bar{p}$-H and $\bar{p}$-D

Run	Gas	Temp	K X-ray yield (%)	3-2 yield (%)	4-2 and above yield (%)
H.1	H_2	300K		7.4 ±2.9	5.6 ±1.5
H.2	H_2	30K	.11±.04	<0.33	.53± .11
D.1	D_2	30K		.16± .05	.36± .06

$\bar{p}$-D K series X-rays is not expected due to the deuterons greater size
compared to the proton, and the increased number of annihilation
channels, making them far too broad and low a yield to be observed.
As a result it provides an excellent background measurement under the
same antiproton stopping conditions as hydrogen. Since no contam-
inant X-rays are observed in the $\bar{p}$-D data in the region of the peak
at 11 KeV in the $\bar{p}$-H data it is reasonable to attribute the peak to
$\bar{p}$-H K X-rays. Cascade calculations suggest at this gas density high
nP-1S transitions will dominate, and hence we identify this peak with
the sum of these transitions. The 2-1 transition, which would be
about 2.5 KeV below this peak, is not observed (as expected from
cascade calculations).

Antiprotonic hydrogen L X-rays are present in runs H.1 (300K
yield of 13%) and H.2 (30K yield of 0.5%). Due to the prompt time
window employed in the histograms of Figure 3 the L X-rays are lost.
Our efficiency at 1.7 KeV is so low we could only set an upper limit
of <0.33% (1 std deV) on the 3-2 transition yield in run H.2. The
$\bar{p}$-D L lines, shifted up in energy by a factor 4/3 (due to the increase in $\bar{p}$ reduced mass) can be seen in run D.1 with a yield of
about 0.5%. The 3-2 transition makes up about 30% of this.

Present work is concentrating on what combinations of input
parameters in the $\bar{p}$-H cascade calculations (e.g. 1S shift and width,
2P width, stark mixing strength) are consistent with the observed L
and K total and component yields.

With these statistics our analysis is obviously incomplete and
requires further data to clarify the situation. However we can
conclude that our results show firm evidence for the existence of
K series X-rays from protonium.

In the future we hope to measure $\bar{p}$-H L and K X-ray yields over
a greater pressure range. At low pressures (below 1/4 atm) the 2-1
transition is predicted to dominate over the high n K series transitions. These measurements should provide a more accurate determination of the $\bar{p}$-p complex scattering length and a thorough investigation of the atomic cascade.

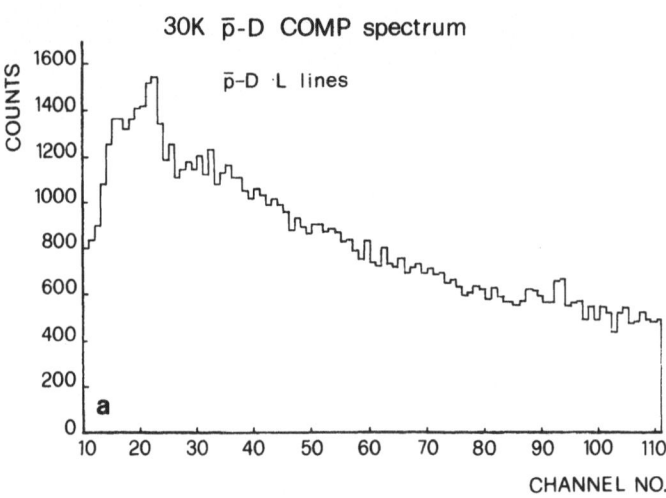

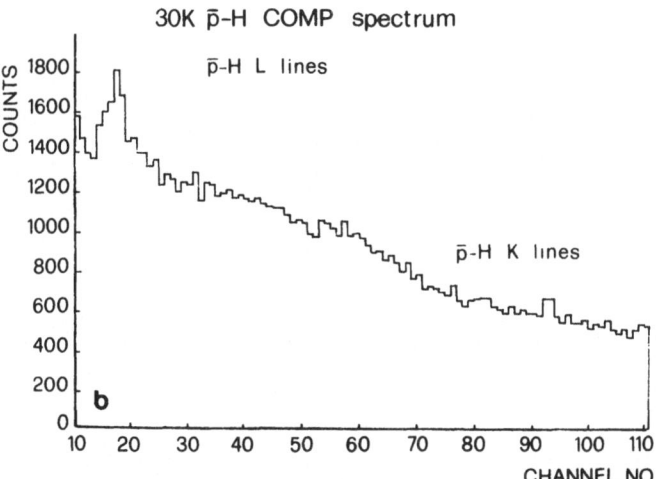

Fig. 3. $\bar{p}$-D (top) and $\bar{p}$-H (bottom) spectra for 1-20 KeV

REFERENCES

1. Richard and Sainio, <u>Phys.Lett.</u>, 110B:349 (1982).
2. Borie and Leon, <u>Phys.Rev.</u>, A21:1460 (1980).
3. Landua and Klempt, <u>Phys.Rev.Lett.</u>, 48:1722 (1982).

ANTIPROTON X-RAY SPECTROSCOPY IN THE CYCLOTRON TRAP:

PRELIMINARY RESULTS

John Missimer

Schweizerisches Institut fuer Nuklearforschung
CH-5234 Villigen, Switzerland

The principle of the experiment designed to measure the X-ray spectrum of protonium was verified by the Karlsruhe group* during the December, 1983 run at LEAR. The experiment relies on the cyclotron trap, whose characteristics are summarized in the discussion of its application to neutral current experiments in muonic atoms[1] and in contributions to previous ERICE proceedings[2,3]. The cyclotron trap yields high stopping densities at low gas pressures. During the December run, about 15 % of the beam of antiprotons at 300 MeV/c could be accepted by the trap at pressures between 400 and 600 mbar. A stopping density of $2 \cdot 10^5$ antiprotons per gram per second was achieved. This density represents an increase by a factor of almost 300 compared to alternative techniques for stopping antiprotons in gases.

A comparison of the $\bar{p}$-helium spectra observed by the Basel-Karlsruhe-Stockholm collaboration[4] (top) and by the Karlsruhe group (bottom) in December is shown in Fig. 1. The most striking feature of the bottom spectrum is the absence of contamination lines. The relative intensities of the 4-3 and 3-2 transitions should also be noted. The dominance of the 4-3 line in the bottom spectrum reflects the increased yields of transitions between circular orbits. This increase is expected at lower gas pressures, since Stark mixing decreases with target density. The reduction of Stark mixing is the

---

* The Karlsruhe group consists of P. Blum, M. Droge, D. Gotta, W. Kunold, D. Rohmann, M. Schneider, L.M.Simons

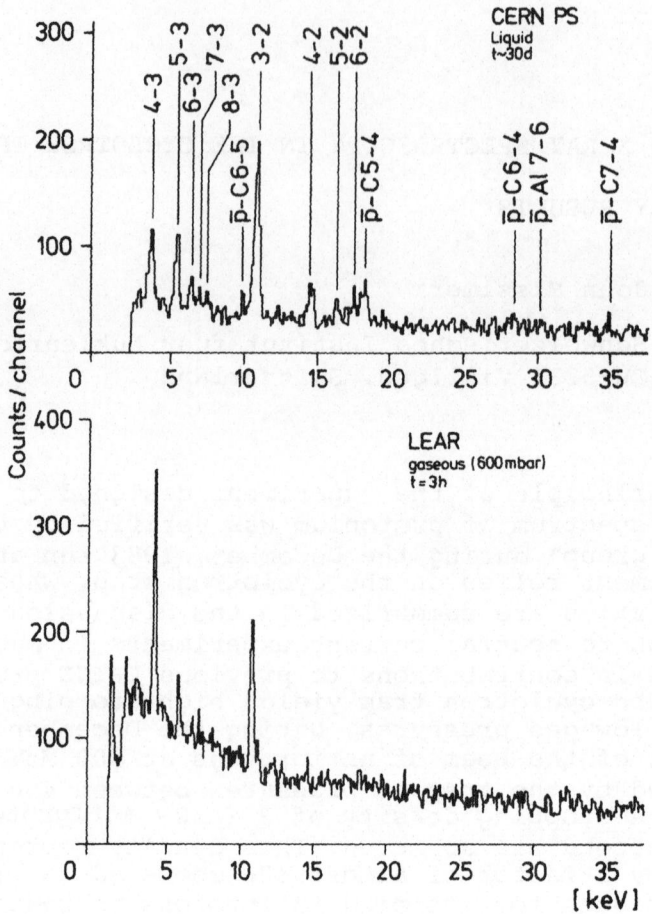

Fig. 1. Spectra of $\bar{p}^4$He. The top spectrum was observed in liquid helium in about 30 days by the Basel-Karlsruhe-Stockholm collaboration[4] using conventional techniques. Contamination lines are indicated. The bottom spectrum was observed in gaseous helium at 600 mbar by the Karlsruhe group in 3 hours.

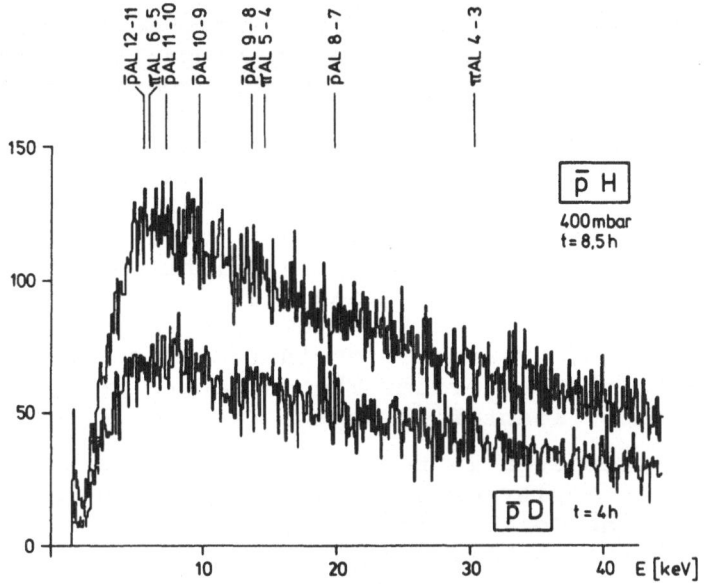

Fig. 2. X-ray spectra cf $\bar{p}$-hydrogen (top) and $\bar{p}$-deuterium (bottom) gases at a pressure of 400 mbar. The spectra were measured in runs of 8 hours and 4 hours respectively. Positions of the most likely contamination lines are indicated.

motivation for measuring the protonium spectrum in gases at low pressures.

The yield of the 3-2 transition implies an annihilation width

$$\Gamma_{3D} = 2.6 \pm 0.5 \text{ meV}$$

which has been calculated without the Fried-Martin correction[6]. This width is consistent with that obtained by the Mainz-Daresbury-TRIUMF collaboration[7] and therefore contradicts the width obtained by the Basel-Karlsruhe-Stockholm collaboration[4].

The prompt spectra of $\bar{p}$-hydrogen and $\bar{p}$-deuterium atomic states are displayed in Fig. 2; comparison permits an analysis of systematic effects in the experiment. Again, the most striking feature of these spectra is the absence of contamination lines. Careful examination of the spectra suggest the presence of structure in the hydrogen spectrum which is absent in the deuterium spectrum. A line-fitting program confirms the structure. It uses the fact that the energy differences between (Lyman $(n \to 1)$ lines are determined to an accuracy of better than 1 eV by quantum electrodynamics for $n \geq 3$. The analysis indicates three lines corresponding to the Lymann series in the hydrogen spectrum and none in the deuterium. The structure at approximately 9 keV can be identified as the K_α-line.

The structures seen in the hydrogen spectra are not prominent; the intensity of the proposed K-lines is only a three standard deviation effect:

$$N_{n \to 1} = 153 \pm 50$$

If the line at 9 keV is the K_α-line, it displays a shift,

$$\Delta E_{1S} = 428 \pm 60 \text{ eV}$$

For the width, only an upper limit

$$\Gamma_{1S} < 500 \text{ eV}$$

can be extracted. These magnitudes are consistent with the wide range of theoretical expectation. If the structures have been correctly identified, the yields of the K-series,

$$Y_K = (8 \pm 4) \cdot 10^{-3}$$

and of the K_α-line

$$Y_{K_\alpha} = (3.2 \pm 1.5) \cdot 10^{-3}$$

are also consistent with the cascade calculation of Borie and Leon[5].

In conclusion, the December run at LEAR, has demonstrated that the cyclotron trap does yield high stopping densities of antiprotons at the low pressures required to suppress Stark mixing. The spectra measured display little contamination and the spectrum of protonium suggests the observation of the K-series. An improved experiment is planned for the summer. The lower antiproton beam energy of 200 MeV/c and a magnetic yoke should increase the acceptance of antiprotons. In addition, improved detectors and more beam time should permit an unambiguous identification of the K-series, and precise determination of its characteristics.

References

1. J. Missimer, Neutral currents in muonic atoms, contribution to this meeting
2. D. Gotta, X-ray spectroscopy of $\bar{p}$-hydrogen in the cyclotron trap, LEAR Workshop, Erice (1982)
3. L.M. Simons, The cyclotron trap: status of preparation and planned experiments, LEAR Workshop, Erice (1982)
4. H. Poth, R. Abela, G. Backenstoss, P. Blum, W. Fetscher, R. Hagelberg, M. Izycki, H. Koch, A. Nilsson, P. Pavlopoulos, L. Simons and L. Tauscher, PL 76B:523 (1978)
5. E. Borie and M. Leon, PR A21:1460 (1980)
6. R. Welsh, private communication
7. Mainz-Daresbury-TRIUMF Collaboration NP A384:306 (1982)

PS184: A STUDY OF ANTIPROTON-NUCLEUS INTERACTIONS AT LEAR

D. Garreta, P. Birien, G. Bruge, A. Chaumeaux, D.M. Drake[1],
S. Janouin, D. Legrand, M.C. Mallet-Lemaire, B. Mayer,
J. Pain and J.C. Peng[1]

DPHN/ME, CEN, Saclay, France

M. Berrada, J.P. Bocquet, E. Monnand, J. Mougey and
P. Perrin

DRF, CEN, Grenoble, France

E. Aslanides and O. Bing

CRN, Strasbourg, France

.A. Erell, J. Lichtenstadt and A.I. Yavin

Tel Aviv University[2], Tel Aviv, Israel

(Presented by D.M. Drake)

The PS184, a collaboration among experimental groups from
Saclay, Strasbourg, Grenoble and Tel Aviv, is one of the experiments
that has been approved for the study of nuclear physics with the
antiproton beam provided by the Low-Energy Antiproton Ring (LEAR) at
CERN. We use standard nuclear physics techniques centred around a
magnetic spectrometer, SPES II[1], which has a momentum acceptance of
±18%, a subtended solid angle of 30 msr, and a maximum momentum of
800 MeV/c.

---

1) Permanent address: Los Alamos National Laboratory, New Mexico,
   USA. Supported in part by the US Department of Energy.
2) Supported by the Israel Fund for Basic Research.

We have proposed and partially completed four types of experiments which should provide data for building and testing models of $\bar{p}$-nucleus interactions. These are: 1) $A(\bar{p},\bar{p})A$ elastic scattering; 2) $A(\bar{p}\bar{p}')A^*$ inelastic scattering; 3) $A(\bar{p}p)A_{Z-1,\bar{p}}$ knockout; and 4) $A(\bar{p},x)X$ annihilation; of these, (3) and (4) have no exact analogy in normal nuclear physics.

Very little $\bar{p}$-nucleus data existed before the summer of 1983, and most of these were concerned with total reaction cross-sections[2]. Theoretical models[3] were based on the elementary amplitudes of $\bar{N}N$ interactions folded over the nuclear volume. More pragmatic[4] optical models were built on data which came from $\bar{p}$-atomic X-rays and $\bar{p}$ absorption. The energy shift and width of the innermost X-rays can be interpreted in terms of V and W of the nuclear potential constrained by an assumed nucleon distribution. Potentials deduced from these data differ by factors as large as 3, have large errors, and are not unambiguous.

Our primary effort has been directed toward measuring elastically scattered $\bar{p}$ angular distributions at 300 and 600 MeV/c and extracting optical model parameters therefrom[5]. Figures 1 and 2 illustrate two examples: Fig. 1 shows $^{12}C(\bar{p}\bar{p})$ experimental cross-sections at 300 MeV/c, several theoretical predictions, and our coupled channel optical model fit; Fig. 2 is for $^{208}Pb(\bar{p}\bar{p})$ elastic scattering at 600 MeV/c. Table 1 goes with Fig. 1, giving several sets of optical model parameters which fit the data reasonably well.

This table is typical of the fits for all the data in that χ^2/N is a slowly varying function of potentials. A common feature of these fits is that specific values of V and W are always found at a particular radius, $R_{1/2}$, almost independent of the fixed values of V_0 and W_0.

One can search for the existence of $\bar{p}$ nuclear states by detecting protons emitted in the forward hemisphere from nuclei bombarded by $\bar{p}$'s. Since the elementary amplitude for $\bar{p}p$ scattering at 180°

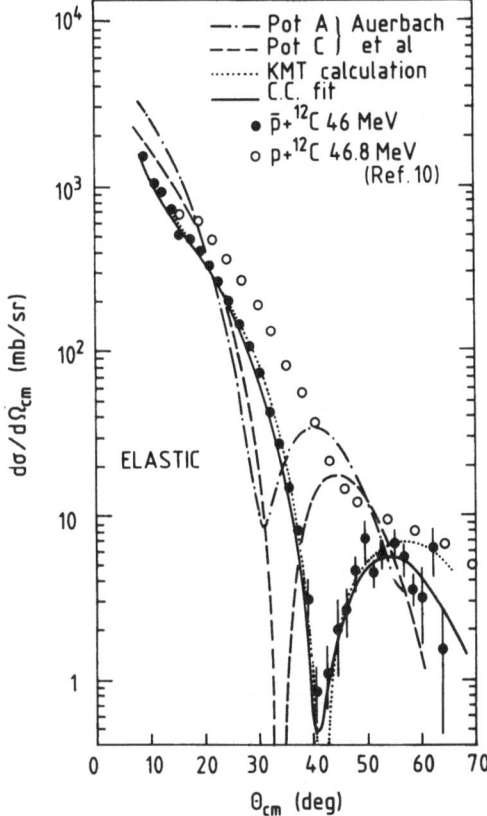

Fig. 1. Angular distribution for $\bar{p}$-^{12}C elastic scattering at 46.8 MeV.

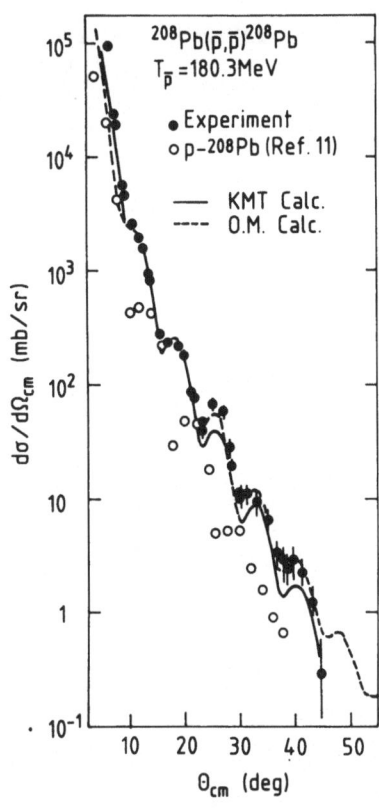

Fig. 2. Angular distribution for $\bar{p}$-^{208}Pb elastic scattering. The two curves represent: an optical model calculation (dashed line) with parameters $V_0 = 30$ MeV, $W_0 = 176.4$ MeV, $r_V = 1.2$ fm, $a_V = 0.52$ fm, $r_W = 0.95$ fm and $a_W = 0.55$ fm, and a KTM-type analysis (full line).

is known[6], one can construct simple models[7] in which a ($\bar{p}$p) knockout reaction in a nucleus would take place. Because the incident particle is a $\bar{p}$, for which neither the binding energy nor the annihilation probability is known, positive results, i.e. the appearance of a knockout peak above a smooth proton spectrum produced by annihilation, would allow us to extract information about $\bar{p}$-nucleus

Table 1. Sets of optical model parameters which
fit the $\bar{p}$-^{12}C scattering reasonably well.
For different fixed values of V_0 and W_0,
V and W tend to be about 4 and 8 MeV at a
radius $R_{1/2}$ = 3.3 fm.

V_0 (MeV)	W_0 (MeV)	r_0 (fm)	a_0 (fm)	χ^2/N	V $(R_{1/2})$ (MeV)	W $(R_{1/2})$ (MeV)
10.9	20.3	1.55	0.4	0.95	4.3	8.0
12.4	25.6	1.48	0.45	0.88	3.9	8.0
15.3	34.5	1.37	0.50	0.87	3.6	8.0
23.2	52.8	1.22	0.55	0.90	3.5	8.0
43.3	89.9	1.00	0.60	1.00	3.7	8.7

states. Figure 3 shows the proton spectrum from three targets at 0°
for incident $\bar{p}$'s of 600 MeV/c. No obvious peak appears in any of
these spectra. Although the statistical quality of the data does
not allow strict limits on the possible cross-section or width such
a peak might have, our cross-sections are about a factor of 2 lower
than those predicted by the models. These spectra can also be
analysed in terms of the fourth area of our study, $\bar{p}$ annihilation in
the nucleus and its subsequent decay. The spectra can be parametrized in a cascade model by a temperature representing the slope
of the spectra and by a proton multiplicity related to the annihilation cross-section. The temperatures T_0 for our spectra are about
100 MeV, somewhat higher than the experimental results of Shibata[8]
who measured the proton evaporation spectra from nuclei bombarded by
4 GeV/c protons and pions and than the results of cascade model
calculations[9].

If we assume that the absorption cross-section goes as σ_a =
$\sigma_0 A^{0.67}$, and that the cascade spectra of the targets have the same
shape, we can extract relative proton multiplicities of 1.15 and

326

0.93. These numbers represent the ratios of the number of protons emitted per absorption for copper and bismuth to that for carbon.

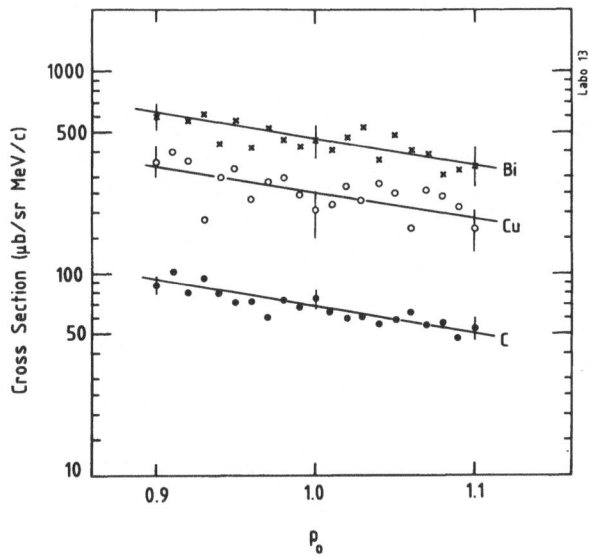

Fig. 3. Cross-sections for ($\bar{p}$p) reactions on carbon, copper, and bismuth as a function of emitted proton momentum; p_0 is 600 MeV/c.

In future experiments we hope to study the A dependence of the $\bar{p}$ A potential; the effect of targets with spin; scattering from different isotopes of the same element, e.g. ^{16}O, ^{18}O. We plan to study the knockout reaction with much better statistics and a lighter target such as lithium for which the cascade background should be smaller and the nuclear protons more accessible.

REFERENCES

1. J. Thirion and P. Birien, Le spectromètre II, Rapport DPHN/ME (1978).
2. H. Aihara et al., Nucl. Phys. $\underline{A360}$:291 (1981).
3. A. Bouyssy and S. Marcos, Phys. Lett. $\underline{114B}$:397 (1982).
   C.B. Dover and J.M. Richard, Phys. Rev. $\underline{C21}$:1466 (1980).
   E.H. Auerbach et al., Phys. Rev. Lett. $\underline{46}$:702 (1981).
   J.A. Niskanen and A.M. Green, Nucl. Phys. $\underline{A404}$:495 (1984).
   A.M. Green and S. Wycech, Nucl. Phys. $\underline{A377}$:441 (1982).
4. P.D. Barnes et al., Phys. Rev. Lett. $\underline{29}$:1132 (1972).
   M. Poth et al., Nucl. Phys. $\underline{A294}$:435 (1978).
   R.J. Abrams et al., Phys. Rev. $\underline{D4}$:3235 (1971).
   C.J. Batty, Nucl. Phys. $\underline{A372}$:418 (1981).
5. D. Garreta et al., Phys. Lett. $\underline{135B}$:266 (1984).
6. M. Alston-Garnjost et al., Phys. Rev. Lett. 43:1901 (1979).
7. H. Heiselberg et al., Phys. Lett. 132B:279 (1983).
8. T.A. Shibata et al., Nucl. Phys. $\underline{A408}$:525 (1983).
9. M. Cahay et al., Nucl. Phys. $\underline{A393}$:237 (1983);
   M.R. Clover et al., Phys. Rev. $\underline{C26}$:2138 (1982).
10. H. Willmes et al., Particles and Nuclei 4:192 (1972).
11. A. Johansson et al., Ark. Fys. $\underline{19}$:541 (1961).

PERSPECTIVES OF ANTINEUTRON PHYSICS

T.Bressani[1,2], E.Chiavassa[1,2], S.Costa[1,2], G.Dellacasa[1,2], N.De Marco[1,2], M.Gallio[1,2], F.Iazzi[3,2], B.Minetti[3,2], M.Morandin[4], A.Musso[1,2], G.A Puddu[5,2], C.Sciolla[3,2], S.Serci[5,2], E.Vercellin[1,2], and C.Voci[4].

1 Istituto di Fisica Superiore dell'Università, Torino (Italy)
2 I.N.F.N., Sezione di Torino (Italy)
3 Dipartimento di Fisica Politechnico, Torino (Italy)
4 Dipartimento di Fisica and I.N.F.N., Sezione di Padova (Italy)
5 Dipartimento di Scienze Fisiche, Università di Cagliari (Italy)

1. INTRODUCTION

Experimental data in $\bar{n}$ interactions are up to now rather scarce, mainly due to the lack of suitable beams. With the recent improvements of $\bar{p}$ beams and with the advent of the LEAR facility at CERN, it is now the time of thinking of new experiments in order to study the properties and the interactions of an elementary particle like the $\bar{n}$. In the following we will discuss briefly how we can obtain $\bar{n}$ beams, how we can detect these particles and finally some significant experiments that can be performed.

2. ANTINEUTRON BEAMS

We may distinguish two different approaches for the production of $\bar{n}$ beams of reasonable intensity and energy definition:
i) dumping a proton beam, of at least 10 GeV, into a production target and separating the $\bar{n}$ component by means of time-of-flight (TOF) over large distances;
ii) using the charge-exchange (CEX) reaction $\bar{p}p \rightarrow \bar{n}n$.
Brando et al.[1] followed the first approach at the AGS and succeeded in observing a time-separated $\bar{n}$ beam. The advantage of the method is that it requires a primary proton beam, that can obviously be used for the simultaneous production of several beams of secondary particles, the disadvantage that the resulting $\bar{n}$ beam has a rather low intensity and a poor energy definition.
The second method was foreseen in several options:

a) external CH_2 target
b) external LH_2 target (with or without tagging)
c) internal jet target
The option a) was used at the ZGS by Gunderson et al.[2] in an experiment on the measurement of σ_{tot} on protons and ^{12}C for $\bar{n}$ between 250 and 880 MeV/c and proposed by Lowenstein et al.[3] for an experiment on σ_{ann} by low energy $\bar{n}$ on protons at the AGS. The option b) was studied by Voci[4] (without tagging), Dieterle[5] and Bressani et al.[6] (with tagging). The option c) was studied also by Voci[7]. It is rather difficult to make a detailed comparison between the different options, since each of them has some advantages with respect to the others. The final choice has to be done following considerations related to the features of the $\bar{p}$ beam and mainly to the experiment that has to be performed. An exemple of how the $\bar{n}$ beam can be optimized to a specific experiment will be given in Sec.4.

3. ANTINEUTRON DETECTORS

An $\bar{n}$ detector must be able to discriminate with very good efficiency $\bar{n}$ from other neutral particles (n,γ,K). This objective can be a-chieved if one measures simultaneously the interaction point, the TOF and the energy release of the annihilation products (or the topology of the produced charged particles). Some simplified $\bar{n}$ detectors were built by using sandwiches of scintillators and converters (metallic sheets)[1,8]. We describe here the performances of a more elaborate $\bar{n}$ detector that has been operational at LEAR during 1983 and 1984, parasiting on an external LH_2 target according to the option b) mentioned in Sec.2. The apparatus is made of ten equal modules placed along the $\bar{n}$ path. Each module consists of:
1) an iron slab, 20 x 20 x 1 cm^3 in which the annihilation event may occur;
2) a scintillator, 20 x 20 x 1 cm^3, giving the time information of the event;
3) a scintillator wall, 100x100x1cm^3, segmented in two parts, detecting the charged particles ($\pi\pm$) produced in the annihilation;
4) an iron sheet, 100x100x0.2 cm^3 that stops low energy, highly ionizing particles produced in the annihilation
5) a plane of x-y limited streamer tubes, 96x100 cm^2 with 1 cm pitch, allowing the spatial reconstruction of the event.
The trigger conditions can be chosen according to the number of annihilation prongs that have to be selected. The minimal condition is that at least two subsequent elements of scintillator be activated, corresponding to one prong at least in the final state. The performances of this detector are summarized by Fig.1, showing the TOF separation between $\bar{n}$ and γ produced by $\bar{p}$ of 600 MeV/c and by Fig. 2, where a typical 4 -prongs annihilation event is depicted. The efficiency of the detector was not yet measured directly, but can be estimated around 55%.

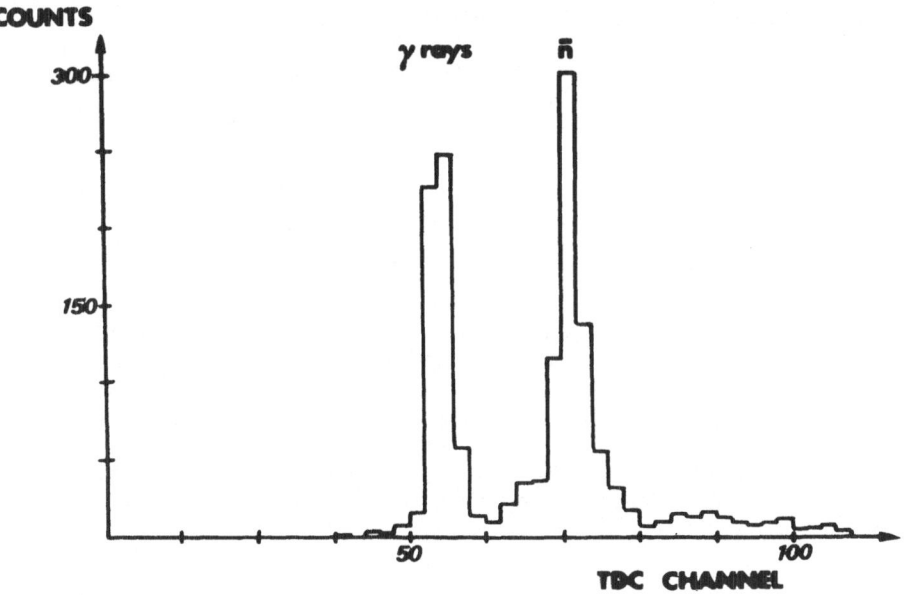

Fig. 1. TOF separation between n̄ and γ produced by p̄ of 600 MeV/c.

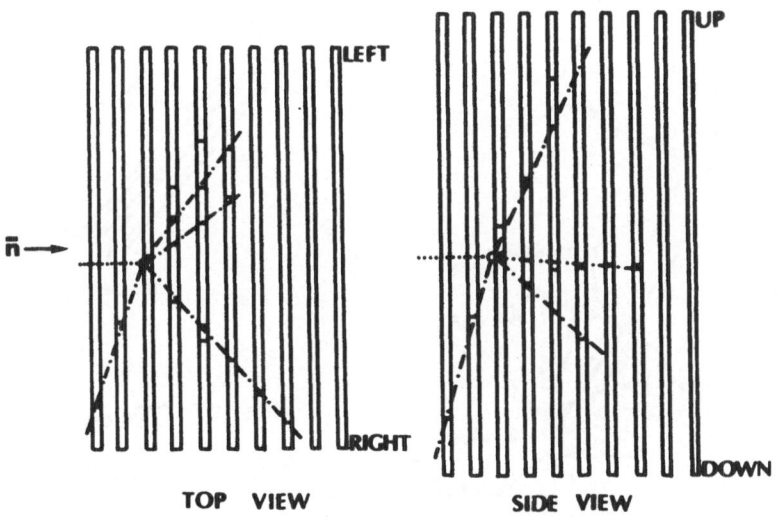

Fig. 2. 4 - prongs annihilation event as measured by the n̄ detector.

4. SOME EXPERIMENTS WITH ANTINEUTRONS

Especially at low energies (<300 MeV/c) there is a complete lack of information on the n̄ interactions. The first measurement that one can try to perform is then that of δ_{tot} on protons and nuclei. The np state is a pure I = 1 state and, at low energy, the interaction is dominated by the s – wave. Therefore one can expect that these measurements could shed light on the N̄N system in a pure I = 1 state. The measurement of δ_{tot}[9] on nuclei could give informations on the relative importance of the real and imaginary parts of the optical potential used to describe the N̄–Nucleus interactions. Fig. 3. shows a set-up that was proposed[6] for the measurement of δ_{tot} with a tagged n̄ beam. The p̄ beam impinges on a LH$_2$ production target, 15 cm long. Up to 300 MeV/c p̄ are brought at rest along the length of the target and produce n̄ via CEX up to 98 MeV/c. The neutron from the CEX reaction is detected by the n hodoscopes, allowing the measurement of its TOF and impact position. The energy and direction of the associated n̄ can then be determined to a good precision.

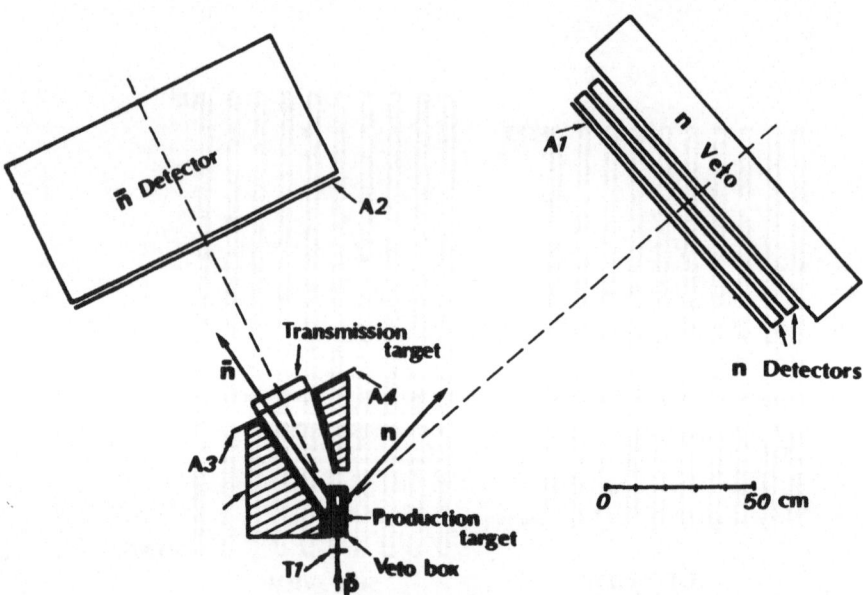

Fig. 3.   Layout for δ_{tot} measurements with tagged n̄. A1,...A4 are anticounters, T_1 the timing scintillator for the start on p̄ of the TOF measurements.

Fig. 4. shows the expected incertitudes, calculated by means of a detailed Monte-Carlo simulation of the experiment. The transmission target (LH$_2$ or nuclear) is located behind a defining Lead collimator and at ⪆50 cm from the n̄ detector. By alternating runs with empty and full target and by profiting of the good localization properties of the n̄ detector one can measure simultaneously:
a) the removal of tagged n̄'s due to interactions in the target;
b) the angular distribution of scattered n̄.

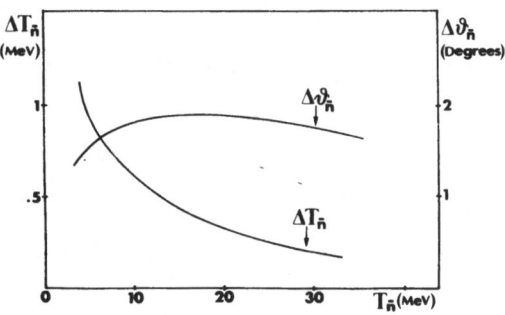

Fig. 4.   Incertitudes on the reconstruction of the energy $T_{\bar{n}}$ and the direction of the tagged n̄ as a function of $T_{\bar{n}}$

In this way one can obtain both $\sigma_{tot.}$, by using the usual extrapolation for $\Omega \to 0$ to determine the rate of n̄ scattered through small angles, and $(d\sigma/d\Omega)_{el}$ from 0° to ~45°. Fig. 5. shows the expected flux of n̄ per bin of 20 MeV/c obtained with p̄ of 300, 400, 500 and 600 MeV/c in the above geometry. The total number of tagged n̄ is reduced by a factor ~ 0.35 taking into account the efficiency of the n̄ detector. It can be seen that measurements to a 5% statistical precision can be obtained in reasonable running times. With the same detectors, in different geometrical configurations, it is possible to measure also the n̄ mass with a precision of 2 keV[10], and the cross-sections for the CEX reaction[4,7]

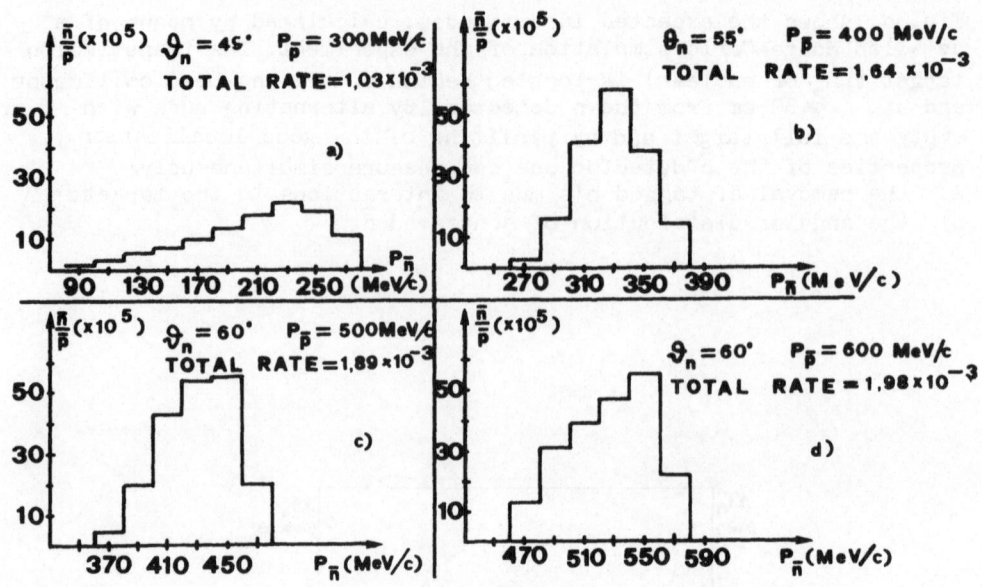

Fig. 5. Calculated flux of the n̄ in the tagged beam.

REFERENCES

1. T. Brando, A. Fainberg, T. Kalogeropoulos, D. Michael and
   G. Tranakos, Nucl. Instrum. Meth. 180 (1981), 461.
2. B. Gunderson, J. Learned, J. Mapp. and D.D. Reeder, Phys. Rev.
   D23 (1981), 587.
3. D. I. Lowenstein, E.V. Hungerford, B.W. Mayes, L.S. Pinsky,
   C. Bromberg, A. Micks, R.A. Lewis, R. Miller, B.Y. Oh, T.Potter,
   G.A. Smith, J. Whitmore and G.S. Mutchler, AGS proposal E-767
   (1981).
4. C. Voci in Proc. Int. School of Intermediate Energy Nuclear
   Physics, Verona, 1981, p.425.
5. B. Dieterle, report LA - 9798 - P (1983), p.162.
6. T. Bressani, M. Caria, E.Chiavassa, S.Costa, G.Dellacasa,
   U. Dosselli, M. Gallio, F. Iazzi, M.P. Macciotta, B. Minetti,
   M. Morandin, A. Musso, G.A. Puddu, S. Serci and C. Voci, proposal CERN/PSCC/84-27/PSCC/P73 (1984).
7. C.Voci, in Proc. of the Joint CERN-KFK Workshop on Physics
   with cooled low energetic antiprotons, Karlsruhe, 1979, p.125
8. W. Bruckner, H. Dobbeling, J. Ciborowski, S. Majewski, B. Pooh,
   R. Ransome, M. Treichel and Th. Walcher, in Physics at LEAR
   with Low-Energy Cooled Antiprotons (Plenum Press, N.Y., 1984),
   p. 437.

9. L. Bracci, G. Fiorentini, G. Mezzorani and P. Quarati, Nuovo Cimento 78A (1983), 306.
10. T. Bressani and B. Minetti, in Physics at LEAR with Low-Energy Cooled Antiprotons (Plenum Press, N.Y., 1984), p. 781.

WEAKLY CHARGED EXOTIC PARTICLES

A. Zehnder

Swiss Institute for Nuclear Research

CH-5234 Villigen, Switzerland

1. INTRODUCTION

Global symmetries and the resulting conservation laws
are the foundations of our understanding of the physical
world. But certain symmetries are locally broken. The
nature of this symmetry breaking and its consequences are
not well understood. Two main concepts are in discussion
today: The dynamic and the spontaneous symmetry breaking.
The present talk is limited to the latter: Its main experimental consequence is the prediction of new particles;
they can be classified in two groups: the "massless"
Goldstone bosons, which would lead to "longrange" interaction between particles, and the massive particles like
the Higgs boson, which could be detected via their production or decay processes. The talk presents an over-
view, followed by a specific discussion of selected
particles with predictions in the standard Weinberg-Salam
model.

Electroweak interaction is described in the standard
W-S-model[1], predicting the gauge particles (γ, Z, W) with
their masses; all of them are experimentally established
today. In addition, the W-S-model postulates a scalar
boson with unknown mass: the Higgs boson, which is still
to be found. By introducing strong interaction, one has
to enlarge the W-S-model: there are two possibilities[2]:
(i) relate particles with the same spin: leads to so-
called grand unified theories (GUT) or (ii) relate
particles with different spin: leads to super symmetric
theories (SUSY).

337

In GUT one is faced with various problems: the explanation of the family hierarchy which obviously is a broken symmetry with its Goldstone particle, the familon; the CP problem which leads to the prediction of an axion, and others. In the case of SUSY theories, there is a SUSY particle with a difference of spin of 1/2 for every existing particle: e.g. one has spin 1/2 photinos, gluinos, etc. and spin 0 sleptons, squarks etc. In addition, the theory predicts a new spin 1 particle: the U-boson. Its mass scale is not fixed, it is therefore interesting to search as well in the low energy region (up to 100 MeV).

2. MANIFESTATION OF WEAKLY COUPLED PARTICLES

a) Coupling

The present discussion is limited to weakly coupled particles, which means that a particle X is coupled to a fermion f with the coupling strength:

$$\alpha_w \equiv g_{xff}^2/4\pi = \frac{G_F \, m_f^2}{2\sqrt{2}\pi} \equiv \frac{m_f^2}{4\pi V^2} \tag{1}$$

G_F is the Fermi coupling constant, expressed more generally by V, the mass scale of the symmetry breaking. Numerically, in the W-S-model we have:

$$V_o = (\sqrt{\sqrt{2} \, G_F})^{-1} = 250 \text{ GeV} \tag{2}$$

Generally

$$V_o \leq V \leq m_{GUT} \simeq 10^{15} \text{ GeV} \tag{3}$$

m_{GUT} being the mass scale of symmetry breaking. A consequence of this large variation in the symmetry breaking mass scale is the possibility to predict observables that are experimentally unreachable, like the decay of the invisible axion (see below).

b) Observables

Exchange potential. For $0^+, 1^-$ particles, there is a Yukawa-type exchange potential:

$$V(r) = \alpha_w \frac{e^{-m_x r}}{r} \tag{4}$$

which leads, among other things to energy shifts and additional terms in scattering amplitudes.

For $0^-,1^+$ particles, specifically for Goldstone bosons, Gelmini et al.[3] showed that, in a non relativistic limit, the coupling must be derivative for the exchange of a Goldstone particle G, using Dirac's equation, one gets:

$$\frac{1}{V} \delta_\mu G \bar{f}_1 \gamma_\mu (a+b\gamma_5) f_2 = \frac{1}{V} G \bar{f}_1 \{a(m_1-m_2) +b(m_1+m_2)\gamma_5\}f_2$$

(5)

if $f_1 = f_2$, only pseudoscalar coupling remains with an amplitude proportional to:

$$\frac{m_f}{V} \bar{f} \gamma_5 f$$

(6)

Therefore the resulting exchange potential has dipole-dipole character:

$$V(r) \propto \left[\vec{S}_1\vec{S}_2-3(\vec{S}_1\hat{r}_1)(\vec{S}_2\hat{r}_2)\right]/r^3$$

(7)

with a $1/r^3$ dependence. As shown below, the experimental limits are rather weak for this type of exchange potential.

Decay. If the mass of the X-particle is larger than $2m_e$, it can decay into two leptons. The decay width is given by[4]:

$$\Gamma (X \to \ell^+\ell^-) = \frac{G_F m_\ell^2}{2\sqrt{2}\pi} m_X \left[1-4m_\ell^2/m_X^2\right]^{3/2}$$

(8)

Since the coupling strength is proportional to the square of the lepton mass, the decay, if kinematically allowed, mostly proceeds via the heaviest lepton decay mode. 10^{-8} sec for $m_X = 2\ m_e$ is a typical lifetime for the X into e^+e^- decay. Above the $2\ m_\pi$ threshold, hadronic decays are dominant, but their decay rate is strongly model-dependent.

If the mass of X is less then $2\ m_e$, only decay into gammas and neutrinos is possible. In the case of spin 0^- particles, decay into neutrinos is kinematically forbidden.

Therefore, the spin 0^- particles decay via a triangle graph into two gammas with a approximate decay width of[4]:

$$\Gamma (X-\gamma\gamma) \approx \sqrt{2}G_F \alpha^2 m_X^3/16\pi^3$$

(9)

The typical lifetime of 10^{-2} sec is much longer than the lifetime of the leptonic decay mode.

Production. X particles are always produced in
competition with known particles, therefore, their
production rate can be estimated via the ratio of the
coupling strength. In a proton beam dump, X-particles
are produced at the approximate ratio $g^2_{XNN}/g^2_{\pi NN} \approx 10^{-7}$
of the π-production. In a electron beam dump, they
are produced via X-bremsstrahlung with a ratio of
$\alpha_w/\alpha \approx 2.2 \ 10^{-4}$ [5].

They also may be produced via nuclear deexitation
or particle decay[6]. For low energy particles, $\approx$ 1 MeV,
the study of nuclear deexcitation is a very powerful
tool, as will be shown in the case of axion research.

In the following chapter, we individually discuss
pseudoscalar (0^-), scalar (0^+), axialvector (1^+), and
vector (1^-) particles.

3. PSEUDOSCALAR PARTICLES

a) Effects of pseudoscalar exchange potentials

As mentioned before, massless pseudoscalars would
manifest themselves via spin dependent, long-range
exchange forces which would contribute to the anomalous
magnetic moment of muons and electrons, and would shift
the hyperfine energy splitting of positronium and myonium.
Table 1 shows the experimental results and their
theoretical predictions.

One sees that only the myonium reaches the sensitivity
which the W-S-model requires to confirm or rule out the
existence of such particles. Therefore, from these experiments no conclusion can be drawn about the existence
of light Goldstone bosons. The experimental situation
is even worse for heavy pseudoscalar particles, because
of the shorter range of exchange forces.

It is clear that gravitational experiments have no
sensitivity for pseudoscalars, because no spin dependent
experiments were ever performed.

b) The Axion

The axion is the best known exotic Goldstone
particle: its properties are described by the W-S-model
and QCD. The proved existence of the axion would confirm
the basic ideas about grand unification, spontaneous
symmetry breaking, and Higgs mechanism. The axion is

Process	Contribution from x	Experimental Uncertainty	Exp. Limit on α_x	Expected W-S-Limit
$\frac{1}{2}(g-2)_e$	$\frac{\alpha_{xee}}{4\pi}$	$3 \cdot 10^{-11}$	$4 \cdot 10^{-10}$	$5 \cdot 10^{-13}$
$\frac{1}{2}(g-2)_\mu$	$\frac{\alpha_{x\mu\mu}}{4\pi}$	$\sim 10^{-8}$	$1 \cdot 10^{-7}$	$2 \cdot 10^{-8}$
$\Delta E(^3S_1 - ^3S_o)$	$\frac{5}{3}\alpha_{xee}\alpha^3 m_e$	$7 \cdot 10^{-10}$ eV	$4 \cdot 10^{-7}$	$5 \cdot 10^{-13}$
$\Gamma(^3S_1 \to x\gamma)$	$2\pi\,\alpha_{xee}\,\alpha\,|\psi_i(o)|^2/m_e^2$	BR $< 10^{-3}$	$\sim 10^{-8}$	$5 \cdot 10^{-13}$
μ^+e^- -hfs	$\frac{4}{3}\sqrt{\alpha_{xe}\,\alpha_{x\mu}}\,\alpha^3 m_e^2/m_\mu$	$0.8 \cdot 10^{-13}$ eV	$1.1 \cdot 10^{-10}$	$1 \cdot 10^{-10}$

accessible in low energy experiments, since its mass is of order 100 keV.

A part of the following chapter is already published in review articles[8], but for completeness, it is included in this talk.

It is experimentally obvious that we live in a world where P and CP are good symmetries at the level of the strong interaction. However, the QCD-Lagrangian could contain a CP-nonconserving term[9], which would give rise to a nonvanishing electric dipole moment of the neutron[10]. To be in agreement with the experimental limit[11], such a term must be extremely small. This makes it difficult to include the CP-violating weak interaction in a combined theory of QCD and electro-weak interaction. A natural solution for this problem was suggested by Peccei and Quinn[12], who postulated a global $U(1)$-symmetry for the total Lagrangian (QCD + electro-weak). In this way any strong CP-violation could be rotated away. As pointed out by Weinberg and by Wilczek[6] independently, one should expect a new particle through symmetry breaking: the axion, a pseudoscalar particle like the π^0 (if isovector) or the η^0 (if isoscalar).

The properties of the axion can be calculated in terms of only one free parameter $X = tg\ \lambda$[13] given by the ratio of the expectation value of the two Higgs fields f_1 and f_2 ($f_1 = \sin \lambda/(\sqrt{2}\ G_F)^{1/2}$, $f_2 = \cos \lambda/(\sqrt{2}\ G_F)^{1/2}$) as introduced by Peccei and Quinn. The coupling strength is given by eq. (1).

The axion-nucleon coupling can, as an order-of-magnitude estimate, be expressed by the pion-nucleon coupling strength[5]

$$g_{aNN}^2 = g_{\pi NN}^2 (f_\pi/V_o)^2 = 1.45 \cdot 10^{-7}\ g_{\pi NN}^2 \qquad (10)$$

Thus the coupling strength of the axion to a nucleon is very much smaller than the corresponding pion-nucleon coupling strength.

The mass of the "standard" axion and the decay rate into two photons are predicted by Bardeen and Tye[13] to be

$$m_a = m_\pi (f_\pi/V_o)\ N(X+1/X)\ \sqrt{Z}/(1+Z) = 75(X+1/X)\ (keV) \qquad (11)$$

and

$$\Gamma(a \to \gamma\gamma) = \Gamma(\pi^0 \to \gamma\gamma) \cdot Z \cdot (m_a/m_\pi)^5 = 1.5 \ (m_a/100 \ keV)^5 (sec^{-1})$$

(12)

The ratio of the up-quark mass to the down-quark mass $Z = m_u/m_d = 0.56$ was taken from ref. 14. N is the number of quark doublets of the theory, for the present discussion we adopt $N = 3$. Because X is a positive number by definition, $(X+1/X) > 2$; therefore the axion mass has a minimal value of $m = 150$ keV corresponding to a maximal lifetime of $\tau(a \to \gamma\gamma) = 90$ msec (see eqs. 12 and 13). The axion coupling strength to 2/3-charge quarks is proportional to X, the coupling strength to -1/3-charge quarks and to leptons is proportional to 1/X, respectively[5]. These facts enable us to design specific axion search experiments, yielding unambiguous results.

The isoscalar and isovector axion-nucleon formfactors are given by[5]

$$\rho^{(0)} = -(N-1) \ (X+1/X) \ F_A^{(0)}/F_A^{(1)}$$

(13a)

$$\rho^{(1)} = X(1-N \cdot (1-Z)/(1+Z))-1/X \cdot (1+N \cdot (1-Z)/(1+Z))$$

(13b)

Note that the isovector coupling vanishes for $X = 3.4$; $F_A^{(0)}/F_A^{(1)} = 3/5$ as predicted by a quark model estimate[15].

We shall call an axion with the above properties a standard axion.

Experimental search for axions. There are two kinds of experiments: Searches for axion production and searches for axion decay. Experiments of the first kind are investigating the two-body decay of certain particles: an axionic event manifests itself by a decay into a known monoenergetic particle plus "nothing" (an undetected axion). In experiments of the second kind axions are identified by their characteristic decay into an electron-positron pair or into two gamma rays. The axion source strength in terms of the competitive production process is given in eq. 10 for hadronic processes and for electro-magnetic processes by[5]

$$\Gamma_a/\Gamma_\gamma = g_{aNN}^2/4\pi\alpha = 2.2 \cdot 10^{-4}$$

(14)

This indicates that axions should be searched for in electro-magnetic processes.

343

<u>Axion decay into $e^+ + e^-$</u>. Axions decay only into an electron-positron pair if their rest mass $m_a > 2\ m_e$, thus a negative result can only set an upper mass limit.

- Calaprice et al.[16] searched for axion production in the isovector M1-transition of ^{12}C (15.1 MeV). The experiment was only sensitive to the decay into e^+e^-. The negative results ruled out axions in the mass range from 1 to 15 MeV, except for small X-values (X<0.3).
- Faissner et al.[17] performed an experiment at the 590 MeV proton beam dump at SIN to look for the decay into e^+e^-. The axion production rate was estimated from the known $\pi^+ + \pi^-$ rate to be $N_a = 2.7 \cdot 10^8$/Clb/m^2 in front of the apparatus for long-lived axions. The negative results led the authors to the conclusion that axions with a rest mass $> 2\ m_e$ are ruled out. However, axions with a laboratory lifetime shorter than 10^{-9} sec would have decayed before reaching the detector. Assuming $E_a = 100$ MeV the experiment only excludes axions of masses up to 25 MeV.

These experiments indicate that axions have to be looked for through the decay into two-gamma or via the ovservation of ψ/J,T or K^+-decay into axions.

<u>Axions produced by ψ/J,T or K^+ decay</u>
- The $K^+ - \pi^+ a_0$ decay search was performed in conjunction with the $K^+ - \pi^+ \nu\bar{\nu}$ search by Asano et al.[18] at KEK (Japan). The experiment required a $K - \pi - \mu - e$ decay decay chain in order to suppress the $K^+ - \mu^+ \nu\gamma$ decay. In addition, a large solid angle lead glass counter system was used to veto the radiative decay. The $K^+ - \pi^+ a_0$ decay would produce monoenergetic pions with a energy of about 120 MeV. Only an upper limit for the branching ratio was found

$$BR(K^+ \to \pi^+ a_0) \leq 3.8 \cdot 10^{-8} \tag{15}$$

The theoretically expected value is $> 10^{-5}$ for pure isoscalar axions and $>2 \cdot 10^{-8}$ for pure isovector axions[19], respectively. Therefore, isoscalar axions can be ruled out safely.
- The $\psi/J - \gamma a_0$ decay search was performed by the Crystal Ball Collaboration at SLAC[20], who was looking for a monoenergetic gamma ray plus "nothing". Its result is

$$BR(\psi/J \to \gamma a_0) < 1.4 \cdot 10^{-5} \tag{16}$$

The search $T \to a_0 \gamma$ was measured at CESR[21] and DORIS[22],

344

leading to the limit (90% CL):

$$B(\Upsilon(1s) \to a_0\gamma) \leq 1.4 \cdot 10^{-4} \tag{17}$$

The product of the two theoretical branching ratios is independent of X, since its contribution for $Q = 2/3$ quarkonium and $Q = -1/3$ quarkonium is reversed.

$$B(J/\psi \to a_0\gamma) \cdot B(\Upsilon - a_0\gamma) = B(J/\psi \to \mu\mu) \cdot B(\Upsilon \to \mu\mu) \cdot \frac{G_F^2}{2\pi^2\alpha^2} m_c^2 m_b^2 \tag{18}$$

using the leptonic branching ratios the expected value of the product is estimated as $(16\pm3) \cdot 10^{-9}$. The product of the measured branching ratios is $4.2 \cdot 10^{-9}$ not a very strong limit, but nevertheless it rules out axions.

<u>Axion decay into two gamma rays</u>. In this kind of experiment the decay of the axion into two gammas outside a well shielded axion source is observed. The axions are produced in an electron or a proton beam dump or via nuclear deexcitation. The two-gamma coincidence rate is given by

$$R_{\gamma\gamma} = R_y \cdot \Gamma_a/\Gamma_y \cdot \ell/v_a \cdot \Gamma(a \to \gamma\gamma)/\gamma \cdot \varepsilon_{tot} \tag{19}$$

where R_y is the source strength for the competing process (gamma emission rate for a specific nuclear transition or π^0 production rate in a beam dump), ℓ is the length of the decay region, v_a the axion velocity, $\gamma = E/m_a$ the Lorentz factor and ε_{tot} the total efficiency of the two gamma detector.

Proton beam dump experiments were performed at high and medium energy beams. The experiment of Jacques et al.[23] was carried out at the BNL 7 ft. bubble chamber looking for e^+e^--pair not associated with any other vertex. The upper limit of the ratio

$$\sigma(pN \to a^0 + \ldots)/\sigma(pN \to \pi^0 + \ldots) \leq 10^{-8} \tag{20}$$

is about an order of magnitude smaller than estimated in eq.10. An energy cut of 0.5 GeV was applied.

In a second experiment at Gargamelle (CERN) Faissner et al.[24,25] analyzed single gamma events with an energy cut of 0.2 GeV < E < 2 GeV. The plot of the angular distribution shows a clear peak in the direction of the beam dump. This could indicate an axion decay process. (Note that the two photons are chiefly emitted in forward direction). Faissner[25] shows also a two-gamma event whose vertex lies clearly inside the bubble chamber. However, a detailed background analysis is not present.

At the 590 MeV proton beam dump at SIN, Faissner et al.[26] performed another axion search with a spark chamber set-up and a lead converter looking specifically for two gamma events. A detailed description of the experiment is given in ref. 26. Again the angular distributions of one- and two-gamma events show a peak away from the beam dump. From the observed events they deduced an axion lifetime of about 7 msec corresponding (see eq. 12) to an axion mass of about 250 keV.

An improved experiment is presently carried out at SIN[27].

Axion production via nuclear deexcitation. Axions can also be produced via nuclear deexcitation as proposed by Donnelly et al.[5]. Because of the axion's pseudoscalar nature its production competes with magnetic gamma transition (except for MO-transitions). Searches for axions in nuclear transitions have several advantages:
- The rate Γ_a/Γ_γ is quite reliably calculable, primarily because of the cancellation of the reduced nuclear matrix elements[28].
- The sum of the energy of the two coincident gammas yields a monoenergetic peak at the known transition energy.
- Low energy nuclear gammas are easy to shield off.

The branching ratio Γ_a/Γ_γ for a single particle transition (neutron n or proton p) is given by[5]

$$\Gamma_a/\Gamma_\gamma = (\alpha_a/\alpha_\gamma) \cdot (L/L+1) \cdot (k_a/k_\gamma)^{2L+1} (\rho_i/2\mu_i)^2$$

(21)

where $\alpha_a/\alpha_\gamma = 2.2 \ 10^{-4}$, k_a and k_γ are the axion and gamma momenta, L is the multipolarity, μ_i (i=n or p) are the magnetic formfactors and ρ_i their axionic analogues. The ρ's may be calculated within the axion model[5]. They are plotted in Fig. 1 versus the axion mass m_a. Note that the axion-neutron coupling vanishes for X = 0.7; the axion-proton coupling, on the other hand, is finite for all X. Inserting eq. (21) into eq. (19) yields the expected two-gamma rate $R_{\gamma\gamma}$.

346

A first experiment was carried out by Zehnder[29], using a ^{137}Ba-source ($4\cdot10^{13}$ decay/sec). The decay width was calculated by Barroso and Mukhopadhyay[28]. ^{137}Ba has 81 neutrons, therefore the 662 keV M4-transition is fairly well described by a single-neutron (hole) transition. The experimental set-up is shown in Fig. 2, it consists mainly of a pair of well shielded NaI-crystals in coincidence. The result excludes axions with a mass > 160 keV if X < 0.7 or > 200 keV if X > 0.7, only 0.45 < X < 1.4 is not ruled out.

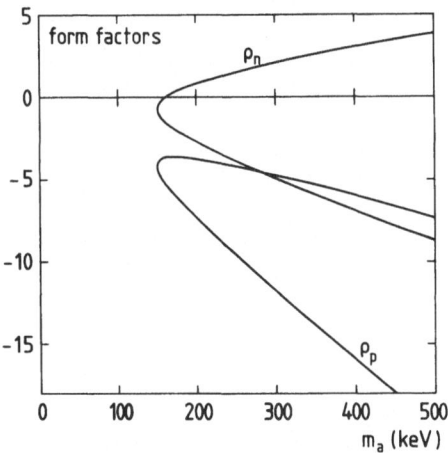

Fig. 1.   Axionic form factors for single-proton and single-neutron transitions versus axion mass. Axions with m_a < 150 keV are excluded in the standard axion model.

From the Ba-experiment one can conclude that axions do not couple to neutrons. However, the beam dump experiment of Faissner et al.[26] is sensitive to the axion-proton as well as to the axion-neutron coupling, since in a beam dump, protons produce axions incoherently on protons and on neutrons. This suggests that the axion

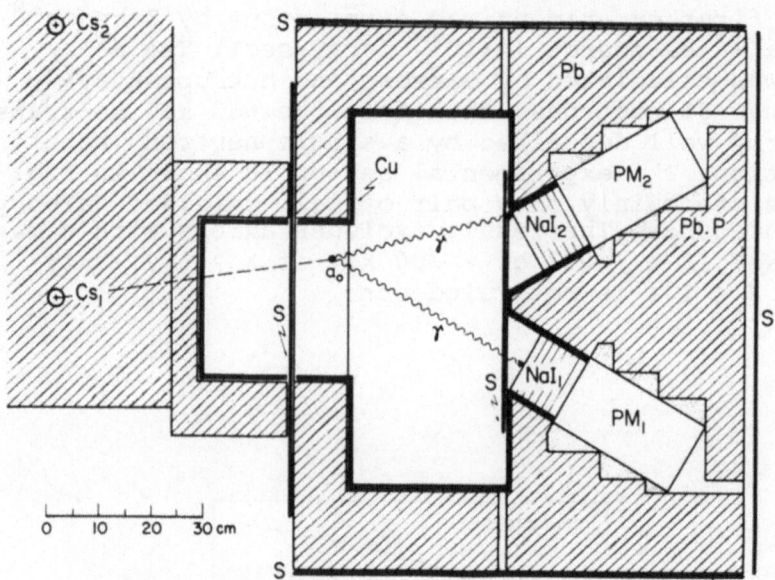

Fig. 2. Plan view of the experimental layout. Container
with two ^{137}Cs sources: Cs_1 = 950 Ci, Cs_2 =
150 Ci. S = plastic scintillators, Pb = lead
shielding, Pb.P = lead pellets, Cu = 5 mm
electrolytic copper to suppress Pb X-rays.

search should also be extended to single-proton
nuclear transitions.

Such experiments were carried out by Zehnder et al.[30]
at the nuclear power reactor in Goesgen (Switzerland),
and by Lehmann et al.[31] using a ^{65}Cu-source. Figure 3
shows the coincident summed-up energy spectrum of the
axion into two gamma sum peaks. Clearly, no visible line
is present. From the upper limits one can deduce a model-
independent quantity Q for the product of axion production and axion decay width

$$Q = (\Gamma_a/\Gamma_\gamma) \cdot \Gamma(a\to\gamma\gamma)/\gamma$$

by substituting the values of the relevant parameters in
eq. (19). The upper limits of Q are listed in Table 2.
For the standard axion one can calculate the theoretically
predicted $R_{\gamma\gamma}$ and compare it with the experimental limits
(see Fig. 4). Obviously, the standard axion is ruled out
by all reactions. The same conclusion is found by
Lehmann et al.[31] and by the results of Cavaignac et al.[32]

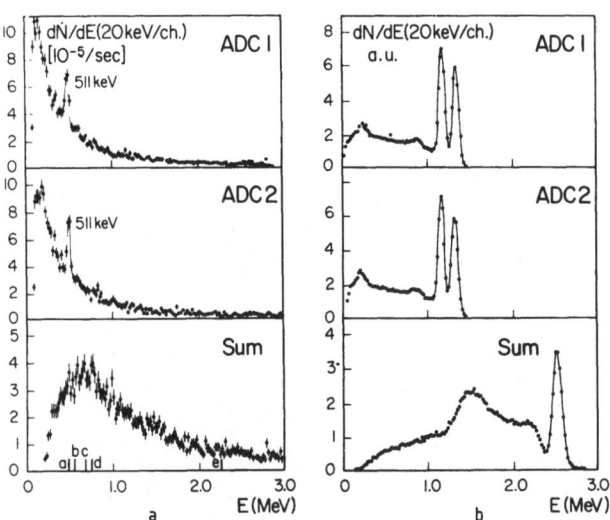

Fig. 3.   (a) Energy spectra in the NaI detectors 1 and
2, and the corresponding sum spectrum (open
configuration). The symbols a through e indicate
the positions of the expected $a_9 \to 2\gamma$ sum peaks
for ^{7}Li: a, ^{91}Y: b, ^{137}Ba: c, ^{99}Nb: d and the
n + p reaction: e. (b) ^{60}Co spectrum with peaks
at 1173 keV and 1333 keV. The sum is a coincidence between these two cascading gamma rays.
Same experimental conditions as (a).

Table 2. List of transitions considered R_γ, $R_{\gamma\gamma}$, Q are explained in the text.

Transition	Energy	Mode	transition rates in the reactor R_γ (sec^{-1})	axion rates "open minus closed" $R_{\gamma\gamma}$ (10^{-5} sec^{-1})	Q (10^{-6} sec^{-1})
$^7\mathrm{Li}(1/2^-) \to {}^7\mathrm{Li}(3/2^-)+\gamma(a)$	478 keV	single proton	$1.0\cdot10^{19}$	1.1 ± 0.7	$\leq0.71(95\%CL)$
$^{91}\mathrm{Y}(9/2^+) \to {}^{91}\mathrm{Y}(1/2^-)+\gamma(a)$	555 keV	single proton	$2.1\cdot10^{18}$	-1.3 ± 1.0	$\leq2.46(95\%CL)$
$^{97}\mathrm{Nb}(1/2^-) \to {}^{97}\mathrm{Nb}(9/2^+)+\gamma(a)$	743 keV	single proton	$5.0\cdot10^{18}$	1.7 ± 1.5	$\leq2.77(95\%CL)$
$^{137}\mathrm{Ba}(11/2^-) \to {}^{137}\mathrm{Ba}(3/2^+)+\gamma(a)$	662 keV	single neutron	$1.8\cdot10^{17}$	-0.2 ± 1.5	$\leq44.8(95\%CL)$
$n+p \to d+\gamma$ (a)	2230 keV	isovector	$7.5\cdot10^{18}$	-0.7 ± 0.9	$\leq1.34(95\%CL)$

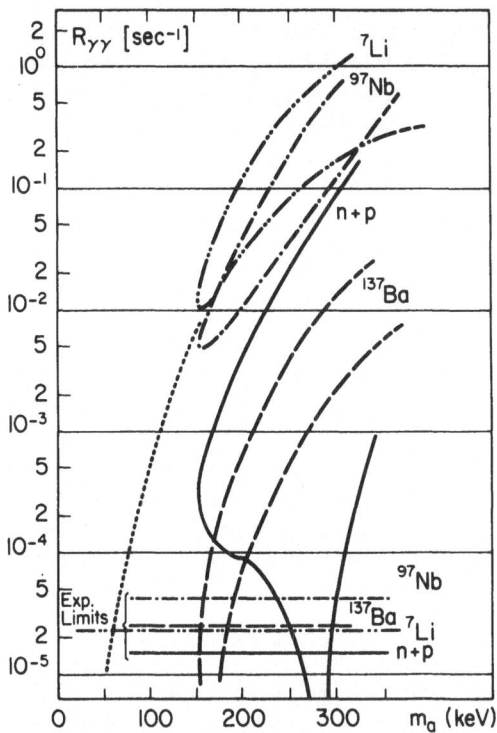

Fig. 4. Expected 2γ-rates from decaying axions versus
mass for the different reactions considered, as
obtained from eq. 19, where the branching ratios
and the axion decay width are calculated from
the standard axion model. Standard axions are
ruled out by the experimental limits for the
^{7}Li and ^{97}Nb transitions, as well as by the
combined results of the n + p reaction and
the ^{137}Ba transition. The dotted line is explained in the text.

Independently of any specific model, using dimensional
arguments only, one can infer that $R_{\gamma\gamma}$ is proportional to
m^6 and to a formfactor ρ_p. Setting $\rho_p = 4$ to its minimal
value we get an upper mass limit for a non-standard axion
of $m_a < 60$ keV.

Conclusions. From different experiments one can
safely conclude that standard axions do not exist. An
obvious way to solve the CP problem of Peccei and Quinn
was proposed by Kim[33] and later refined by Dine, Fischler

and Scrednicki[34], reducing the axion coupling to matter, making it lighter and increasing its lifetime: The axion becomes invisible. The astrophysical consequences are discussed by several authors[35], their main conclusion is that the invisible axion must be coupled extremely weakly, in order to disturb the evolution of stars. The inverse coupling strength V may be as big as 10^{12} GeV (compared to 250 GeV for the standard axion). This large value brings it close to the grand unification scale of 10^{15} GeV; indeed, Wise, Georgie and Glashow[36] have extended the axion model to SU(5) and $V = m_{GUT}$.

Another possibility was proposed by Anselm and Uraltsev[37], postulating two independent symmetries for quarks and leptons. As a result the axions are produced in pairs. This hypothesis would lessen the experimental discrepancy between the beam dump experiment of Faissner et al.[26] and the nuclear transition experiments[29-32]. If two axions are emitted simultaneously, the nuclear transition experiments are less conclusive because:

- the available transition energy may be too small to produce two axions,
- the axions are no longer monoenergetic, and therefore, the present experiments searching for monoenergetic lines are much less sensitive.

The ψ/J,T and K^+-decay experiments are equally unfit to rule out the two-axion hypothesis. For the beam dump experiment none of the above restrictions apply because the energy available is high enough, and one is not restricted to the search of monoenergetic axions.

c) The Majoron

If neutrinos are Majorana particles, the lepton number is broken. Several models are possible to describe this fact. One of them is the assumption that the lepton number is globally broken, thus predicting the existence of a Goldstone boson[38]: the majoron. In this model majorons couple only to neutrinos and with a strong coupling strength h of order 1 to 10^{-2}. The neutral current interaction with fermions is given by the following vertices[38]:

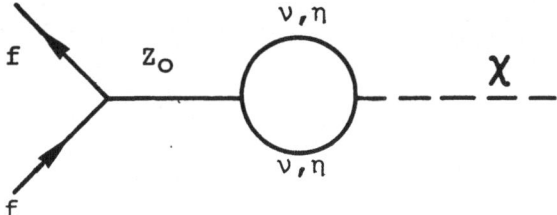

Fig. 5. Coupling of the majoron to fermions through
Z_0-χ mixing. ν, η are the Majorana neutrinos.

the charged current interaction by

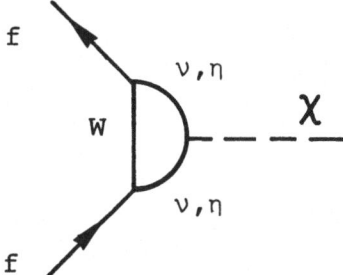

Fig. 6. Coupling of the majoron through W exchange.

the resulting exchange potential is given by

$$V_f(r) = \left(\frac{G_F m_\nu m_f}{16\pi^2}\right)^2 \frac{h^2}{4\pi} [3(\vec{\sigma}_1 \hat{r})(\vec{\sigma}_1 \hat{r}) - \vec{\sigma}_1 \vec{\sigma}_2]/r^3 \qquad (23)$$

$$\approx 10^{-38}/m_f^2 \text{(GeV)} \text{ for } h = 10^{-2} \quad m_\nu = 1 \text{ eV}$$

The best experimental limit is 10^{-6} from hfs of
the H_2 molecules, and obviously not in contradiction
with the above model.

A possible experiment to test long-range, spin-
dependent interaction could be the following: two
polarized macroscopic objects are placed in two superconducting shields at a distance R = 10 cm. Each object
contains N polarized nucleons N $\approx 10^{23}$. The force
resulting from χ exchange is given by:

$$F = \lambda_x^2 \, m_p^2 \, \frac{\mu_N^2}{R^4} \, N^2 \approx 10^{-45} \text{ Newton} \qquad (24)$$

which is hopelessly small. λ_x^2 stands for the coupling constants in eq. (23).

In the model of ref. 39, no additional right handed neutrinos are introduced, but the Higgs sector is enlarged. This leads to a similar prediction as above. The authors also consider the existence of a new bound system: the neutrinium, a $\nu\nu$-"atom" bound by majoron and Higgs exchange. Owing to the rather strong coupling of the neutrinos the system has hydrogen like features. It has a radius of about 10^{-7} to 10^{-4} cm and corresponding binding energies of 10^{-2} eV to 10 keV. Therefore, the neutrinium is a low energy system.

Even without detailed calculation one can predict that most of the cosmic 1.9° K neutrinos (a relic of the big bang, analoguous to the 2.7° K photons) are in bound neutrinium states, thus altering the results of proposed experiments. For details of these experiments see the paper of Langacker et al.[40]. Another consequence of strong neutrino-neutrino coupling should be the cooling of neutrinos which emerge from reactors and stars. No such calculations were done.

d) <u>Familons</u>

Wilczek[41] considered the implication of the spontaneous break down of the family symmetry. He predicts the existence of a Goldstone boson called familon, which couples derivatively to flavor quantum number changing currents. Therefore, the most promissing search is in flavor changing decays like:

$$\Gamma(\mu \rightarrow e+f) = m_\mu^3/16\pi \ v_{\mu f}^2 \qquad (25)$$

Experimentally, the branching ratio is less then $6 \cdot 10^{-6}$, giving a limit of the scale parameter $V_{\mu f}$ (eq. (1)) of $6 \cdot 10^9$ GeV. From K-πf decay search, one has a limit of 10^{11} GeV. Improved experiments, especially $\mu \rightarrow$ fe decay, could push the limit of V close to the GUT-mass of 10^{15} GeV.

4. SCALAR PARTICLES

The most prominent representative of the possible scalar particles is the still missing Higgs particle, which is required in the standard model. Being aware of the existence of strong theoretical arguments[42] for a Higgs mass larger then 7 - 10 GeV, search in the low energy mass range is nevertheless important. As pointed out by several authors, an extention of the minimal model could also produce light Higgs particles[43].

The experimental limit for scalar particles results from experiments sensitive to exchange potential, particle decay and nuclei deexcitation. They will be discussed in detail.

Exchange Potential

The exchange of weakly coupled scalar particles with mass m_x between two fermions with mass m_f leads to a Yukawa-type potential:

$$V(r) = \frac{G_F m_f^2}{2\sqrt{2\pi}} \, e^{-m_x r}/r \qquad (26)$$

it is a $1/r$ potential and would appear as anomalous contribution to the Coulomb or gravitation potential. For a more general discussion see[44] and[45].

Gravitation experiments test the long-range behavior of the exchange potential, therefore, they are sensitive to small m_x masses.

The Eötvös-type experiments[46] measure the force between earth and objects of different materials, but identical masses. Because of the mass defect, the different objects, say hydrogen and copper, have a different number of hadrons: consequently, any additional force resulting from a potential which is proportional to the number of hadrons, should be measurable.

The experiments measure a difference which is smaller than one part in 10^9 between the gravitational forces of copper and of hydrogen with the same mass. From this result, one gets a limit for the mass of scalar (or vector) particles which rules out particles with a mass larger than 10^{-12} eV.

In Cavendish-type experiments, the gravitational forces between laboratory-size objects are measured. These are not nearly as precise as the Eötvös-type experiments, but, owing to the short distance, they are much more sensitive to potentials that fall off faster than $1/r$. Long[47] presents a survey of Cavendish-type experiments which show an actual deviation from a pure Newtonian potential. For the present discussion, we assume the deviation as an upper limit, excluding masses smaller than 10^{-4} eV.

It goes without saying that an experimental clarification of the discrepancy, discussed by Long, is needed.

Additional exchange potentials of the type of eq.(26) would also modify the amplitude of the neutron-nucleus scattering. Recently, Bracci et al.[45] reanalyzed the low energy n - Pb scattering experiment done by Aleksandrov et al.[48] in 1966.

They found no additional contribution and can, in agreement with Barbieri and Ericson[49], rule out scalar particles with a mass less than 15 MeV.

In this context, it is interesting to note, that from an old experiment performed to measure the electric polarisability of the neutron, one can derive such good bounds. Therefore, an improved experiment on n-nucleus scattering could rise the limit considerably.

As discussed by Leisi[50] at this school, precise measurements of μ-atomic X-ray energies also exclude anomalous scalar or vector exchange bosons with masses less than 10 MeV.

Production of scalar particles. Nuclear physics offers a way to search for scalar particles as well. In contrast to axions, scalar particles, in addition to $0^+ \to 0^+$ transition, compete with gamma transition of electric multipolarity. Experiments in $^4He (0^+ \to 0^+$ with $E_\gamma \simeq 20$ MeV)[51] and $^{16}O (0^+ \to 0^+$ with $E \simeq 6$ MeV)[52] were performed. The maximum transition energy sets an upper limit to the detectable mass. Since the decay into e^+e^- is the signature, the lower bound of the mass is $2m_e$. Having reached the required sensitivity and not observing any decay into e^+e^-, the authors[51,52] rule out scalar bosons in the mass range 1 MeV $\leq m_x \leq$ 14 MeV.

The study of the K into e^+e^- decay offers another possible search for scalar bosons. Willey and Yu[53] assumed that this decay can be simulated by the decay sequence K into H (the Higgs boson) and H into e^+e^-. With an analogous quark model calculation which was used to compute the K into $\mu^+\mu^-$ decay, the authors found the following theoretical branching ratio for K into H:

$$B(K^+ \to \pi^+ H) \simeq 1 \cdot 10^{-4} \text{ for } m_H \ll m_K$$

Because $B(H \to e^+e^-)$ is order unity for $2m_e < m_H < 2m_\mu$ and the experimental value for the branching ratio is $B(K \to e^+e^-) = (2.6 \pm 0.5)10^{-7}$, one is tempted to rule out Higgs bosons with a mass $m_H < m_K - m_\pi = 325$ MeV.

However, a direct search for scalar bosons with a mass larger than 15 MeV is still interesting, because

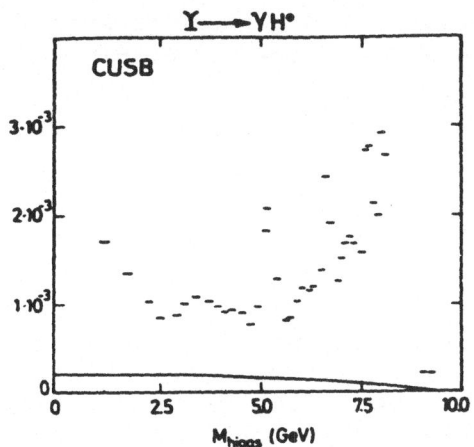

Fig. 7. 90 % C.L. upper limit on the branching ratio
$\Upsilon \rightarrow \gamma H^O$, plotted as a function of the Higgs
mass[52]. The curve is the expected branching
ratio.

the above analysis of the K into e^+e^- decay uses a
particular quark model. As seen from Fig. 7, high energy
experiments such as the search for Higgs bosons in the
Υ-decay[54], do not give any limits for masses below 1 GeV[55].

A feasible experiment was proposed by Ellis et al.[56].
They urged to look for π-capture at rest in nuclei, such
as ^9Be and ^{12}C, where the π^O production is energetically
forbidden. The Higgs boson would be produced via $\pi^- + p + H$
reaction with a branching ratio of order 10^{-6}, and would
decay into e^+e^- in about 10^{-10} sec, yielding a unambiguous
signal with invariant mass.

5. VECTOR PARTICLES

The Decay

Spin 1 vector particles have quite different decay
modes than scalars. In addition, to decay into gammas
and leptons, there is a decay in first order into
neutrinos. In fact, this decay has a width of the same
order as the decay into e^+e^-. The gamma decay is even

slower than the decay into 2 gammas of spin 0 particles, because it proceeds via a 3 gamma emission.

These facts show that
(i) the branching ratio of the gamma decay is less than 10^{-6}.
(ii) even if the mass of X is less than $2m_e$, the decay is still fast $(0(10^{-10}$ sec)) owing to the decay into neutrinos.
(iii) For masses $0(10$ MeV) the lifetime is $0(10^{-10}$ sec) and therefore, if $\gamma = E/m \approx 100$, the decay proceeds close to the production (beam dump). Since the detectors are 100 to 1000 decay lengths away from the beam dump, the search of vector particle becomes very difficult, if not impossible.

Exhange Potential

Vector particle exchange is analogous to the scalar particle exchange and is discussed in section IV.

Production

In analogy to the axion, vector particles are also produced via nuclear deexcitation, but now in competition with electro-magnetic radiation of electric multipolarity. In contrast to photon emission, 0^+ to 0^+ transitions are also possible via massive vector particle emission.

The main difference between vector particle search and axion search is the small branching ratio of the decay into gammas, caused by the allowed decay into neutrinos. Therefore, search for vector particles must be done through looking for missing transition rates, instead of searching for gammas. Missing transitions can be detected knowing the population of the excited states (via the feeding beta or gamma transition, or nuclear reaction), and by an absolute measurement of the decay products (in case of a $0^+ \rightarrow 0^+$ transition, they are: two photons or e^+e^-, or internal conversion).

No such experiments have been done, but would be worthwhile doing. The fact that the branching ratio for vector particle emission is quite large reduces their difficulty. Watson[57] did specific calculations plotted in Fig. 8 for the supersymmetric U-particle which are also valid for other weakly coupled vector particles.

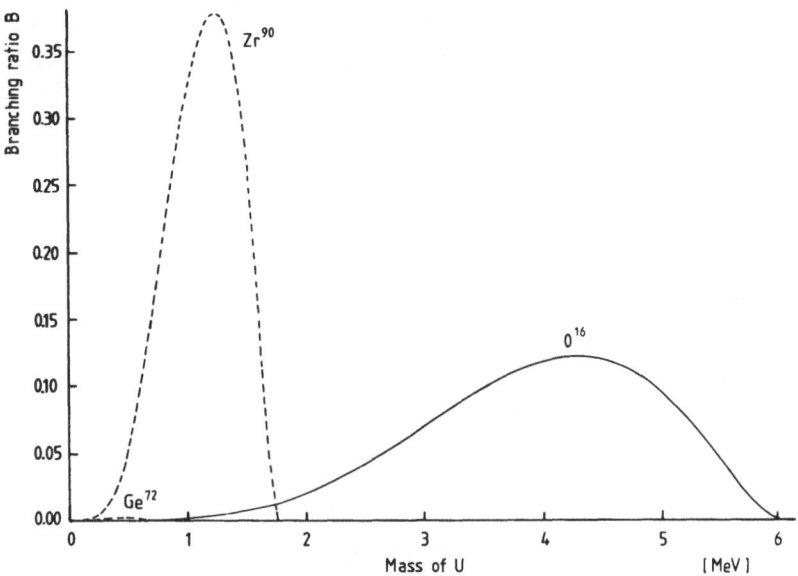

Fig. 8. Branching ratios for the emission of a vector particle U in selected nuclear transitions versus the mass of the U-particle. The calculations were done by Watson[57].

In conclusion, it is safe to say that from beam dump experiments one has rather poor limits for masses of vector particles, owing to their short lifetime of order 10^{-10} sec. If the mass is below $2m_e$, the observation of their decay is not possible, they would manifest themselves through "missing" nuclear gamma transition.

6. AXIAL VECTOR PARTICLES

The limits for axial vector particles are also poor because of the following facts:
- The exchange potential is $1/r^3$ and spin dependent.
  As shown in section III, the experimental limits do not

reach the sensitivity required.
- The decay for $m_x < 2m_e$ proceeds via the decay into 3 gammas, no experiments were done like the one described in section III, requiring 3 instead of 2 gammas in coincidence.

7. CONCLUSION

The present talk has shown that existing limits for low mass weakly coupled, exotic particles are not very good. Only the standard axion can be ruled out, but essentially, no limits exist for other light pseudo-scalar (Goldstone) bosons and axial vector particles. Direct search for scalar particles, like the Higgs boson, excludes them with masses below 15 MeV. Other limits are using debatable theoretical arguments.

A number of experiments is suggested to improve the yet unsatisfactory situation.

I am greatful to U. Zehnder for the critical reading of the manuscript.

References

1. S. Weinberg, Phys. Rev. Lett. 19:1264 (1967)
   A. Salam, Proc. 8th Nobel Symp. ed. N. Svartholm (Stockholm 1968)
   S. Glashow, J. Iliopoulos and L. Maiani, Phys. Rev. D2:1285 (1970).
2. P. Fayet, Proc. l'Ecole d'Eté Physique des Particules 1982, Gif-sur-Yvette, France to be published.
3. G.B. Gelmini, S. Nussinov and T. Yanagida, Nucl. Phys. B219:31 (1983).
4. L. Resnick, M.K. Sundaresan and P.J.S. Watson, Phys. Rev. D8:172 (1973).
5. T.W. Donnelly et al., Phys. Rev. D18:1607 (1978).
6. S. Weinberg, Phys. Rev. Lett. 40:223 (1978)
   F. Wilczek, Phys. Rev. Lett. 40:279 (1978).
7. M. Yoshimura, Proc. of Workshop on grand unified theories and early universe. KEK 83-13 Japan, 140 (1983).
8. A. Zehnder, Proc. l'Ecole d'Eté de Physique des Particules 1982 Gif-sur-Yvette France, to be published.
9. G.T. Hooft, Phys. Rev. Lett. 37:8 (1976), and Phys. Rev. D14:3432 (1976).

10. V. Baluni, Phys. Rev. $\underline{D19}$:2227 (1979), R. Crewther,
    P. diVecchia, G. Veneziano and E. Witten, Phys.
    Lett. $\underline{89B}$:123 (1979).
11. Altarev et al., Phys. Lett. $\underline{102B}$:13 (1981).
12. R.D. Peccei and H.R. Quinn, Phys. Rev. Lett. $\underline{38}$:1440,
    (1977).
13. W.A. Bardeen and S.H.H. Tye, Lett. $\underline{74B}$:229 (1978).
14. R. Dashen, Phys. Rev. $\underline{183}$:1245 (1969).
15. S.L. Adler et al., Phys. Rev. $\underline{D11}$:3309 (1075).
16. F.P. Calaprice, R.W. Dunford, R.T. Kouzes, M. Miller,
    A. Hallin, M. Schneider, and D. Schreiber, Phys.
    Rev. $\underline{D20}$:2708 (1979).
17. H. Faissner, E. Frenzel, W. Heinrigs, A. Preussger,
    D. Samm, and U. Samm, Phys. Lett. $\underline{96B}$:201 (1980).
18. Y. Asano et al., Phys. Lett. $\underline{107B}$:159 (1981)
19. J. Kandaswamy, P. Salamonson, and J. Schechter,
    Phys. Lett. $\underline{74B}$:377 (1978).
20. C. Edwards et al., Phys. Rev. Lett. $\underline{48}$:903 (1982).
21. M. Sivertz et al., Phys. Rev. $\underline{D26}$:717 (1982)
    M.S. Alam et al., Phys. Rev. $\underline{D27}$:1665 (1983).
22. LENA collaboration, Z. Physik $\underline{C17}$:197 (1983).
23. P.F. Jacques et al., Phys. Rev. $\underline{D21}$:1206 (1980).
24. J. Blietschau et al., Nucl. Phys. $\underline{114B}$:189 (1976),
    H. Wachsmuth and F.J. Hasert in ref.
25. H. Faissner, Invited Talk given at the International
    Neutrino Conference 1982 at Balatonfuered (Hungary)
26. H. Faissner, W. Heinrigs, A. Preussger and U. Samm,
    Phys. Rev. $\underline{D28}$:1198 (1983).
27. Proposal SIN R-81-06.2.
28. A. Barroso and N.C. Mukhopadhyay, Phys. Rev. $\underline{C24}$:2382
    (1981).
29. A. Zehnder, Phys. Lett. $\underline{104B}$:494 (1981).
30. A. Zehnder, K. Gabathuler, and J.L. Vuilleumier,
    Phys. Lett. $\underline{110B}$:419 (1982).
31. P. Lehmann, E. Lesquoy, A. Muller, and S. Zylberajch,
    Phys. Lett. $\underline{115B}$:270 (1982).
32. J.F. Cavaignac et al., Phys. Lett. $\underline{121B}$:193 (1983).
33. J.E. Kim, Phys. Rev. Lett. $\underline{43}$:103 (1979).
34. M. Dine, W. Fischer, and M. Scrednicki, Phys. Lett.
    $\underline{104B}$:199 (1981).
35. P. Sikivie, Phys. Rev. Lett. $\underline{48}$:1156 (1982),
    J. Preskill, M.B. Wise, and F. Wilczek, Phys. Lett.
    $\underline{120B}$:127 (1983), L.F. Abbott and P. Sikivie,
    Phys. Lett. $\underline{120B}$:133 (1983), M. Dine and W. Fischer,
    Phys. Lett. $\underline{120B}$:137 (1983).
36. M. Wise, H. Georgi, and S.L. Glashow, Phys. Rev.
    $\underline{47}$:402 (1981).

37. A.A. Anselm and N.G. Uraltsev, Phys. Lett. 114B:39 (1982).
38. Y. Chikashige, R.N. Mohapatra, and R.D. Peccei, Phys. Lett. 98B:265 (1981).
39. G.B. Gelmini and M. Roncadelli, Phys. Lett. 99B:411 (1981).
40. P. Langacker, J.P. Leveille, and J. Sheiman, Phys. Rev. D27:1228 (1983).
41. F. Wilczek, Phys. Rev. Lett. 49:1549 (1982).
42. J. Ellis, Lectures presented at the Les Houches Summer School (1981) Ref: CERN TH-3174.
43. R.S. Willey, Phys. Rev. Lett. 52:585 (1984), R.M. Barnett, G. Senjanovic, L. Wolfenstein, and D. Wyler, Phys. Lett. 136B:191 (1984).
44. G. Feinberg and J. Sucher, Phys. Rev. D20:1717 (1979).
45. L. Bracci, G. Fiorentini, and R. Tripiccione, Nucl. Phys. B217:215 (1983).
46. R.V. Eötvös, D. Pekar, and E. Fekele, Ann. d. Phys. 86:11 (1922).
47. D.R. Long, Phys. Rev. D9:850 (1974).
48. Y.A. Aleksandrov et al., JEPT Lett. 4:134 (1966).
49. R. Barbieri and T.E.O. Ericson, Phys. Lett. 57B:270 (1975).
50. H.-J. Leisi, talk given at this school.
51. S.J. Freedman et al., Phys. Rev. Lett. 52:240 (1984).
52. D. Kohler, B.A. Watson, and J.A. Becker, Phys. Rev. Lett. 33:1628 (1974).
53. R.S. Willey and H.L. Yu, Phys. Rev. D26:3287 (1982).
54. CUSB Collaboration paper C-188 submitted to the 1983 International Symposium on Lepton and Photon Interactions at High Energies Cornell University (1983).
55. S. Yamada, DESY report 83-100, Nov. 1983.
56. J. Ellis, M.K. Gaillard, and D.V. Nanopoulos, Nucl. Phys. B106:292 (1976).
57. P.J.S. Watson, private communication.

SEARCH FOR MUON-HADRON INTERACTIONS FROM MUONIC X-RAYS

H.J. Leisi

Institut für Mittelenergiephysik der ETH-Z
c/o SIN
CH-5234 Villigen

INTRODUCTION

I would like to report on a number of muonic X-ray experiments which were designed to investigate the muon-nucleus interaction. In earlier days similar experiments were used as a test of the vacuum polarization effect of QED. With the advent of the unified electroweak interaction we expect muon-nucleon forces in addition to those of QED. One, of course, is the "heavy-light" interaction mediated by the vector boson Z^O. Due to the large mass of the Z^O the range of this force is, however, too short to result in a measurable energy shift in muonic atoms. An additional force would be mediated by the Higgs boson which also has definite coupling g to the leptons and to the quarks. In the minimal version of the theory (one Higgs doublet) we have

$$g^2 = m_f^2 \sqrt{2} \, G_F \quad , \tag{1}$$

where m_f is the fermion mass and G_F the Fermi coupling constant. The mass of the Higgs boson has a lower bound (under rather general assumptions) of about 10 GeV. This force is also extremely weak. One can enlarge the Higgs sector of the theory. In fact, most GUT's and all supersymmetric models need at least two Higgs doublets. In such schemes there can be Higgs bosons much lighter than 10 GeV and their couplings to the fermions can be

363

enhanced[1]. It is therefore very important to search for scalar bosons in the whole mass range, also much below 10 GeV. This and other considerations led us to generalize the aim of our experiments still further by asking: Are there <u>any</u> muon-nucleus interactions (of scalar, vector or axial vector type) beyond QED?

EXPERIMENTS

The experiments are listed in table 1. They were all performed with the bent-crystal spectrometer located at the muon channel I at SIN[6].

Experiment 1

This is an improved version of the experiment published in ref. 6. The Rydberg constant of the muonic atom (in wavelength units) is obtained from the Rydberg constant of electronic hydrogen and the muon-electron mass ratio. Experimentally, one determines the wavelength ratio λ_x/λ_γ, where λ_γ is the wavelength measured by Delattes et al.[7] of a near-by nuclear γ ray. Since non-s states are investigated the nuclear structure effects are very small. The corrections for the Mg transition are listed in table 2 together with the combined uncertainties for both transitions. The radiative corrections have negligible errors.

Table 1.   Characteristics of X-ray experiments

Exp.	muonic atom	transition measured	references
1	^{24}Mg, ^{28}Si	3d-2p	2
2	^{12}C, ^{13}C	2p-1s	3, 4, 5
3	^{7}Li	2p-1s	4

364

Table 2.   Experiment 1: main uncertainties (in ppm) and
           corrections for the Mg transition.

	corrections, Mg transition	uncertainties (ppm)
Finite nuclear size	-0.89(3) eV	
electron screening	-0.49(12) eV[a]	
nuclear polarization	0.17(3) eV	1.8
m_e/m_μ	±0.5 ppm	
statistics		2.1
source-hight correction	-10.4(5) ppm	1.0
λ_γ	±0.9 ppm[b]	
total		3.0

[a]Preliminary; an improved evaluation based on the
  measured 4f-3d transition energy is in progress.
[b]Preliminary; improved experiments are currently
  performed by the NBS group.

Comparing the measured transition wavelength with
the calculated one we find, averaged over both
transitions:

$$\frac{\lambda_{exp} - \lambda_{QED}}{\lambda_{QED}} = (0.5 \pm 3.0) \times 10^{-6} \quad \text{(preliminary)} \quad (2)$$

An additional (isoscalar) muon-nucleon interaction
can be parametrized by a neutral boson exchange. For a
scalar or vector interaction this leads to the Yukawa
potential

$$V(r) = - A \cdot \frac{g_N \cdot g_\mu}{4\pi} \cdot \frac{e^{-mr}}{r} \quad , \quad (3)$$

where g_N and g_μ describe the boson coupling to the
nucleon and to the muon, respectively, and m is the

boson mass. Combining the energy shift of (3) with (2) we can deduce limits for $g_N \cdot g_\mu$ as a function of m. For any muon-nucleon interaction mediated by a scalar or vector boson we find

$$g_N \cdot g_\mu < 10^{-6} \times e^2 \quad , \tag{4}$$

in the mass range m $\lesssim$ 1 MeV. The result can also be characterized by the coupling (1): Any interaction with the strength of a (one-doublet) Higgs boson coupling would have been seen if the mass of the exchanged boson were $\lesssim$ 10 MeV.

The result of experiment 1 has two more alternative interpretations:

- The experiment is thusfar the most accurate test of the vacuum polarization effect. Provided there is no additional interaction and $m_{\mu^-} = m_{\mu^+}$ (CPT theorem), then (2) is a 0.09 % test of the vacuum polarization effect.

- If there are no additional interactions and the QED calculations are complete, then (2) represents the most precise direct measurement of the mass of the negative muon.

Experiment 2

In order to investigate muon-nucleon interactions of shorter range (larger m) we have measured transitions involving s states. The idea is to deduce from the measured nuclear finite-size shift the nuclear r.m.s. charge radius and compare it with the same quantity deduced from electron-nucleus scattering. A difference between the two would indicate the presence of an additional interaction.

In a very careful procedere we have subtracted from the finite-size shift the small contributions from charge moments other than $<r^2>$ by using moment ratios from a charge-model independent analysis of electron scattering data by Sick[8]. The calculation of the nuclear polarization shift by Rosenfelder is used[9], and the radiative corrections were calculated with the code MURKS[10]. We then obtain from the measured finite-size shift[3]:

$$\langle r^2 \rangle_\mu^{1/2} = 2.4832(18) \text{ fm.} \tag{5}$$

This value is to be compared with $\langle r^2 \rangle$ as determined from electron scattering[8,11]:

$$\langle r^2 \rangle_e^{1/2} = 2.4705(50) \text{ fm.} \tag{6}$$

There is a difference (2.4 σ) between the values (5) and (6). Being unable to find another explanation we tentatively ascribe this difference to an additional muon-nucleon interaction (3) with

$$\frac{g_N \cdot g_\mu}{e^2 \cdot m^2} = (-0.134 \pm 0.056) \text{ GeV}^{-2} \quad , \tag{7}$$

where m is around 1 GeV. Such an interaction could be due to a light scalar meson, or to a vector boson with a different coupling to the muon and to the electron.

We performed a similar experiment with muonic ^{13}C[5]. The r.m.s. radius of ^{13}C is found to be

$$\langle r^2 \rangle_\mu^{1/2} = 2.4621(36) \text{ fm.} \tag{8}$$

Comparing this again with electron scattering data and combining it with the ^{12}C result we obtain a 2.5 standard deviation error for the quantity (7).

Experiment 3

In this experiment we have observed the magnetic hyperfine splitting of the 1s ground state of μ-^7Li. The splitting is experimentally well-resolved and is found to be

$$\Delta E = 4684(49) \text{ meV.} \tag{9}$$

A comparison with the h.f.s. derived from the electromagnetic interaction yields bounds on a muon-nucleus
interaction mediated by an axial-vector boson. Information about the magnetization distribution (Bohr-
Weisskopf effect) is obtained again from electron
scattering. No axial vector interaction was found. The
limits, strongly improved by the present experiment,
can be expressed by

$$\frac{g_N \cdot g_\mu}{e^2 \cdot m^2} = (0.029 \pm 0.052) \text{ GeV}^{-2} \quad ; \tag{10}$$

3 MeV $\lesssim$ m $\lesssim$ few GeV.

I hope to have convinced you that precision
experiments on the beautifully simple muonic atom system
(hydrogen-like) provide interesting results for particle
physics. There are three reasons for this:

1. The muon is closer to the nucleus than electrons in
   a normal atom and therefore more sensitive to a short-
   range interaction.

2. Certain interactions (the Higgs coupling and the
   interaction discussed by Barshay[12]) are proportional
   to the lepton mass and hence 200 times stronger than
   in electronic systems.

3. A clean experiment at zero momentum transfer is a
   very important complement to high-energy experiments.

REFERENCES

1. R.S. Willey, Phys. Rev. Lett. 52:585 (1984);
    R.M. Barnett, G. Senjanović, L. Wolfenstein, and
        D. Wyler, Phys. Lett. 136B:191 (1984).
2. I. Beltrami, "New precision measurements of the
        muonic $3d_{5/2}$-$2p_{3/2}$ X-ray transition in ^{24}Mg and
        ^{28}Si: Vacuum polarization test and search for
        muon-hadron interactions beyond QED", Thesis
        ETH Zurich, Nr. 7062 (1982) (unpublished), and
        to be published.

3. W. Ruckstuhl, B. Aas, W. Beer, I. Beltrami,
    F.W.N. de Boer, K. Bos, P.F.A. Goudsmit,
    U. Kiebele, H.J. Leisi, G. Strassner, A. Vacchi,
    and R. Weber, Phys. Rev. Lett. 49:859 (1982).

4. W. Ruckstuhl, "Search for new muon-nucleon inter-
    actions in s states of muonic atoms", Thesis
    ETH Zurich, Nr. 7061 (1982) (unpublished), and
    to be published.

5. F.W.N. de Boer, B. Aas, P. Baertschi, W. Beer,
    I. Beltrami, K. Bos, P.F.A. Goudsmit, U. Kiebele,
    H.J. Leisi, W. Ruckstuhl, G. Strassner, A. Vacchi,
    and R. Weber, to be published.

6. B. Aas, W. Beer, I. Beltrami, P. Ebersold,
    R. Eichler, Th. von Ledebur, H.J. Leisi,
    W. Ruckstuhl, W.W. Sapp, A. Vacchi, J. Kern,
    J.-A. Pinston, W. Schwitz, and R. Weber, Nucl.
    Phys. A375:405 (1982).

7. R.D. Delattes, E.G. Kessler, W.C. Sauder, and
    A. Henins, Ann. Phys. (N.Y.) 129:378 (1980).

8. I. Sick, Phys. Lett. 116B:212 (1982), and private
    communication.

9. R. Rosenfelder, Nucl. Phys. A393:301 (1982);
    J. Bernabéu, and T.E.O. Ericson, Z. Phys. A309:213
    (1983).

10. W. Ruckstuhl, "MURKS, ein Programm zur Berechnung
    von Korrekturen zu den Bindungsenergien in
    myonischen Atomen", LHE-report, ETH Zurich
    (1982).

11. W. Reuter, G. Fricke, K. Merle, and H. Miska,
    Phys. Rev. C26:806 (1982).

12. S. Barshay, Phys. Rev. D7:2635 (1973), and Phys.
    Lett. 37B:397 (1971).

EXPERIMENTAL SEARCH FOR STRONG VAN DER WAALS FORCES

P.F.A. Goudsmit

Institut für Mittelenergiephysik der ETH-Z
c/o SIN
CH-5234 Villigen

INTRODUCTION

In the electromagnetic interaction we know about the presence of a long range - van der Waals - force acting between neutral atoms. This interaction is caused by the electric dipole polarizability of atoms. In second order perturbation calculation one can show that the v.d. Waals potential is proportional to the product of the polarizabilities of the atoms and that it is proportional to the inverse sixth power of the interatomic distance. Due to retardation effects this radial behaviour changes to r^{-7} for large distances. It has been suggested repeatedly that a similar phenomenon may be found in strong interactions: a long range force acting between colour neutral objects.

In this seminar I would like to present some ideas about experimental determination of such strong van der Waals interactions.

Following the discussion of Feinberg and Sucher[1] let us define a strong van der Waals potential as

$$V_N(r) = \lambda_N \, \hbar c \cdot \frac{1}{r} \left(\frac{r_o}{r} \right)^{N-1} , \quad \text{valid for } r \gg r_o \qquad (1)$$

where the radial scale of the interaction is given by r_o

- taken as typically 1 fm - and λ_N is the strength. From a phenomenological point of view there are no restrictions to N. The cases N < 4 are discussed in ref. 1 and 2. I will restrict myself to N = 4, 5 and 6. The main topic of this talk will be a discussion on the problem of making a clear separation between the short range and the long range interaction effects. Since this problem becomes dominant for higher N-values I will mainly concentrate on the N = 6 case.

In the electromagnetic case nature provides us with an excellent playground for v.d. Waals interactions. In many substances the average interatomic distances are significantly larger than the atomic dimensions but, at the same time, small enough not to make the van der Waals interaction become unmeasurably small. This is the reason that in daily life the electric van der Waals force plays such an important role.

With a strong van der Waals force of the type just described the situation is completely different. No natural system exists where a strong v.d. Waals shift - say with an r^{-6} behaviour - is dominant. In order to observe its effects one has to rely on precision experiments on hadron-hadron scattering or on transition energies in exotic atoms where the orbiting particle is a hadron. The experimental set up of fig. 1 cannot be realized in the laboratory; information on a n-nucleus van der Waals force can, however, be obtained from low-energy n-nucleus scattering experiments as has been shown in ref. 2.

No experiments have been carried out thus far with the specific purpose of a search for such interactions. In a series of review articles[1,2,3,4], however, existing experiments have been analyzed for the presence of strong van der Waals interactions and upper limits for their strength have been given. The search for van der Waals effects through the observation of level energies in exotic atoms has the advantage that due to the centrifugal barrier effect one can, in principle, place the orbiting particle at an appropriate radius. In this way the short range interaction can be strongly reduced while a long range component remains visible.

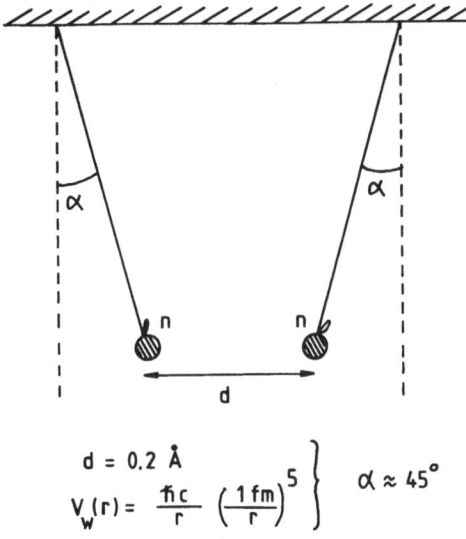

$$d = 0.2 \text{ Å}$$

$$\left. V_w(r) = \frac{\hbar c}{r} \left(\frac{1\,\text{fm}}{r}\right)^5 \right\} \quad \alpha \approx 45°$$

Figure 1.   Idealized version of an experimental deter-
            mination of a long range neutron-neutron
            interaction.

In a recent SIN research proposal[5] our group has
suggested to measure specific pionic atom transition
energies with very high precision in order to further
investigate the possible presence of strong van der
Waals forces in the π-nucleus system.

PIONIC X-RAY EXPERIMENTS - A PROPOSAL

Before discussing our proposal let me mention a
rather technical point which, nevertheless, has led to
misinterpretation of experimental data.

The phenomenological potential (1) contains a
strong short range part. Before a comparison between
predictions of this potential and experiments can be
made one has to make sure that this short range part
does not play a significant role in the comparison.
This has been done in different ways e.g. by cutting
the potential to zero or to a constant value below a
certain radius (typically m_π^{-1}). I will point out in the
next section that this procedure is not adequate in all
cases.

A program for a search for a strong v.d. Waals
interaction - say between pions and nucleons - should
look as follows

(a) it should include all essential observables
    available from (π-nucleon (nucleus)) scattering
    experiments.

In the exotic (pionic) atom section it should contain

(b) measurements of transitions in which a level shift
    due to a v.d. Waals interaction originates from a
    region well outside the nucleus and
(c) measurements of transitions that are mainly sensi-
    tive to the short range component of the strong
    interaction.

The total data set then should be analyzed in a common
fitting procedure of the optical-potential parameters
and the parameters λ of the v.d. Waals interaction.
Our program[5] includes the 5-4 transition in pionic ^{90}Zr
and the 4-3 transition in ^{40}Ca (point (b)), a series of
3d-2p transitions[6] in ^{12}C, $^{16,18}O$, $^{24,26}Mg$ and $^{28,30}Si$
and existing s-state shifts for point (c).

        In order to get a rough impression of the sensi-
tivity of the experiment before the data are actually
available and the common fit can be made we have chosen
a special short range approximation of the van der Waals
potential (1), suggested to us by F. Lenz. We again cut
the potential at 1.4 fm but at the same time we add a
constant negative part $-V_O$ in the region between 0 and
1.4 fm, choosing V_O such that the π-nucleon s-wave
scattering length for low q becomes zero (i.e.
$\int r^2 V(r)dr = 0$). This potential is then folded with the
nuclear matter distribution and first order perturbation is used to calculate the long range shifts.

        The choice of this short range behaviour of (1)
will automatically result in a reduction of the "double
counting" of short range effects. The reason being that
the s-wave part is essentially eliminated and in our
preliminary analysis we can treat the p-wave and higher
contributions independently of the (empirical) s-wave
effects both for the pionic atom case and for the
π-nucleon scattering.

Let us check now if the measurement of the 4f-state in pionic ^{90}Zr really probes a region well outside the nucleus. In fig. 2 I have plotted the integrand of the expression for the 4f-energy shift $V(r) \cdot r^2 \cdot Y^2(r)$ for a r^{-6} potential as a function of r; $Y(r)$ being the pionic radial wave function calculated with a conventional optical potential code. Indeed the energy shift originates mainly from a region of typically 15 fm, considerably large than the nuclear radius.

A similar curve for the 3d-state in pionic ^{40}Ca shows that also here (although less pronounced) the sensitive region lies outside the nucleus.

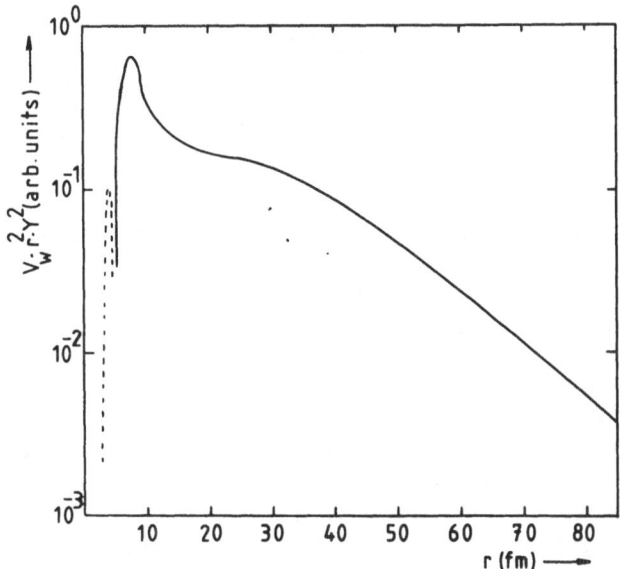

Figure 2. The radial distribution of the contribution to the energy shift by a r^{-6} potential of the 4f-state of pionic ^{90}Zr. The points below $\approx$ 5 fm correspond to the negative part of the potential.

The measurements of the pionic transition energies
will be performed with the πKS crystal spectrometer at
SIN (also the 3d-2p trantisions, mentioned before, were
measured with this instrument).The upper limits expected
from our work on the basis of the rough analysis
sketched above is given below.

$$\lambda_4 < 5 \cdot 10^{-4}, \qquad \lambda_5 < 2 \cdot 10^{-2}, \qquad \lambda_6 \lesssim 0.4$$

As a first part of our van der Waals measurement
we are at the moment engaged in a precision measurement
of the wavelength of the 4-3 transition in pionic ^{24}Mg.
This transition is extremely insensitive to as well
the short range conventional strong interaction as to
van der Waals forces described above. We expect from
this measurement to obtain a new experimental value of
the mass of the negative pion with a precision of
2-3 ppm. This mass value is needed since the existing
value for m_π is too inaccurate for a proper evaluation
of the ^{90}Zr and ^{40}Ca measurements.

COMMENTS ON OTHER DETERMINATIONS OF λ_N

Let me now say a few words about some other experiments used for determinations of upper limits of λ.
Again, I will concentrate on the problem of unraveling
the short range and the possible long range components
of the interaction. As mentioned before several authors
have cut the potential (1) at typically 1.4 fm to get
rid of its short range component. A convolution of the
resulting potential with the nuclear matter distribution, however, yields a potential that, for small radii,
looks rather similar to the nuclear matter distribution
itself. Let us take the case of the 1s-state in pionic
^{16}O (included in the analysis of ref. 4). Fig. 3 shows
the convolution of a r^{-6} potential, cut at 1.4 fm (see
insert) with the nuclear matter distribution. If the
resulting potential is used in an analysis of s-state
shifts[4] the computer may have a hard time in deciding
between a short range and a van der Waals interaction.
The very low upper limits deduced in ref. 4 from the
transitions to inner pionic states loose some of their
significance due to the fact that no clear description
is given of how the short and long ranged parts of the
interaction were separated.

376

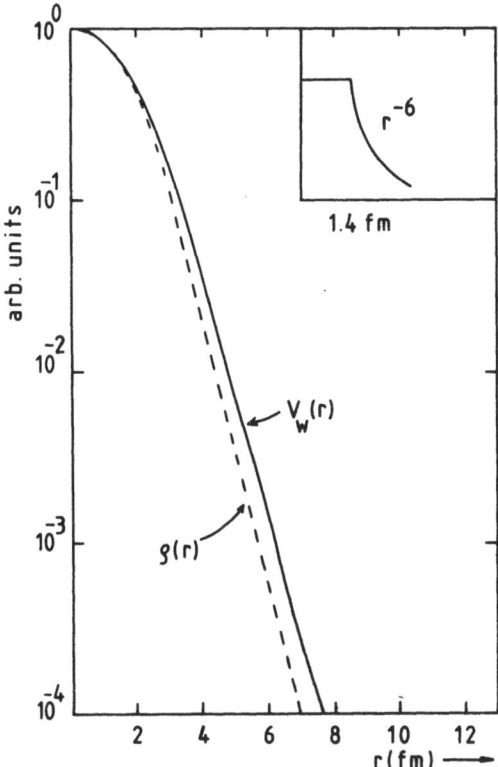

Figure 3.   The convolution $V_w(r)$ of a r^{-6} potential
cut at 1.4 fm - as shown in the insert -
with the nuclear matter distribution $\rho(r)$
of ^{16}O.

To show this point in a different way I have
plotted in fig. 4 ratios between the van der Waals
shift ΔE_w (calculated once more with the potential
described in the preceeding section) and the short
range shift ϵ_o, calculated with a standard set of
optical model parameters. Evidently, even with our van
der Waals potential, designed to suppress the low range
effects the van der Waals effect becomes a constant
fraction of the short range effect for pion orbits that
are close to the nucleus (large Z, small n). The ^{90}Zr
point and to a lesser degree that of ^{40}Ca satisfy the
requirement of being significantly sensitive to a long
range effect. In the same figure we have indicated also
some points used in the analysis of other authors[1,3,4].
The figure illustrates the higher sensitivity of the
^{90}Zr point.

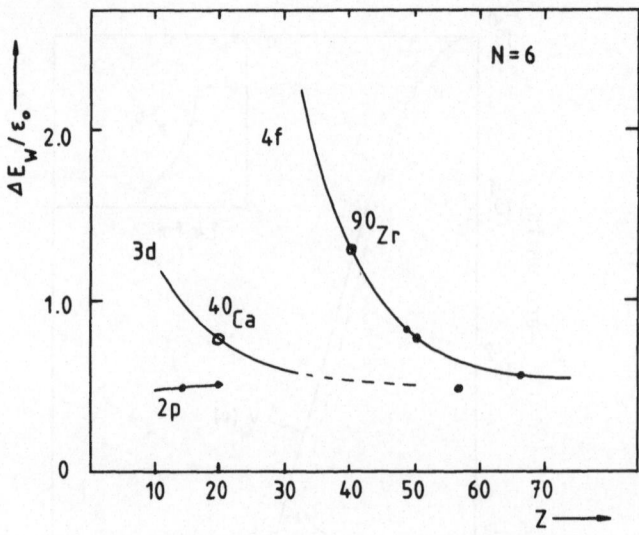

Figure 4.   Ratios of long range and short range energy
shifts for several pionic levels. For the
long range potential we took N = 6, λ_6 = 1.

Let us turn now to scattering experiments. Bracci
et al.[2] analyzed a low energy $n-^{208}Pb$ scattering experiment by Alexandrov et al.[7] and showed that this experiment is suited for the determination of upper limits
on λ. In our own analysis of the data we used the same
short range behaviour of the r^{-6} van der Waals potential
as described above. The radial range of sensitivity of
the experiment was investigated in a way similar as
that described for the ^{90}Zr case. The difference of
the integrands occuring in the expression for the $\ell = 1$
scattering amplitude between the extreme values of q
used in the analysis was plotted as a function of r.
This difference is directly related to the long range
effects. The resulting plot is rather similar to that
of fig. 2, showing that the experiment is indeed sensitive to long range effects. For N = 6 we obtain $\lambda \lesssim 0.6$
from the same set of data used in ref. 2. This value is
higher than that of ref. 2 by a factor 3; the difference
being mainly due to the difference in the short range
part of the potential (1). For N = 4 where the short
range part plays a minor role only, our values agree
with each other. We think that our value 0.6 is a
slightly more realistic upper limit for long range
contributions.

Finally let me mention an analysis by Lyth[3] of the π^+p scattering experiments of Bertin et al.[8]. Since here we are no longer dealing with an extended nucleus this experiment is sensitive in a range of smaller radii. A plot for this experiment similar to that described for the ^{208}Pb case shows a major sensitivity between 1.4 and ≈ 5 fm. The analysis with our potential resulted in an upper limit ($\lambda_6 \leq 0.8$) that is higher than that given in ref. 4 (0.14). The discrepancy is partly - but not completely - due to the difference in the short range behaviour of the potentials used and to our slightly more conservative fit to the data points. The sensitivity of the π^+p experiment for N < 6 seems to be significantly lower than that of the pionic atom case[3].

I hope to have convinced you that precision measurement of selected transitions in exotic atoms can improve the existing limits on a strong van der Waals interaction. Bracci et al.[2] in their review article mention the possibility of obtaining very interesting limits on λ in antiprotonic atoms. Also here it seems that a crystal spectrometer measurement would be the most suitable choice in order not to let the instrumental error become dominant. Such a measurement, however, will be possible only when the intensities at LEAR will be approximately one order of magnitude higher than at present.

I would like to thank B. Jeckelmann for evaluating numerous convolutions of a van der Waals potential and nuclear densities.

REFERENCES

1. G. Feinberg and J. Sucher; Phys. Rev. D20:1717 (1979).
2. L. Bracci, G. Fiorentini, and R. Tripiccione; Nucl. Phys. B217:215 (1983).
3. D. Lyth; Z. Phys. C15:177 (1982).
4. C.J. Batty; Phys. Lett. 115B:278 (1982).
5. H.J. Leisi et al; SIN proposal R-82-10.1.
6. G. de Chambrier et al.; to be published.
7. Yu.A. Aleksandrov et al.; JETP Lett. 4:134 (1966).
8. P.Y. Bertin et al.; Nucl. Phys. B106:341 (1976).

PHOTON-PHOTON INTERACTION DETECTION VIA THE VACUUM BIREFRINGENCE

INDUCED BY A MAGNETIC FIELD: STATUS OF THE EXPERIMENT

E. Iacopini

Dipartimento di Fisica dell'Università di Pisa, Italy
Sezione INFN di Pisa, Italy
CERN, Geneva, Switzerland

1. INTRODUCTION

The light-by-light scattering is a process which cannot be described by classical electrodynamics, because of the linear structure
of the Maxwell equations. The linearity of the equations is a consequence of the fact that electromagnetic waves do not possess any
electromagnetic structure, such as electric charge, magnetic moment,
etc. This is not so in the case of gravity waves, for example, where
the "charge role" is played by the energy-momentum density tensor,
which is carried also by the wave itself; and this is why Einstein's
equations are intrinsically non-linear.

In the framework of the theory of quantum electrodynamics (QED),
the photon remains without electromagnetic structure, which means
that photons cannot scatter directly. However, they can scatter via
the interaction with virtual charged pairs, i.e. because of the vacuum polarization process.

In terms of Feynman graphs, the diagrams describing the lowest
order contribution to the photon-photon scattering are shown in
Fig. 1, for virtual electron-positron pairs. The amplitude corresponding to such Feynman diagrams is the so-called vacuum polariza-
tion tensor of fourth rank $G_{\mu\nu\rho\sigma}(1234) = G_{\mu\nu\rho\sigma}(k_1,k_2,k_3,k_4)$.

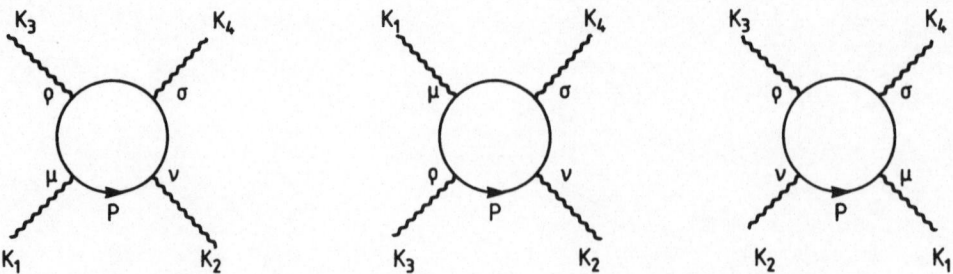

Fig. 1. Three of the lowest order Feynmann graphs contributing to the photon-photon interaction (the other three differ only in the arrow direction in the closed loop).

The general structure of the tensor $G_{\mu\nu\rho\sigma}$ has been studied by Karplus and Neuman[1] and, more recently, also by De Tollis and collaborators[2] using Mandelstam variables and double-dispersion techniques. On a general basis, by using only relativistic invariance, gauge invariance, and symmetry properties, Karplus and Neuman showed that the tensor $G_{\mu\nu\rho\sigma}$ can be written in terms of only five independent, scalar amplitudes A^{2143}, A^{2341}, A^{2111}, A^{2121}, and A^{2311}

$$
\begin{aligned}
G_{\mu\nu\rho\sigma}(1234) = \sum \{ & A^{2143}(1234) I^{(1)}{}_{\mu\nu\rho\sigma}(1234) \\
& + A^{2341}(1234) I^{(2)}{}_{\mu\nu\rho\sigma}(1234) + A^{2111}(1234) I^{(3)}{}_{\mu\nu\rho\sigma}(1234) + \\
& + A^{2121}(1234) I^{(4)}{}_{\mu\nu\rho\sigma}(1234) + A^{2311}(1234) I^{(5)}{}_{\mu\nu\rho\sigma}(1234) \} ,
\end{aligned} \qquad (1)
$$

where the sum Σ refers to simultaneous permutations of the four arguments k_1, k_2, k_3, k_4 and of the tensor indices, and we have put

$$
I^{(1)}{}_{\mu\nu\rho\sigma} = (1/32) T_{\mu\alpha\beta}(k_1) T_{\nu\beta\alpha}(k_2) T_{\rho\gamma\delta}(k_3) T_{\sigma\delta\gamma}(k_4) ,
$$

$$
I^{(2)}{}_{\mu\nu\rho\sigma} = (1/8) T_{\mu\alpha\beta}(k_1) T_{\nu\beta\gamma}(k_2) T_{\rho\gamma\delta}(k_3) T_{\sigma\delta\alpha}(k_4) ,
$$

$$
I^{(3)}{}_{\mu\nu\rho\sigma} = - \left[1/4(k_3 \cdot k_4) \right] T_{\mu\alpha\beta}(k_1) T_{\nu\beta\alpha}(k_2) k_{1\gamma} T_{\rho\gamma\delta}(k_3) T_{\sigma\delta\epsilon}(k_4) k_{1\epsilon} ,
$$

$$
I^{(4)}{}_{\mu\nu\rho\sigma} = - \left[1/4(k_3 \cdot k_4) \right] T_{\mu\alpha\beta}(k_1) T_{\nu\beta\alpha}(k_2) k_{2\gamma} T_{\rho\gamma\delta}(k_3) T_{\sigma\delta\epsilon}(k_4) k_{1\epsilon} ,
$$

$$I^{(5)}_{\mu\nu\rho\sigma} = \left[1/3(k_2 \cdot k_4)\right]k_{2\alpha}T_{\sigma\alpha\beta}(k_4)T_{\mu\beta\gamma}(k_1)$$

$$\left[T_{\nu\gamma\delta}(k_2)T_{\rho\delta\epsilon}(k_3) - T_{\rho\gamma\delta}(k_3)T_{\nu\delta\epsilon}(k_2)\right]k_{1\epsilon} ,$$

the tensor $T_{\alpha\beta\gamma}(k)$ being defined as

$$T_{\alpha\beta\gamma}(k) = \delta_{\alpha\beta} \, k_\gamma - \delta_{\alpha\gamma} \, k_\beta . \qquad (2)$$

The dynamics of the process is only contained in the five amplitudes A^{2143}, A^{2341}, A^{2111}, A^{2121}, and A^{2311}. Since in the low-energy limit the last three amplitudes vanish, only A^{2143} and A^{2341} contribute to the photon-photon scattering process far from the pair-production threshold. In this limit ($\omega \ll mc^2$) the photon-photon cross-section due to spinor (σ_{sp}) and scalar (σ_{sc}) virtual charged pairs reads

$$\sigma_{sp} = (973/10125\pi)\alpha^4(\hbar/mc)^2(\omega/mc^2)^6 ,$$
$$\sigma_{sc} = (119/10125\pi)\alpha^4(\hbar/mc)^2(\omega/mc^2)^6 , \qquad (3)$$

where ω is the photon energy in the centre-of-mass reference frame and m is the mass of the virtual charged particle. For visible light ($\omega = 2.5$ eV) we obtain

$$\sigma_e \approx 2.0 \times 10^{-63} \text{ cm}^2 , \qquad \sigma_\pi \approx 6.4 \times 10^{-84} \text{ cm}^2 . \qquad (4)$$

Below pair-production threshold, the photon-photon scattering can be described without using the virtual pair field by only modifying the linear structure of Maxwell equations. Euler, Heinsenberg and Weisskopf[3,4] have shown that, for slowly varying fields*

$$(\hbar/mc)|\text{grad}(\vec{A})| \ll |\vec{A}| , \qquad (\hbar/mc^2)|\partial\vec{A}/\partial t| \ll |\vec{A}| , \qquad (5)$$

---

*A stands for the electric (E) or magnetic (B) field amplitude.

and far from the critical value

$$|\vec{A}_{cr}| = m_e^2 c^3/e\hbar , \qquad (6)$$

i.e.

$$|\vec{E}_{cr}| \approx 1.3 \times 10^{18} \text{ V/m} , \qquad |\vec{B}_{cr}| \approx 4.5 \times 10^{13} \text{ G} , \qquad (7)$$

the real part of the photon-photon scattering amplitude can be accounted for by introducing in the free electromagnetic Lagrangian suitable higher order terms in the two relativistic invariants

$$F = E^2 - B^2 , \qquad G = \vec{E} \cdot \vec{B} . \qquad (8)$$

If the only contribution comes from electron-positron virtual pairs, then the e.m. field is described (to the lowest order) by the Lagrangian

$$8\pi \mathscr{L} = F + 2R \left[F^2 + 7G^2 \right] , \qquad (9)$$

where

$$R = (\alpha^2/90\pi)\hbar^3/mc^2 . \qquad (10)$$

According to Schwinger[5], if the virtual pair is made of spinless particles (e.g. pions), the Lagrangian becomes

$$8\pi \mathscr{L} = F + (R/8)\left[7F^2 + 4G^2 \right] . \qquad (11)$$

The constant R, in the two above-mentioned cases, takes the values

$$R_e \approx 1.3 \times 10^{-32} \text{ cm}^3/\text{erg} , \qquad R_\pi \approx 2.2 \times 10^{-42} \text{ cm}^3/\text{erg} . \qquad (12)$$

The highest continuous electromagnetic field amplitudes which can be produced in vacuum are

$$E_{max} \approx 3 \times 10^2 \text{ sV/cm} , \qquad B_{max} \approx 10^5 \text{ G} . \qquad (13)$$

They correspond to e.m. energy densities of the order of

$$u_{E_{max}} \approx 10^4 \text{ erg/cm}^3 , \qquad u_{B_{max}} \approx 10^9 \text{ erg/cm}^3 . \qquad (14)$$

They are to be multiplied by R to obtain the order of magnitude of
the linearity violation of Maxwell's equations that one can get in
the laboratory*. It is then obvious why Jauch and Rohlrich[7] affirm:

"The non-linear corrections to the linear classical Maxwell
theory are extremely small, so that the basic principle of
superposition is not violated to an observable degree".

2. INDIRECT OBSERVATIONS

The photon-photon interaction contributes, in general, to the
higher order radiative corrections of any electromagnetic process,
but it turns out that these corrections generally give rise to small
contributions to the main process, which would be present also in
the absence of γ-γ scattering.

In some sense, the situation is similar to the one found in the
detection of gravitational waves. The observation of the period time-
derivative in the binary pulsar PSW1913+16 shows[8] a power loss of the
system which fits very well with the one due to the emission of gravity waves. However this finding, although of primary relevance,
constitutes effectively an indirect observation of gravity wave
emission.

The two more relevant experiments that I will discuss in connection with the indirect observation of the γ-γ scattering are the
g_μ-2 experiment[9,10] and the Delbruck scattering[11].

2.1 g_μ-2 Experiment

The g_μ-2 experiment is one of the most stringent and precise
tests of QED. It is, therefore, not surprising that it is able to
say something also concerning the photon-photon interaction. Accor-

---

*Near neutron stars, where magnetic fields up to 10^{12} G could be
present, the QED non-linear corrections to Maxwell's equations become very important[6] and they explain the depolarization observed
in the X-rays emitted near the surface of some pulsars.

ding to Ref. 9, the value of the muon anomaly $a = (g_\mu - 2)/2$ experimentally found so far is

$$a = (1165924 \pm 8.5) \times 10^{-9} . \tag{15}$$

By assuming that the muon is a structureless particle described by a Dirac field, in the framework of QED, the radiative corrections[10] to the fundamental Feynman graph shown in Fig. 2a are responsible for an anomaly

$$a_{QED} = 0.5(\alpha/\pi) + 0.76578223(\alpha/\pi)^2 + (24.452 \pm 0.056)(\alpha/\pi)^3$$
$$+ (135 \pm 63)(\alpha/\pi)^4 + (420 \pm 30)(\alpha/\pi)^5 + \ldots$$
$$= (1165852.0 \pm 1.9) \times 10^{-9} . \tag{16}$$

Vacuum polarization terms involving hadrons add a contribution of

$$a_{hadr} = (66.7 \pm 8.1) \times 10^{-9} , \tag{17}$$

and the virtual emission of Weinberg-Salam weak bosons gives

$$a_{weak} = (2.1 \pm 0.2) \times 10^{-9} . \tag{18}$$

Therefore, the theoretical value for the muon anomaly amounts to

$$a_{th} = (1165921 \pm 8.3) \times 10^{-9} , \tag{19}$$

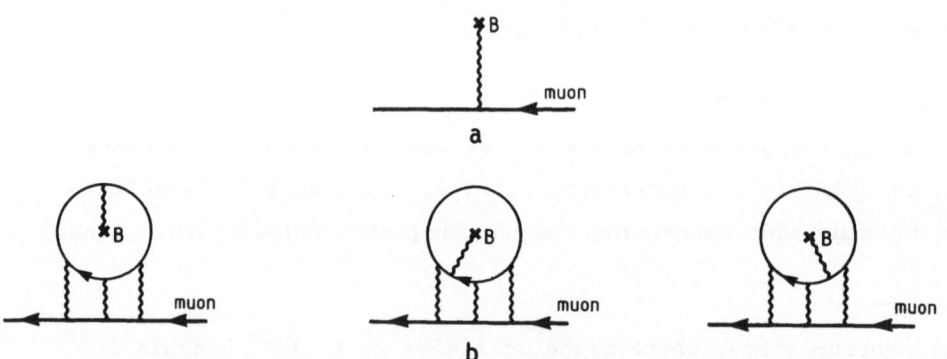

Fig. 2. Feynman graphs concerning the $g_\mu - 2$; a) fundamental diagram; b) three of the lowest order corrections to the muon anomaly, due to the photon-photon interaction (the other three are obtained by reversing the arrow direction in the closed loop).

the uncertainty being essentially due to the calculations in the hadron sector. The agreement between values (15) and (19), i.e. between QED and experiment is very striking and is the reason why the experiment is one of the most important confirmations of QED.

The photon-photon interaction contributes (to the lowest order) to the anomaly a_{QED} via the Feynman graphs shown in Fig. 2b. According to Ref. 10 the photon-photon contribution is

$$a_{\gamma\gamma} = 21.69 \pm 0.05)(\alpha/\pi)^3 = (271.8 \pm 0.6) \times 10^{-9} . \qquad (20)$$

As a consequence, given the errors quoted in Eqs. (15) and (19), the $g_\mu-2$ experiment tests the photon-photon interaction at the 5% level, besides the fact that its contribution is only 2.3×10^{-4} of the whole anomaly.

This test must, however, be considered "an indirect observation" of $\gamma-\gamma$ interaction since it is based on an interpretation of the experimental result which requires the use of QED and some hypothesis on the muon interactions, and on the absence of any other mechanism or particle able to influence the anomaly calculations.

2.2 Delbruck Scattering

The Delbruck scattering[11-13] is a radiative correction to the Compton scattering of a photon from a nucleus (See Feynman diagrams shown in Fig. 3a). In the Thomson region, the nuclear field can be assumed to be a static Coulomb field. The scattering is, in this case, elastic and is sometimes referred to as "potential scattering of light" (see Fig. 3b).

According to Refs. 2 and 12, the two differential cross-sections for no photon spin-flip (σ_{++}) and photon spin-flip (σ_{+-}) read, respectively,

$$
\begin{aligned}
d\sigma_{++}/d\Omega &= (Z\alpha)^4 \ r_0^2 |M_{++}|^2 \\
&= (Z\alpha)^4 \ r_0^2 \{(73/72)(1/32)(\omega/m_e c^2)^2 \cos^2 \theta/2\}^2 \\
d\sigma_{+-}/d\Omega &= (Z\alpha)^4 \ r_0^2 |M_{+-}|^2 \\
&= (Z\alpha)^4 \ r_0^2 \{(5/8)(1/32)(\omega/m_e c^2)^2 \sin^2 \theta/2\}^2 ,
\end{aligned}
\qquad (21)
$$

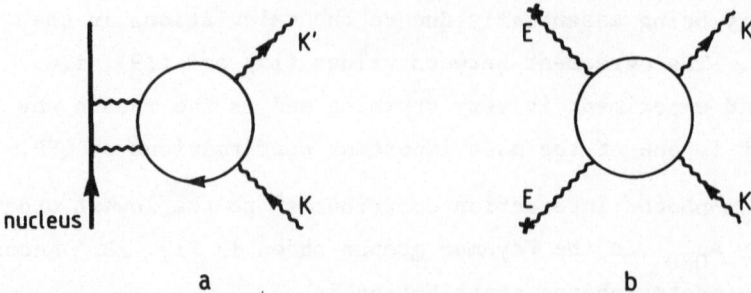

Fig. 3. Delbruck scattering Feynman graphs; a) one of the diagrams describing the Delbruck scattering (the others are obtained by permutations of the photon legs); b) Feynman diagram concerning the electric potential scattering of light.

in both cases a factor $Z^4\alpha^2$ higher than in the pure γ-γ case.

In the forward direction, the scattering amplitude is proportional to the second power of the photon energy ω ($\omega \ll m_e c^2$). As a consequence, in the presence of a macroscopic electric field in vacuum, one should expect[14] an index of refraction of vacuum such that

$$n - 1 \propto 1/\omega^2 \, M(\theta=0) \, , \qquad (22)$$

i.e. independent from the photon energy. This result is in fact what the Lagrangian (9) predicts for an electric field. This is also what does occur for the magnetically induced vacuum birefringence (see Section 3), because of the symmetry in E and B of Lagrangian (9).

Above pair-production threshold there is, of course, a non-zero imaginary part in the scattering amplitude, which is related, via the optical theorem, to the total cross-section for real-pair creation. However, only the real part of the Delbruck scattering is connected to virtual-pair creation, i.e. to the vacuum polarization phenomenon. It is, therefore, the experimental observation of this real part that should be proved in order to confirm in this way the existence of a photon-photon interaction.

In general, when a photon scatters on a nucleus, four coherent processes take place:

i)   Thomson scattering;
ii) Delbruck scattering;
iii) Rayleigh scattering;
iv) Nuclear resonance scattering.

The four amplitudes add together coherently, and this is why it is quite difficult (apart from when $\omega \gg m_e c^2$) to prove the presence of a Delbruck real amplitude.

The best evidence of the phenomenon, to my knowledge, has been obtained by Kahane et al.[15] by using ^{181}Ta (Z = 73) and photons having an energy of 7.9 and 9.0 MeV. Figure 4 is taken from Ref. 15 and shows the comparison between theoretical calculations and experimental data. On this basis, one can conclude that the real amplitude of the photon scattering from a Coulomb potential has been tested at the 20% level.

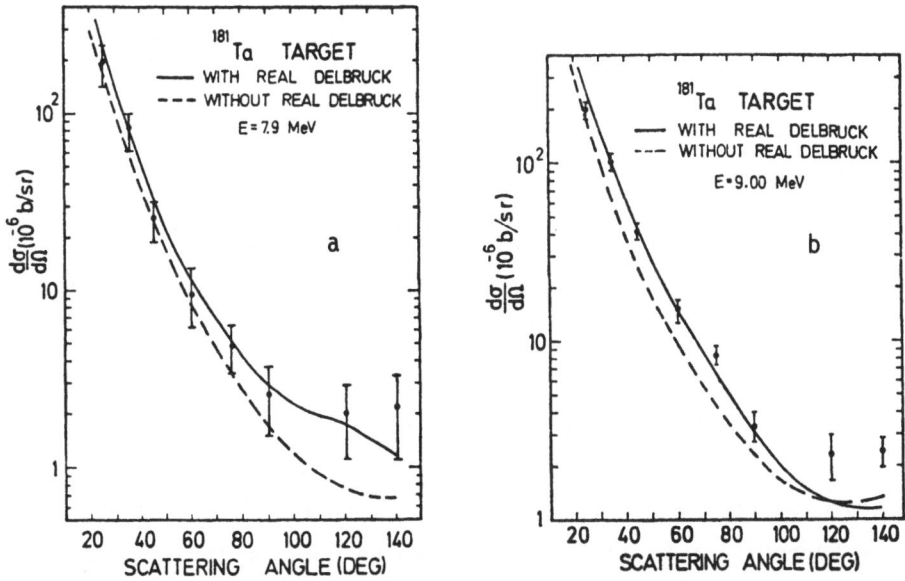

Fig. 4. Experimental results (taken from Ref. 15) concerning the Delbruck scattering on tantalum: a) 7.9 MeV of photon energy; b) 9.0 MeV of photon energy.

3. DIRECT OBSERVATION BY VACUUM BIREFRINGENCE DETECTION

The idea of the experiment[16],[17] is to observe, by optical means, the birefringence induced in vacuum by a magnetic field via the electron-positron quantized field. As a matter of fact, from the Lagrangian (9), one obtains that the vacuum, in the presence of a magnetic (electric) field, behaves, in some ways, as a material medium and becomes polarized.

The electric and magnetic susceptivity tensors ε_{ij} and μ_{ij} defined, in terms of the auxiliary fields

$$\vec{D} = 4\pi \, \partial\mathcal{L}/\partial\vec{E} \quad \text{and} \quad \vec{H} = -4\pi \, \partial\mathcal{L}/\partial\vec{B} \; , \tag{23}$$

as

$$D_i = \varepsilon_{ij}(E,B)E_j \quad \text{and} \quad B_i = \mu_{ij}(E,B)H_j \; , \tag{24}$$

read

$$\varepsilon_{ij} = \delta_{ij} + 2R\big[2(E^2 - B^2)\delta_{ij} + B_i B_j\big] \; , \tag{26}$$

$$\mu_{ij} = \delta_{ij} - 2R\big[2(E^2 - B^2)\delta_{ij} - E_i E_j\big] \; . \tag{27}$$

The structure of ε_{ij} and μ_{ij} shows that in the presence of a static magnetic field, the vacuum is no longer isotropic for the propagation of light. It can be shown[18] that the indices of refraction n_L and n_T of vacuum in a magnetic field B_0, for light linearly polarized respectively parallel and transverse to the plane defined by $\vec{B}_0$ and the light propagation vector $\vec{k}$, are

$$n_L = 1 + 7R_e B_0^2 \sin^2 \theta \; , \tag{28}$$

$$n_T = 1 + 4R_e B_0^2 \sin^2 \theta \; , \tag{29}$$

where θ is the angle between $\vec{B}$ and $\vec{k}$.

Clearly, a measurement of a non-zero vacuum birefringence induced by a magnetic field is a direct, genuine observation of the violation of the linearity of Maxwell's equations, i.e. of the existence of a photon-photon interaction. A measurement of the value

$$\Delta n_{QED} = n_L - n_T = 3R_e B_0^2 \sin^2 \theta = (\alpha^2/30\pi)(\lambdabar_e^3/m_e c^2)B_0^2 \sin^2 \theta \tag{30}$$

would show that such a violation is really due to the presence of virtual electron-positron pairs, as affirmed by QED.

It should be noted that if the electron were a spinless particle, according to the Lagrangian (11), one would have obtained

$$n_L = 1 + (R/4)B_0^2 \sin^2 \theta \ ,$$

$$n_T = 1 + (7R/4)B_0^2 \sin^2 \theta \ , \tag{31}$$

i.e. a value for the magnetically induced vacuum birefringence of

$$\Delta n_{spinless} = -(3R/2)B_0^2 \sin^2 \theta \ , \tag{32}$$

of opposite sign with respect to value (30).

Finally, it may be interesting to note that Born-Infeld's non-linear theory of electrodynamics[19] predicts no vacuum birefringence induced by a magnetic (electric) field, as is pointed out in an earlier paper[20].

The first tentative of observing a violation of the linearity of Maxwell's equations by studying the propagation of light in a magnetic field (10 kG) in vacuum is, to my knowledge, that of Watson[21], performed in 1929. He wanted to measure a possible magnetic moment of the photon which, in a magnetic field, would change its energy and, therefore, its frequency. He tried to observe changes in the fringes pattern of a Fabry-Perot interferometer placed in a transverse magnetic field, and he established an upper limit for the photon magnetic moment of

$$|\mu| < 0.015 \ \mu_B \ . \tag{33}$$

More recently, another experimental proposal[22] has been published to observe the departure from unity of the index of refraction given by Eq. (28) with a Michelson interferometer having in the two arms modulated transverse magnetic fields. From Eqs. (28) and (29) the signal is 7/3 higher in this case than for the vacuum birefringence expressed by Eq. (30). There are, however, many experimental difficulties that we have considered and that have convinced us that the "birefringent method" is experimentally easier.

3.1 The Method

When light propagates through a transverse magnetic field $\vec{B}_0$ in vacuum, the two linear polarization states, parallel and perpendicular to the $\vec{B}_0$ direction, propagate with different phase velocities. As a consequence, light originally linearly polarized at an angle β with respect to $\vec{B}_0$ emerges from the magnetic field region (of length L) elliptically polarized, with an ellipticity ψ_{QED} such that

$$\psi_{QED} = \pi \Delta n_{QED}(L/\lambda) \sin(2\beta) = 3\pi R_e B_0^2 (L/\lambda) \sin(2\beta) , \qquad (34)$$

where λ is the light wavelength.

In our case, by considering $B_0 = 80$ kG, from Eq. (30) we get

$$\Delta n_{QED} = 0.25 \times 10^{-21} \qquad (35)$$

and therefore we shall have to measure very small ellipticities.

The intensity transmitted by an analyser prism nearly crossed with respect to the original light polarization vector is quadratic in the total light ellipticity ψ (to the lowest order), and reads

$$W = W_0(\sigma^2 + \psi^2 + \phi^2) , \qquad (36)$$

where W_0 is the light power prior to the analyser, ϕ is the residual (misalignment) angle between the prism axis and the light polarization vector, and σ^2 is the extinction factor[*] of the system. In order to obtain a linear dependence of the signal on ψ_{qed} we have introduced an external modulation $\Phi_F(t)$ in the light polarization state. We use a glass Faraday modulator which maintains an extinction factor of the order of 10^{-7}, necessary to approach the ultimate sensitivity represented by the shot-noise limit[**] (see Section 3.3)

---

[*]It measures the quality of the light polarization state of the light beam. Using good crystal polarizer prisms, one can reach $\sigma^2 \approx 10^{-8}$.

[**]For interferometer experiments, the quantity equivalent to the extinction factor is the fringe contrast, which for Michelson interferometers is of the order of 10^{-2}.

with a light power in the watt region. The Faraday modulator requires an ellipticity-to-rotation conversion which is obtained by reflection on a gold mirror. A very schematic diagram of the apparatus is shown in Fig. 5.

The light intensity after the analyser prism is then

$$W(t) = W_0\{\sigma^2(t) + \left[\psi_{QED}(t) + \Phi_F(t) + \phi(t)\right]^2 + \chi(t)^2\} , \qquad (37)$$

where χ is the light ellipticity after reflection on the gold mirror.

The current from a photodiode detector placed after the analyser prism A (see Fig. 5), to the lowest order in ψ, ϕ, and χ, reads

$$i(t) = W_0\eta\left[\sigma^2 + (\Phi_{0F}^2/2) - (\Phi_{0F}^2/2) \cos 2\omega_F t\right.$$
$$\left. + 2\Phi_{0F}\psi_0 \sin \omega_F t \sin 2\omega_M t + 2\Phi_{0F}\phi(t) \sin \omega_F t\right] , \qquad (38)$$

where η is the photodetector efficiency and

$$\Phi_F(t) = \Phi_{0F} \sin \omega_F t , \qquad (39)$$

$$\psi_{QED}(t) = \psi_0 \sin 2\omega_M t . \qquad (40)$$

The useful signal, proportional to $W_0\eta\Phi_{0F}\psi_0$ is present, buried in noise, at the two frequencies $\omega_F \pm 2\omega_M$, in the presence of a d.c. level proportional to $W_0\eta(\sigma^2+\Phi_{0F}^2/2)$ and a pure modulation signal at $2\omega_F$ proportional to $W_0\eta\Phi_{0F}^2/2$. We have found it extremely important to have the possibility of adjusting the frequency ω_F in order to place the signal in a frequency region with a minimum of disturbances. Moreover, in order to overcome the low-frequency noise on the polarization state described by $\phi(t)$, the signal frequency needs to be as high as possible ($f_M = \omega_M/2\pi > 25$ mHz). According to Eq. (34), the modulation of the signal can be obtained by pulsating the current in the magnet or by rotating the magnetic field itself.

The signal is recovered by Fourier analysis of the photodetector current sampled during the period T.

In test experiments we have proved[23] the possibility of detecting ellipticities smaller than 10^{-10} by integrating the signal over sampling periods of one day.

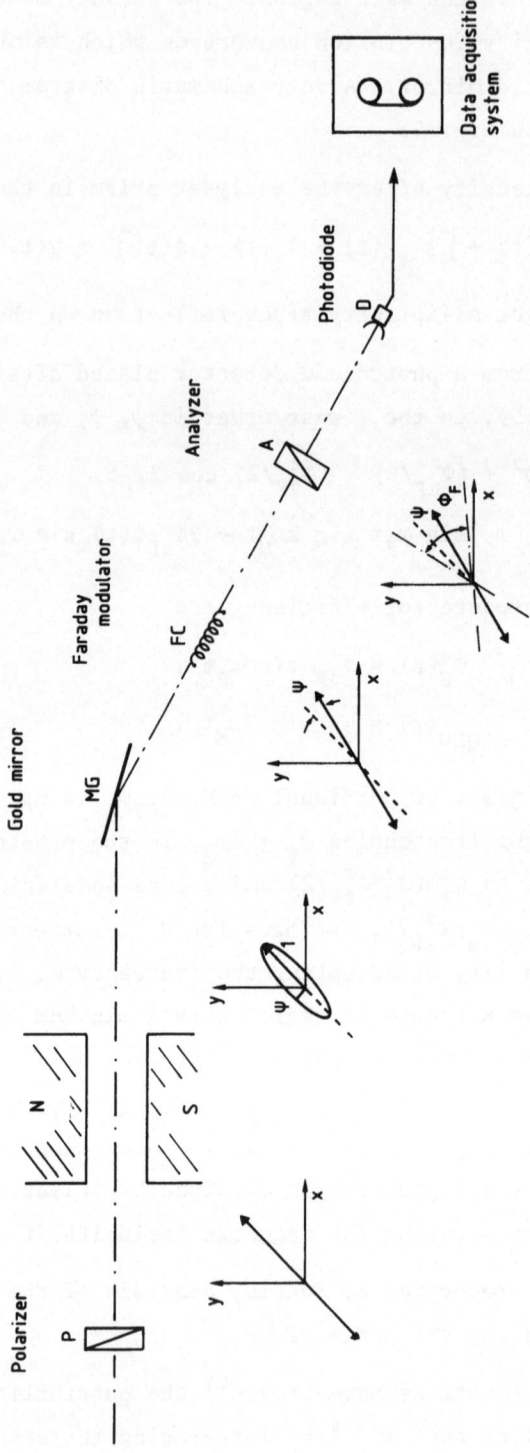

Fig. 5. Schematic diagram of the apparatus.

The test measurements have also been used to study possible sources of systematic errors. For instance, the stray field from the magnet on the cavity mirrors (see Section 3.2) must be smaller than 1 μG, otherwise the Faraday effect in the dielectric multi-layers[24] would generate systematic errors of the order of the effect to be measured. Also, the residual pressure in the vacuum pipe inside the magnetic field must be kept below 10^{-8} Torr in order to eliminate the effect due to the magnetically induced gas birefringence[25,26] (Cotton-Mouton effect).

However, to get an ellipticity ψ_0 of the order of 10^{-11}, the optical path L required in the magnetic field is of the order of several kilometres. The method to achieve this condition is explained in the next section. ·

3.2 Optical Cavity

To obtain L ≈ some kilometres we intend to use an optical delay line[27,28], consisting of two concave spherical mirrors, placed at a distance of about 7 m from one another. The central cavity region (6 m) is traversed by a magnetic field with an intensity of B ≈ 80 kG. In this way, with 500 reflections, i.e. L = 3 km, we will get

$$\psi_{QED} = 0.46 \times 10^{-11} . \tag{41}$$

An optical delay line having these characteristics and a diameter of 7 cm has been realized without magnetic field using interferential spherical mirrors having a reflectivity of 99.5%. The problems studied experimentally were the depolarizatio⁻ ᶠ the beam produced by the reflections and the cavity stability, which influences the noise in the polarization state.

The light source used was the same as planned for the final experiment, i.e. a CW argon-ion laser (λ = 514.5 nm) used at an output power of up to 2 W. The use of laser light is only dictated by the necessity of having the maximum source brilliance, i.e. the maximum power light in the minimum beam phase space (no coherence properties are used).

It is easy to show[27] that, in the Gauss approximations, a light beam in the cavity hits the mirror surfaces in points disposed on an ellipse, which is centred at the intersection point between the mirror surface and the optical axis defined by the two mirror surface centres. The beam spot positions are given by

$$x_n = A \cos (n\Theta + \xi) \ ,$$
$$y_n = B \cos (n\Theta + \zeta) \ , \tag{42}$$

with Θ such that

$$\cos \Theta = 1 - d(f_1 + f_2 - d/2)/f_1 f_2 \ , \tag{43}$$

where d is the mirror distance, f_1 and f_2 are the two mirror focal lengths, and A, B, ξ, and ζ are constants depending upon the initial conditions. To make better use of the mirror surface, we have introduced an astigmatism in the optical delay line by mechanically deforming one of the two mirrors. Instead of ellipses one then obtains Lissajous patterns which cover almost all the mirror surface, as shown in Fig. 6. The beam enters and comes out from the delay line by the same hole drilled in one of the two mirrors, which guarantees a minimum of ellipticity produced by the reflections.

We have matched the TEM_{00} laser beam to the fundamental mode of the cavity, using a suitable telescope. The mode-matching eliminates spot-size oscillations on the mirrors and, since the constant-phase beam surfaces have the same radius of curvature as the mirrors when the beam is reflected, the spread in the incidence angle is minimized, which reduces the beam depolarization produced by the delay line.

Another important consideration has been to orient properly the mirrors with respect to the light polarization direction. In agreement with what was also found by Bouchiat[29], it turns out in

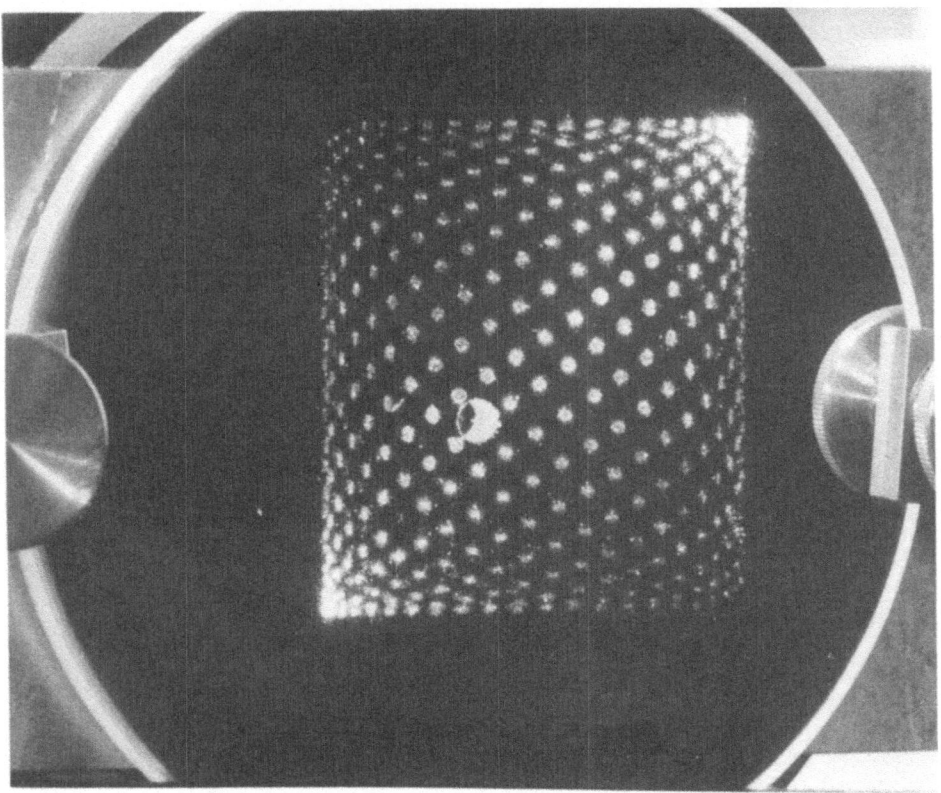

Fig. 6. Photograph of a Lissajous pattern on the cavity entrance mirror.

fact that interferential mirrors present a birefringence axis
($\psi \approx 10^{-4}$ per reflection).

With all these precautions, we have been able to maintain the extinction factor in the 10^{-7} region after up to 600 reflections.

3.3 Sensitivity

Several sources of noise and instabilities determine the sensitivity of the apparatus. We have performed various test measurements to establish the sensitivity we are able to reach. It has been determined by generating a known Faraday rotation (0.67×10^{-8} rad) with an air Faraday cell, and by comparing this with the noise during the measurements. The experimental conditions chosen are summarized in Table 1 and are essentially those planned for the final experiment.

Table 1. Experimental Conditions

Laser power W_0 prior to the analyser	$\sim$ 100 mW
Laser wavelength	514.5 nm
Number of reflections	$\sim$ 500
Cavity attenuation	$\sim$ 12
Optical path	$\sim 3.5 \times 10^5$ cm
Extinction factor σ^2	$\sim 10^{-7}$
Faraday modulation amplitude Φ_F	3.0×10^{-4} rad
Experimental sensitivity ψ_{sens}	$6.8 \times 10^{-11} \sqrt{day}$

The shot-noise in the photocurrent of the photodiode (efficiency $\eta = 0.2$ A/W) is given, from Eq. (38), by

$$\left\langle i_{shot} \right\rangle = \sqrt{2e\eta W_0 (\sigma^2 + \Phi_{0F}^2/2)\ \Delta\nu} \approx 3 \times 10^{-14}\ \sqrt{\Delta\nu}\ A \qquad (44)$$

and would correspond to a sensitivity of

$$\psi_{shot} = 3.5 \times 10^{-9}/\sqrt{Hz} = 1.2 \times 10^{-11}\ \sqrt{day}\ . \qquad (45)$$

The Johnson noise in the 400 MΩ feedback resistor R on the current-to-voltage photodiode preamplifier is

$$\left\langle i_{thermal} \right\rangle = \sqrt{(4kT/R)\ \Delta\nu} \approx 0.7 \times 10^{-14}\ \sqrt{\Delta\nu}\ A\ . \qquad (46)$$

The equivalent input noise of the preamplifiers has been measured to be

$$\left\langle i_{electr} \right\rangle \approx 1.0 \times 10^{-14}\ \sqrt{\Delta\nu}\ A\ . \qquad (47)$$

398

Whilst all these noise sources are "white", the laser intensity
fluctuations show a spectral density increasing towards 0 Hz.
Figure 7a shows the Fourier spectrum of the laser intensity between
0 and 1 Hz taken with a low-frequency spectrum analyser. At 0.050 mHz
the laser intensity noise is 35 dB higher than the statistical limit
for the same value of d.c. voltage output. The Faraday modulation,
acting as a carrier at f_F = 370 Hz, shifts the signal to a region in
which the excess noise of the laser is only 4 dB higher than the
statistical limit. Figure 7b shows the laser light intensity spectrum between 0 and 500 Hz. Above 250 Hz, in agreement with Refs. 23
and 30, we observe

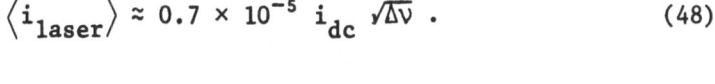

$$\left\langle i_{laser} \right\rangle \approx 0.7 \times 10^{-5} \; i_{dc} \; \sqrt{\Delta\nu} \; . \tag{48}$$

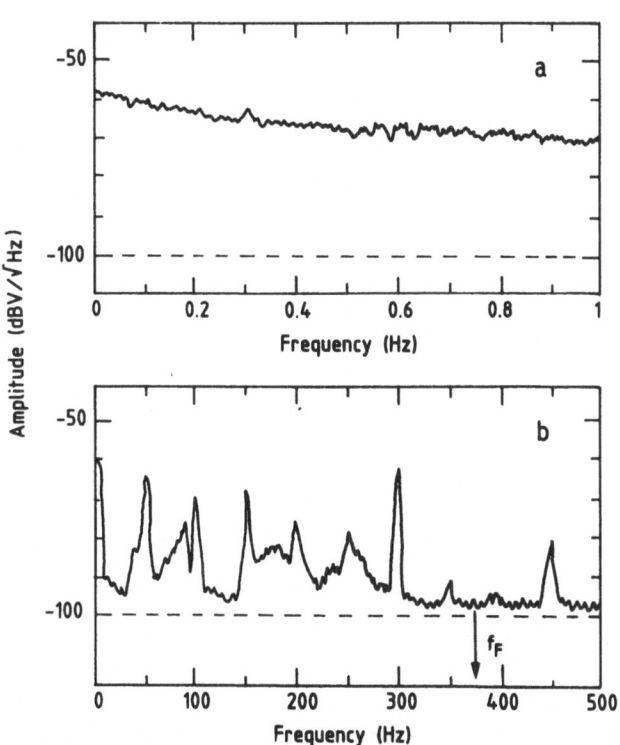

Fig. 7. Laser intensity noise, measured after the prism analyser A,
with a light power ∿ 15 nW. The dashed line represents the shot-
noise limit. a) Fourier noise spectrum taken between 0 and 1 Hz.
b) Fourier noise spectrum taken between 0 and 500 Hz. The peaks
correspond to harmonics of the 50 Hz mains.

The excellent extinction factor of 10^{-7} reached in the apparatus reduces the light power on the detector to such values that the laser noise becomes comparable with the shot noise; in fact we have

$$\left\langle i_{laser} \right\rangle \approx 2.1 \times 10^{-14} \sqrt{\Delta\nu} \text{ A .} \tag{49}$$

The sum of the above-mentioned noise contributions gives

$$\left\langle i_n \right\rangle \approx 3.8 \times 10^{-14} \sqrt{\Delta\nu} \text{ A ,} \tag{50}$$

only a factor of 1.3 higher than the shot-noise limit.

However, in the test measurements so far performed, the sensitivity obtained above 50 mHz has been

$$\psi_{sens} = 6.8 \times 10^{-11} \sqrt{day} \text{ ,} \tag{51}$$

a factor 5.7 worse than the shot-noise limit. This is because the Faraday modulation Φ_F, according to Eq. (38), also acts as a carrier for the fluctuations of the misalignment angle $\phi(t)$ which contribute to the low-frequency noise in the signal. These fluctuations are mainly due to stress birefringence induced by thermal gradients in the optical elements and by mechanical instabilities of the multi-pass optical cavity. To reduce this noise, we are at present mounting another optical cavity, in which much more care has been given to the mechanical mountings. Moreover no windows (see Fig. 8) are used in the polarized "zone", in order to reduce the ellipticity due to thermal gradients in the optics. With this improved cavity we will also study the possibility of reducing its cross-section to 4 cm, in order to provide for the possible future use of a bending magnet developed for 10 TeV proton accelerators. This would possibly lessen the cost problem concerning the magnet.

With the improved apparatus, we hope to reach a sensitivity of

$$\psi_{sens} \approx 2.5 \times 10^{-11} \sqrt{day} \text{ .} \tag{52}$$

In this case, the magnetically induced vacuum birefringence due to the photon-photon interaction could be seen, with a signal-to-noise

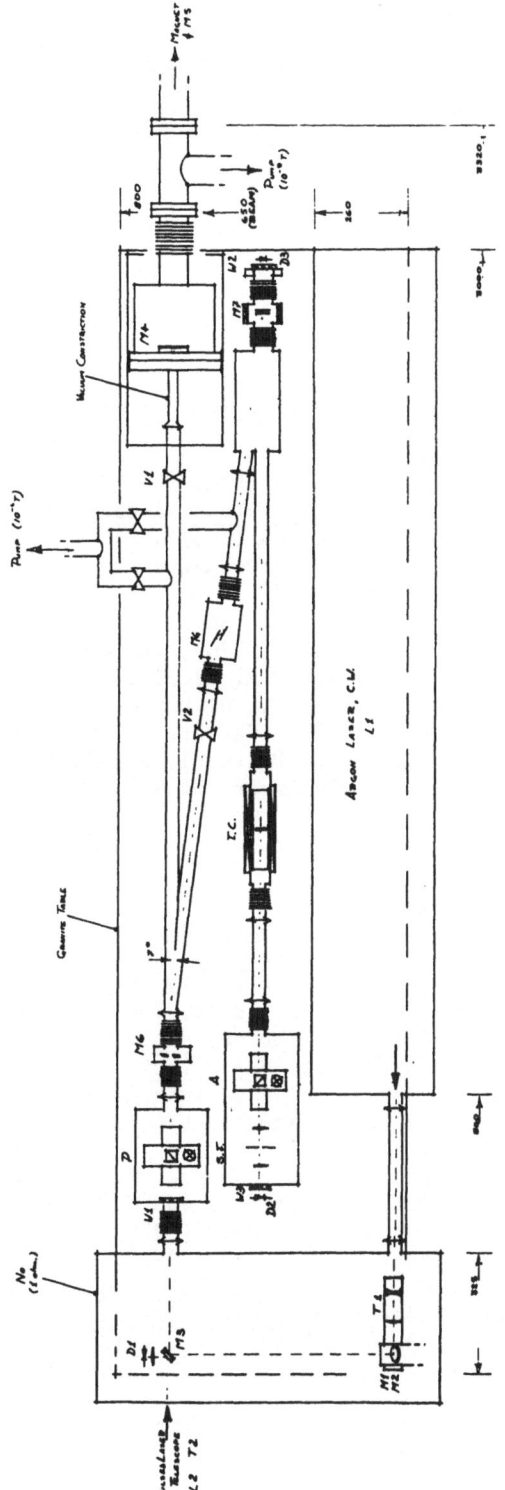

Fig. 8. Layout of the experimental apparatus proposed in order to observe the vacuum birefringence induced by a magnetic field.

ratio of one, by taking data for one month, on the hypothesis of a 6 m long, 80 kG magnet, of 500 reflections and of a signal modulation frequency $\omega_M > 25$ mHz.

REFERENCES

1. R. Karplus and M. Neuman, Phys. Rev. 80:380 (1950).
2. V. Costantini, B. De Tollis and G. Pistoni, Nuovo Cimento 2A:733 (1971).
3. H. Euler and W. Heisenberg, Z. Phys. 98:718 (1936).
4. V.S. Weisskopf, Mat.-Fys. Medd. Dan. Vidensk. Selsk. 14:6 (1936).
5. J. Schwinger, Phys. Rev. 82:664 (1951).
6. R. Novick, M.C. Weisskopf, J.R.P. Angel and P.G. Sutherland, Astr. J. 215:L117 (1977).
7. J.M. Jauch and F. Rohrlich, The theory of photons and electrons (Springer-Verlag, New York, 1976), 2nd edn., p. 298.
8. J.H. Taylor, L.A. Fowler and P.M. McCulloch, Nature 277:437 (1979).
9. J. Bailey, K. Borer, F. Combley, H. Drumm, C. Eck, F.J.M. Farley, J.H. Field, W. Flegel, P.M. Hattersley, F. Krienen, F. Lange, G. Lebée, E. McMillan, G. Petrucci, E. Picasso, O. Runolfsson, W. von Rüden, R.W. Williams and S. Wojcicki, Nucl. Phys. B150:1 (1979).
10. J. Calmet, S. Narison, M. Perrottet and E. de Rafael, Rev. Mod. Phys. 49:21 (1977).
11. M. Delbruck, Z. Phys. 84:144 (1933).
12. B. De Tollis and G. Pistoni, Nuovo Cimento 42A:499 (1977).
13. P. Papatzacos and K. Mork, Phys. Rep. 21:81 (1975).
14. J.M. Jauch and F. Rohrlich, The theory of photons and electrons (Springer-Verlag, New York, 1976), 2nd edn., p. 466.
15. S. Kahane, T. Bar-Noy and R. Moreh, Nucl. Phys. A280:180 (1977).
16. E. Iacopini and E. Zavattini, Phys. Lett. 85B:151 (1979). E. Iacopini and E. Zavattini, Experimental project to detect the vacuum birefringence induced by a magnetic field, preprint CERN-EP/78-162 (1978).
17. S. Carusotto, E. Iacopini, P. Lazeyras, M. Morpurgo, E. Polacco, F. Scuri, B. Smith, G. Stefanini and E. Zavattini, CERN Proposal D2 (1980), and Addendum (1983).
18. S.L. Adler, Ann. Phys. (NY) 87:559 (1971).
19. M. Born and L. Infeld, Proc. R. Soc. London Ser. A 143:410 (1934).
20. E. Iacopini and E. Zavattini, Nuovo Cimento 78B:38 (1983).
21. W.H. Watson, Proc. R. Soc. London Ser. A 125:345 (1929).
22. A.M. Grassi Strini, G. Strini and G. Tagliaferri, Phys. Rev. 19D:2330 (1979).
23. E. Iacopini, B. Smith, G. Stefanini and E. Zavattini, Nuovo Cimento 61B:21 (1981).

24. E. Iacopini, G. Stefanini and E. Zavattini, Appl. Phys. 32A:63 (1983).
25. S. Carusotto, E. Iacopini, E. Polacco, G. Stefanini and E. Zavattini, Opt. Commun. 42:104 (1982).
26. S. Carusotto, E. Iacopini, E. Polacco, G. Stefanini, F. Scuri and E. Zavattini, preprint CERN-EP/83-181 (1983), to be published in J. Opt. Soc. Amer.
27. D.R. Herriott, H. Kogelnik and R. Kompfner, Appl. Opt. 3:523 (1964).
28. D.R. Herriott and J.J. Schulte, Appl. Opt. 4:883 (1965).
29. M.A. Bouchiat and L. Pottier, Appl. Phys. B29:43 (1982).
30. J.H. Cole, Appl. Opt. 19:1023 (1980).

POTENTIAL OF SUPERCONDUCTING TUNNEL JUNCTIONS

AS DETECTORS IN NUCLEAR AND PARTICLE PHYSICS

A. Barone, S. De Stefano, R. Vaglio and S. Vitale

Istituto di Cibernetica, CNR, Arco Felice; Istituto di
Fisica, Università, Napoli; Dipartimento di Fisica
Università di Salerno, Salerno; Dipartimento di Fisica
Università di Genova, Genova; INFN, Sezione di Genova

1. INTRODUCTION

This lecture is devoted to a brief outline of the possibilities offered by superconductivity to the field of radiation detection in different contexts. The idea of investigating such a perspective is quite old and several more or less unsuccessful experiments were reported many years ago. The well developed studies of
superconducting tunnel junctions, supported by the high standard
technology reached for these devices, may open new possibilities
both for energy spectroscopy and for fast detection. Far from being conclusive such a claim requires a great deal of further work
and feasibility studies.

In the next two sections some basic ideas concerning the underlying physics of superconducting junctions are given. Possibilities and drawbacks concerning the application of quasiparticle tunneling in energy spectroscopy are outlined in Section 4. In Section 5 detection processes based on the fast switching of Josephson
junctions is breafly discussed.

2. SUPERCONDUCTING TUNNEL JUNCTIONS

Some of the essential differences between normal metals and
superconductors can be clarifyed by the energy-momentum diagrams
E-k sketched in Fig. 1. For the crude analysis, outlined here, for
superconductive tunneling we shall refer to this simple representation [1].

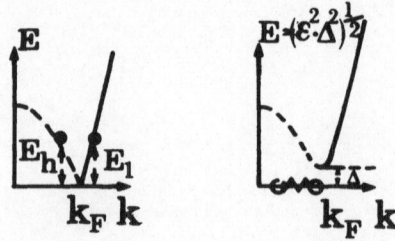

Fig. 1. Energy-momentum diagrams for a normal metal (left) and
for a superconducting metal (right) in which an energy
gap is present.

A wide family of reliable superconducting devices has been developed on the basis of tunneling junction structures [2].

Basically, a superconducting junction device consists of a
sandwich superconductor-dielectric-superconductor (S-I-S). In practice to realize such a structure a metal film is first deposited
under vacuum conditions on a suitable substrate, than an oxidation
process of such a base layer is performed leading thereby to an insulating (oxide) layer which constitutes the tunneling barrier. A
second metal layer is finally deposited to complete the junction
structure.

Let us assume a dielectric barrier film $\backsim$ 50 Å thick; what
do we expect as far as the electrical behavior of such a junction
is concerned? Let us refer first to the case in which the two metal films are normal (i.e. T > Tc). Obviously, as long as the two
metals are "macroscopically" separated they cannot influence each
other, however when they are brought together at a distance of a
few tens of angstroms than, due to quantomechanical tunneling,elec-
trons can travel through the barrier. We can describe the behavior of such a junction by referring to its current-voltage (I-V)
characteristics.

In a rather simplified approach we can get the main results
by a straightforward application of the Fermi "Golden rule". Our
normal metals will be characterized by density of states $N_1 (N_2)$
and Fermi factors $F_1 (F_2)$ (subscripts denotes the two metals). The
current from metal 1 to metal 2 will be given by

$$I_{1 \to 2} = \int_{-\infty}^{+\infty} |T|^2 N_1 F_1 N_2 (1 - F_2) \, dE$$

where T is the tunneling matrix element. Analogous expression
will hold for $I_{2 \to 1}$. For a voltage V across the junction (see
the energy-momentum diagrams of Fig. 2) the expression of the net
current will be

$$I_N = I_{1 \to 2} - I_{2 \to 1} = const \int_{-\infty}^{+\infty} N_1 (E + eV) N_2 (E) (F_1 - F_2) \, dE$$

where the tunneling matrix element has been considered as a constant (let us recall that the electron energy is of the order of
1meV with respect to that of the barrier of the order of 1eV).

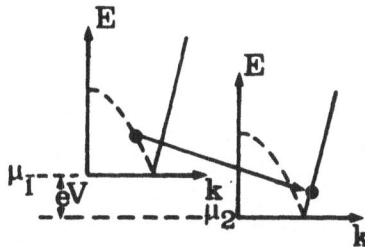

Fig. 2.   Energy-momentum diagrams representing the electron tunnel-
process between two normal metals.

If we make the further assumption of constant density of states
(indeed well verified for energies near the Fermi level) it follows

$$I_N = const \times V$$

the constant identifying a "normal" tunneling conductance G_{NN}.
Thus in the case of two normal metals separated by a tunneling
barrier a ohmic relation is found between the current flowing
through the junction and the corresponding voltage across it
(dashed line of Fig. 3).

Let us now consider the case when the two metals are superconductors (T<Tc). The densities of states have to be replaced
by their expressions in the superconductive state (BCS theory).
The energy-momentum curves representing the superconductive tunnel-

ing are sketched in Fig. 4. It is easy to show that in this case

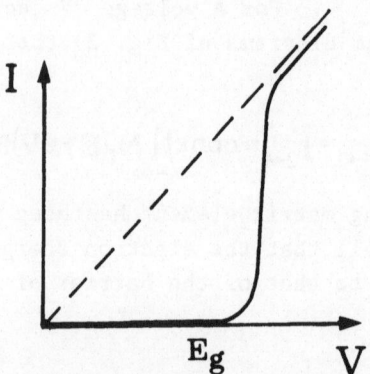

Fig. 3. Sketch of the current-voltage characteristics of a superconductive tunneling junction (equal superconductors).

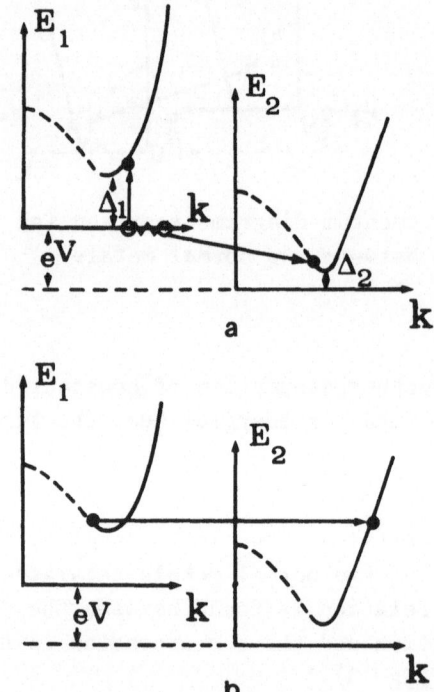

Fig. 4. Energy-momentum diagrams representing tunneling between
superconductors, involving Cooper pair breaking (a) and
direct tunneling of thermally excited quasi-particle (b).

the I vs V dependence becomes that of Fig. 3 where clearly appears
the superconducting gap structure.

3. JOSEPHSON JUNCTIONS

Let us procede in the study of superconducting tunneling referring to the essential phenomenology. We shall confine ourselves
to simple euristic arguments adressing the reader to a proper literature [2]) for more formal aspects of theory and fundamental implications.

In the previous Section we have briefly seen the behavior of
two superconductors divided by a distance (dielectric barrier) of
about 50 Å. What happens when such a distance is further reduced
down to say ∽10 Å? To give an oversimplified picture of the junction behavior in such a new situation let us refer again to the
current voltage characteristics (Fig. 5). It can be seen that, in

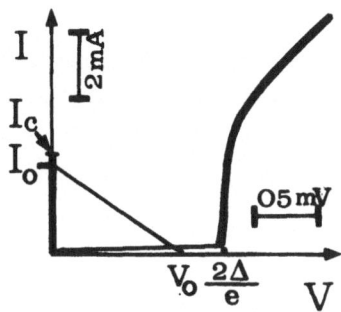

Fig. 5.   Current-voltage characteristics of a Josephson tunnel
          junction (Sn-Sn$_x$O$_y$-Sn). It is also indicated a load line
          suitable for fast switching detector.

addition to the finite voltage branch corresponding to the single
electron tunneling previously discussed, a current at V=0 occurs
with a maximum value I_c.

Let us briefly summarize the basic phenomenological relations
(Josephson equations) which accounts for this behavior.

Let us first recall that a superconductor is described as a
whole by a macroscopic wave function of the type $\Psi = \rho^{\frac{1}{2}} e^{i\varphi}$
where ρ is the density of paired electrons (Cooper pairs) and
$\varphi = \varphi(\underline{r}, t)$ the macroscopic phase. With this assumption it is
$|\Psi|^2 = \rho$. When the two superconductors are at a distance of

$\sim$ 10 Å the two wave functions Ψ_1 and Ψ_2, which characterize the two superconductors 1 and 2 respectively, have a nonvanishing overlap through the barrier (Fig. 6); such a correlation results in a resistanceless flow of Cooper pairs via tunneling effect (see also Fig. 7).

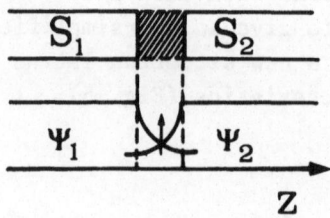

Fig. 6. Sketch of a superconducting junction and qualitative representation of the overlap of the two macroscopic wave functions.

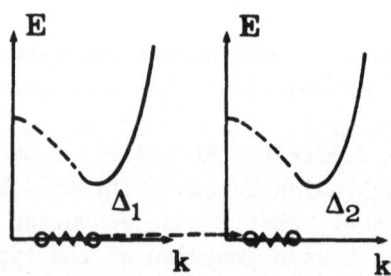

Fig. 7. Energy-momentum diagram representing the tunneling of Cooper pairs.

The corresponding pairs current I_J is related to the relative phase $\varphi = \varphi_1 - \varphi_2$ as follows:

$$I_J = I_c \sin\varphi \qquad (1)$$

there is also a time and space modulation of φ by voltage and magnetic field respectively:

$$\frac{\delta\varphi}{\delta t} = \frac{2e}{\hbar} V \qquad (2)$$

$$\nabla\varphi = \left(\frac{2ed}{\hbar}\right) H \times \underline{n} \qquad (3)$$

where d is the sum of the London penetration depths, λ_1, and λ_2, of the two superconductors and the barrier thickness t (i.e. $d = \lambda_1 + \lambda_2 + t$); n is a unit vector normal to the plane of the junction. Eqs. (1-3) summarize the Josephson effect. By inspection of Fig. 5 we see that a current flows with zero resistance up to a critical value I_c; when such a value is exceeded the junction switches on the single electron tunneling branch. Indeed if we set in (2) $V = 0$, φ is consequently constant and a current I_J given by (1) will flow in the junction. This is the essence of the (d.c.) Josephson effect.

From Equation (3) it can be easily shown that the dependence of the Josephson current vs. applied magnetic field can be cast in a form of a diffractive Fraunhofer-like pattern:

$$I_c(\Phi) = I_c(0) \frac{\sin\pi\frac{\Phi}{\Phi_0}}{\pi\frac{\Phi}{\Phi_0}}$$

where Φ is the magnetic flux through the junction and $\Phi_0 = \frac{hc}{2e}$ is the flux quantum (2.07×10^{-7} G cm^2).

Equations (1-3) combined with the Maxwell equations lead (with a proper normalization of space and time coordinates) to the Sine-Gordon equation

$$\Delta_2\varphi - \frac{\delta^2\varphi}{\delta t^2} = \frac{1}{\lambda_J^2}\sin\varphi$$

where

$$\lambda_J = \left(\frac{\hbar c^2}{8\pi ed J_1}\right)^{1/2}$$

is the socalled Josephson penetration depth. When the junction dimensions L(lenght) and W(width) are W, L < λ_J the junctions are called "small" whereas for W, $L > \lambda_J$ are classified as "large". In the latter case it should be taken into account the effect of the magnetic field associated with the currents flowing into the junction. This circumstance leads to a screening effect with a confinement of the current through the barrier within a distance of $\sim \lambda_J$ to the edges of the junction.

In a useful, though oversimplified approximation, a Josephson junction can be represented by the equivalent circuit of Fig. 8 and the corresponding current balance equation

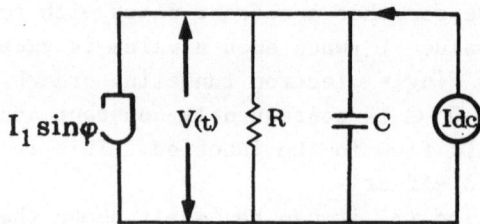

Fig. 8.   Equivalent circuit of a Josephson junction in the approximation of a resistively shunted model.

$$C \frac{dV}{dt} + \frac{V}{R} + I_c \sin\varphi = I_{dc}$$

where it is assumed that 1) the circuit is current biased, which reflects the circumstance that usually junction impedance is small compared to source impedance; 2) only a d.c. current I_{dc} is present; 3) no spatial variations of φ occur.

For superconductive junctions various kinds of materials have been successfully experienced such as Sn, Pb, In, Al as well as "hard" metals such as Nb, V and various alloys. Therefore, in the frame of possible applications as detectors, the choice of the materials is rather wide although is hardly conchievable a superconductor film thickness exceding a few μm. Junctions can be considered in which one of the electrodes could be in a bulk form and therefore made quite thick; in this case however the definition of the actual "sensitive" volume of the detector is not increased.

4. NUCLEAR SPECTROSCOPY WITH SUPERCONDUCTING TUNNEL JUNCTIONS

The idea of using superconductors for this purpose is quite old. It steems from the low value of the energy gap in superconductors ($\backsim$ 1meV) with respect to that of more usual semiconductors ($\backsim$ 1eV). Correspondly this allows, at least in principle, to predict a value ε of the average energy necessary to produce a pair of excitations (ion pair, electron-hole pair etc.) much lower than that of semiconductors. For a given amount of energy say ΔE lost in the detector the number of excitations will be $N = \Delta E/\varepsilon$. Thus larger N (for a given energy loss) in the case of superconductors would imply lower statistical fluctuation and ultimately a better energy resolution.

For a lack of time we shall not review here early interesting attempts using superconductor bulk and films nor the recent stimulating research carried out by Waysand and Coworkers on metastable superconductors in a form of small spheres. Rather we outline the possibilities offered by superconducting junctions.

In the case of semiconductors actual detectors can be made by realizing surface barriers (e.g. Si barrier devices) and large "depletion" depths obtained by compensation (e.g. Ge lithium drift devices). These devices have been successfully used since many years. For the superconductors the possible detection mechanism is somewhat analogous. The large number of quasi particles produced as a consequence of the energy E lost by the ionizing radiation via Cooper pairs breaking in a superconductor can be actually detected in a superconducting tunnel junction. Under proper bias conditions the excess quasi particles can tunnel through the junction. The tunneling of such quasi particles produces a current pulse with exponential decay. The time constant of the decay is related to an effective life-time of the quasi particle in the superconductor. The last quantity is function

of the specific superconducting material as well as of temperature
and other parameters. Problems of quasi particle recombination as
well as pair breaking mechanism and competitive effects due to
phonons fall into the wide and open field of nonequilibrium superconductivity[3,4]. Indeed it is such a three-fluid system whose
components are pairs, quasi particles and phonons, which represents
the quite intricate picture underlying the intrinsic mechanism of
radiation interaction in superconductors.

Simple models based on thermal effects in the sample have
been considered in the first significant attempts of realizing
proportional detectors based on superconducting junctions [5,6].
Further studies have pointed out the central role of nonequilibrium superconductivity and the implications of such theoretical
approach [7,8].

The basic ideas involved in the operation of a superconducting tunnel junction as a radiation detector can be understood by
inspection of Fig. 9. It is shown a family of I-V curves each

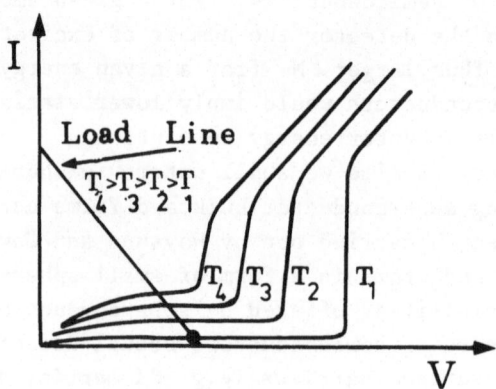

Fig. 9. I-V curves of a superconductive tunnel junction at increasing temperatures. Load line intersects the operating
point.

corresponding to a given sample temperature. If we assume the
crude picture of a heating effect such curves for increasing
values of T can be regarded as corresponding to increasing energies released by the incident radiation. It is worth noting that
in this mode of operation (energy spectroscopy) single electron
tunneling (Gievier type) is considered. However, due to the required high tunnel probability, very thin insulating layers are
considered so that in practice Josephson junctions are actually

employed where the zero voltage current (d.c. Josephson supercurrent) is quenched by a suitable externally applied magnetic field. In the experiments of ref. 5, 6 a Sn-Sn$_x$O$_y$-Sn junction was bombarded by α particles of 5.1 MeV. The total junction thickness and area were of 4-10^3 Å and 7x10^{-4} cm^2 respectively. The resistance at 4.2K was of 77mΩ . In this investigation it was considered that energy lost by the ionizing particles, after a relaxation time of the order of a nanosecond, resulted in a "heat pulse" which in turn produced a detactable current pulse. A suitable electronic chain provided the necessary amplification, shaping and analysis of the pulses. As a preliminary check a "causality correlation" between the occurrence of pulses and the interaction of the α particle with the junction was unambiguously established by a suitable mechanical shutter which could allow or avoid such an interaction. The pulses were recorded just on an oscilloscope and for a lack of a multichannel analyzer, a pulse height spectrum was roughly reconstructed in a histogram. In this experiment preliminary indications confirmed the low energy, ε , necessary for creating a pair of excitations.

More recent experiments, still using α particle source, have contributed to throw more light on the problem of superconducting junctions as potential spectrometers [9]. In these investigations Nb-Nb$_x$O$_y$-Pb high quality junctions were employed. These structures guarantee reliable properties of thermal cyclability and stability. Measurements were performed using also multichannel analysis. Far from being conclusive these investigations provided quite significant indications (see also interesting results reported at this School by a group working at SIN in Zurich).

First of all when a cross junction geometry was employed a rather "flat" spectrum; was observed such as that which could be inferred from previous investigations [5,6]. Could such effect be ascribed to quasi particle diffusion in the junction arms? It appears unrealistic since does not correspond to realistic values of the diffusion lenght. In any case a contribution to the spectrum from these regions appeared to be confirmed by a subsequent experiment using a suitable junction geometry. The sample was patterned, using photolithographic technique, to reduce drastically the ratio between the areas of barrier region and junction arms. In this case a α particle spectrum on the multichannel analyzer with a rather clear peak was obtained. Thus, from this work at first glance it may be concluded that in order to obtain an efficient and uniform charge collection two conditions have to be satisfied:

a) Tunneling probability should be high, that is the insulating layer has to be thin and uniform; b) junction geometry has to be properly designed in order to avoid diffusion of quasiparticles far from the junction barrier region. The apparent diffusion length $l = \sqrt{D\tau} = 100\,\mu m$ (D diffusion coefficient) inferred in the experimental conditions remains to be explained.

In these experiments 5×10^7 electronic charges were collected per MeV lost in the superconductor leading to an energy loss per collected electronic charge $\mathcal{E}$ as low as $\sim$ 20 meV. These results appear to be encouraging.

Among others, the main problem which deserves attention and a great deal of further work is that of the associated electronics. Indeed the request for a thin insulating layer results in high capacitance and low resistance of the device. This circumstance poses serious limits for the design of a low noise amplifier. Thus a new very low input impedance preamplifier is needed to reach performances appropriate to the ultimate intrinsic resolution of these devices. Finally it is also to be stressed that limits on physical dimensions, thickness and environmental constraints will confine the applicabilityof these counters to low energy radiation (i.e. X rays, Auger electrons, low energy β).

5. JOSEPHSON JUNCTION AS FAST SWITCHING DETECTORS

A further perspective of using a superconducting junction for nuclear and particle physics is that based on the Josephson effect [10-12]. In this case one can explore the possibility of having a new discriminator-like detector ("on-off" device) based on the extremely fast switching (order of picoseconds) occurring between a zero voltage state (d.c. Josephson current) and a finite voltage state (quasi particle branch). The junction should be polarized at a current value I_B just below the critical I_c. In fig. 5 it is shown the I-V characteristics and the proper loading (load line at $I = I_o$ and $V = V_o$). The operation principle should be the following. The radiation interacts with the junction producing a break of a number of Cooper pairs along a roughly cilindrical region coaxial with the radiation track. This reduction of Cooper pairs produces in turn a reduction of the Josephson critical current down to a value say $I_{CR} < I_B$. As a consequence a switch occurs from the zero voltage state to another stable point on the finite voltage branch of the I-V characteristics. The intrinsic switching time τ_s as estimated from measurements performed on the context of computer elements [2], is of the order of about 1-10 psec. This approach to the problem follows essentially the same track of the operation of superconducting tunneling "cryotrons". In that case the switching is produced by the magnetic field associated with a suitable control current which can be driven in a thin film overlying the junction. It can be shown that a voltage value say V_1 exists such that if it is chosen $V_o < V_1$ the junction switches back recovering the original zero-voltage as soon as the magnetic field is removed; in our case when the "effect" of the radiation is over.

To conclude whether such a detection operation is effective it is important to see to what extent the effect of radiation is actually equivalent to that of applied magnetic field in superconducting computer elements and whether the time involved for the pair breaking mechanism induced by the radiation and the actual switching is still of the order of τ_s. Also in this case a deep analysis on the basis of the modern concepts of nonequilibrium superconductivity is mandatory. Roughly speaking we can say that the energy release should proceed as follows. The radiation transfers almost all the energy directly to the superfluid component in a time of the order of $10^{-12} - 10^{-11}$ sec, the superelectrons system than relax the energy to the phonon system and a thermalization occurs in a time of the order of nanoseconds. It is indeed the former step of the process, the fastest, which should trigger the switching of the junction into the finite voltage state of the I-V curve.

The first and, to our knowledge, the only experimental results reported so far on the switching mode operation of a Josephson junction induced by nuclear radiation are due to a group of the Naval Research Laboratory in Washington [11, 12].

Let us summarize and analyze their results. The junctions used were Nb-Si-Nb type (let observe that silicon at liquid helium temperature behaves like a dielectric), the radiations employed were α particles produced by Am^{241} source. Nb and Si layers thickness was about $1\,\mu m$ and $80\,\text{Å}$ respectively; junctions areas A_o were ranging within $8-110\,\mu m^2$. The dependence of the efficiency upon I_B, A_o and energy loss was obtained by single counting rates. It was observed a drop of the efficiency for $A_o = 8\,\mu m^2$ at $I_B = 0.80\,I_c$ and for $A_o = 21\,\mu m^2$ at $I_B = 0.92\,I_c$. Moreover, selecting angles of incidence and degrading the energy of the particles by mylar absorbers, an increase of the efficiency with energy loss was evaluated. The explanation proposed by the authors to interpret their results is based on a heating model. The main points can be summarized as follows: 1) The energy loss due to the interaction of the radiation with the superconductor drives a portion, say A, of the junction area A_o into the normal state; 2) switching occurs whenever is $\frac{I_B}{I_c} > 1 - \frac{A}{A_o}$; 3) It is assumed that $A = K \frac{dE}{dx} \frac{1}{\cos\theta}$ with a value of $K \approx 3.5 \, \mu m^2/MeV$ per μm of Nb. On the basis of such considerations a reasonable fit of the data is obtained. Thus such investigations have produced a number of interesting results giving also the condition to get a "one-to-one" correspondence between a α particleimpinging the sample and a switching event.

Experimental results concerning time jitter and time delay of the switching as a response to radiation interaction with the junction are not yet available. As previously mentioned, such time response expected to be extremely short (say order of pseconds) requires a careful investigation in the light of nonequilibrium superconductivity. Such a potential ultrafast detection would require an associated special purpose superconducting circuitry such as that designed for the superconducting computer.

In principle one could immagine also, in the frame of a heating model to recover informations on the energy of the interacting radiation via a measurement of the $I_J(T)$ dependence. However more realistic it appears in this case to use just the single particle tunneling as discussed in the previous section, resorting to the Josephson regime operation only in those problems where events have to be detected in the shortest possible time independently of their energy.

6. CONCLUSIONS

What conclusion can we drawn from the simple ideas outlined in this lecture? It can be safely assumed that the whole topic of the potential of superconductivity in radiation detectors deserves great attention. Together with the stimulating perspectives we have stressed possible drawbacks as well. Among these we recall the necessary cryogenic environment, the extreme sensitivity of the Josephson current upon magnetic fields. Problems of geometrical efficiency certainly would arise in several contexts. However also this last point should be considered in the perspective of making junction arrays, telescope structures and other suitable geometrical configuration. The choice of the material will also represent an important point for the superconducting parameters (critical temperature, energy gap, etc.) and will involve further technological efforts.

REFERENCES

1. Tinkham, M., 1975, "Introduction to superconductivity", Mc. Graw-Hill, New York.
2. Barone, A. and G. Paternò, 1982, "Physics and Applications of the Josephson Effect", J. Wiley and Sons.
3. See Proc. NATO ASI, 1981, "Nonequilibrium Superconductivity, Phonons and Kapitza Boundaries", (K.E. Gray ed.) Plenum.
4. Rothwarf, A. and B.N. Taylor, 1967, Phys. Rev. Lett. 19, 27.
5. Wood, A. and B.L. White, 1969, Appl. Phys. Letters, 15, 237.
6. Wood, A. and B.L. White, 1973, Can. J. Phys., 51, 2032.
7. Kurakado, M. and H. Mazaki, 1980, Phys. Rev. B22, 168.
8. Kurakado, M. and H. Mazaki, 1981, Nuclear Instr. and Meth. 185, 141; 185, 149.
9. Darbo, G., G. Gallinaro, S. Vitale, A. Barone, A. Siri, S. De Stefano and R. Vaglio, 1984, Report INFN/TC - 84/2, January.
10. Barone, A. and S. De Stefano, 1982, Nucl. Instr. Meth., 202, 513.
11. Magno, R., M. Nisenoff, R. Shelby, A.B. Campbell and J. Kidd, 1982, IEEE Trans. Nucl. Sci. NS28, 3994; 1983, IEEE Trans. MAG, MAG19, 1286.

DEVELOPMENT OF A HIGH RESOLUTION SUPERCONDUCTING

DETECTOR FOR KEV RADIATION AT SIN

D. Twerenbold and A. Zehnder
Schweizerisches Institut fuer Nuklearforschung
(SIN), CH-5234 Villigen, Switzerland

V. Zacek
Technische Universitaet Muenchen
D-8046 Garching, Federal Republic of Germany

Since the resolution of a detector is limited by the statistics of the number of produced free charge carriers, a detector with a small minimal ionization energy per free charge is most favorable. The thousand times smaller gap of a superconductor (order of meV) compared to a semiconductor detector ($\Delta_{eff}(Si) = 2.94$ eV, $\Delta_{eff}(Ge) = 3.6$ eV)[1] offers the possibility for particle detectors with an improved energy resolution[2,3,4].

The decisive charge producing mechanism in a superconductor is the breaking up of Cooperpairs, which are the bound states of the conduction electrons responsible for the superconducting effect. The minimal ionizing energy of this detector would be of the order of the binding energy of a Cooperpair, which is twice the gap of the superconductor.

A solid state detector of this type would be an interesting device for measuring, with good precision, the 18 keV endpoint region of the β-spectrum of the tritium decay[5]. The deviation from the linear behaviour of the Kurie plot at the endpoint energy would measure the anti-electron-neutrino mass. Implanting tritium in a superconductor, similar to the experiment by Simpson,[6] offers, in addition to the high resolution, various advantages:

- no problems with source geometry (source=detector)
- no problems with source chemistry (atomic tritium would be deposited at an inter lattice point)
- no problems with excited final Helium-3 state (deexcitation of this state is measured also, because the deexcitation energy is much larger than the gap energy).

In the ideal case (no losses) the deposition of 10 keV energy would produce a charge of 10^7 electrons (assuming a gap of 1 meV), resulting in a resolution of 7.5 eV FWHM.

The energy loss mechanism of ionizing radiation in a superconductor proceeds in two steps: in the first step the energetic charged particle ionizes the atoms along its trajectory. In the case of x-rays a photo electron will be emitted (the photo cross section is dominant for large Z and energies below 20 keV). In the second step the excited electrons loose their energy by breaking up Cooperpairs and by exciting phonons (lattice vibrations). In contrast to semiconductor detectors, phonons play an important role in the superconducting detector, since the maximum phonon energy (Debyeenergy is of the order of 10 meV, i.e. larger then the superconducting gap. This implies that phonons with $E >= 2 \Delta$ are capable of breaking up Cooperpairs; they are thus not a priori lost as in semiconductor detectors. According to reference[7] the energy of the incident radiation ends up in electron and phonon states with energies close to the gap in about 1 psec.

Having produced the "free" charge carriers, the problem arises how these electrons can be extracted from the superconductor. One possible solution is the use of a tunneling junction, where the superconductor is separated from a (not necessarily superconducting) counterelectrode by a thin insulating barrier. Since it is then possible to apply a voltage between the superconductor and the counterelectrode, the "free" charge carriers can be collected by the quantum mechanical tunneleffect. In Fig. 1 the energy vs. density of states diagram of a S-I-N (superconductor-insulator-normalconductor) junction is shown. At a finite temperature excited electron states (quasiparticles) are populated due to the thermal breaking of Cooperpairs. Applying a proper voltage ($U < \Delta/e$) across the tunneling junction a net dc-current i_{th} will flow through the barrier. Similarly the ionizing radiation produces an excessive quasiparticle population giving rise to a current pulse δi_{signal}. The time de-

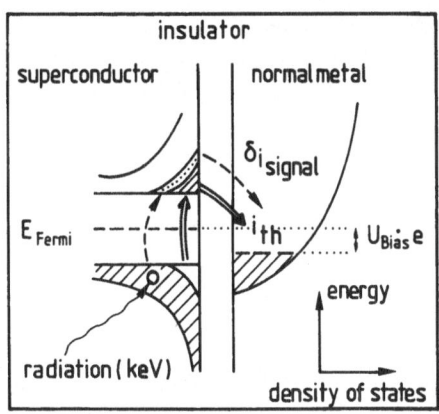

Fig. 1.   Energy diagram of a superconducting-insulator-
          normalconducting junction.

pendence of the signal is determined by the total decay
rate of the excessive quasiparticle population: i.e. the
sum of the tunneling rate, the recombination rate of quasiparticles to Cooperpairs, the phonon loss rates and other
possible rates yet to be determined. In Fig. 2 the current-
voltage characteristic of a Sn/Sn-oxide/Sn junction at a
temperature of 1.2 K is shown. The current i_{th} at a bias
voltage less than Δ/e is strongly temperature dependent.
For a voltage larger than the gap a higher order, temperature independent, current is superimposed[8]. Finally,
at voltages larger than twice the superconducting gap the
superconductivity is quenched and the current-voltage
characteristic changes into the normal linear (ohmic)
characteristic.

At SIN superconducting tunneling junctions of the
following composition have been fabricated:
Sn/Sn-oxide/Sn, Sn/Sn-oxide/Pb,
Aℓ/Aℓ-oxide/Sn and Aℓ/Aℓ-oxide/Pb.
The junctions are of the crossed type, evaporated from
heated boats onto quartz substrates using mechanical
masks. The oxides were grown thermally.

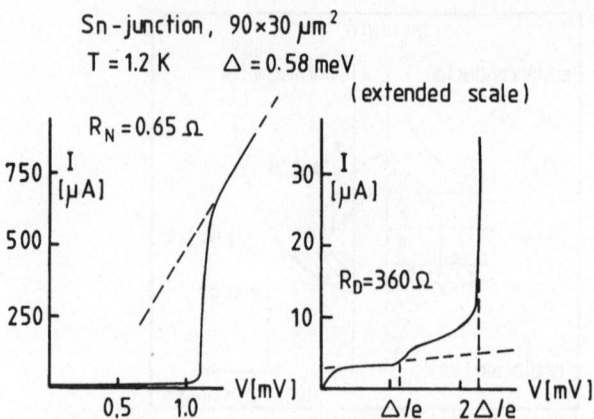

Fig. 2.   Current-voltage characteristic of a
           Sn/Sn-oxide/Sn junction at T = 1.2 K.

The best results were obtained with
Sn-junctions having the following dimensions:

    thickness   = .3 μm
    junction area   = 30 μm x 90 μm
    oxide thickness = approx. 20 Ångstroem

The junctions were exposed to two radioactive sources:
an Am-241 5 MeV alpha source and a Fe-55 5.9 keV x-ray
source. In the case of the alpha source, the 5 MeV alpha
particle looses about 70 keV of energy passing through the
thin (0.3 μm) layer of tin. The excess charge produced by
the ionizing radiation was amplified by a preamplifier[9],
and after appropriate pulseshaping it was analyzed in a
multichannelanalyzer. In Fig. 3 the pulseshape (output of
the preamplifier) and the pulse height spectrum of the
Am-241 source is shown. The peak of the spectrum (70 keV
energy loss) corresponds to a collected charge of roughly
10^7 electrons.

The low energy tail of the alpha spectrum is not due to electronic noise. We believe that it is caused by the inefficient charge collection due to the diffusion of charge carriers into the crossed tin stripes. Backscattering due to the bombardment of 5 MeV alpha particles into the substrate may also be the cause.

We detected also the 5.9 keV x-rays of a Fe-55 source, where $4 \cdot 10^5$ charge carriers were collected. The amplitude of the 5.9 keV signal is in good agreement with the amplitude of the 70 keV energy loss of the alpha source. From these results one can calculate an intrinsic effective gap of 15 meV, which is two orders of magnitude smaller than in a semiconductor detector. The result is also in agreement with the work of Darbo et al.[4]. Because of the thickness of the detector (0.3 μm), only part of the x-ray energy is absorbed, not allowing to see the photopeak. No effort was made to reduce the noise, which corresponds in our apparatus to about 10^5 electron noise equivalent.

We have also started to investigate the loss mechanisms of this detector using the alpha source. The effective collected charge Q_{eff} is given by the following formula:

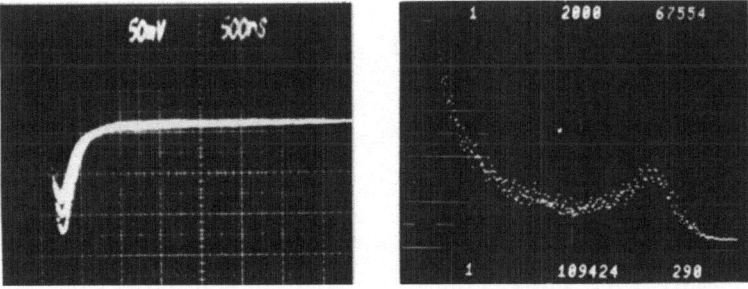

Fig. 3.   Pulseshape and pulse height spectrum with an Am-241 5 MeV alpha source.

$$Q_{eff} = Q_0 \frac{\gamma_T}{\gamma_T + \gamma_R + \gamma_x}$$

where Q_0 = E/Δ
E = deposited energy
Δ = gap
γ_T = tunneling rate
γ_R = recombination rate
γ_x = temperature independent loss rates
(e.g. phonon loss rates, quasiparticle loss
rates)

The tunneling rate can be calculated[10] from the normal-conducting quantities of the junction:

$$\gamma_T = \frac{1}{R_N \cdot e^2 \cdot N_o \cdot A \cdot d} \qquad (1)$$

where R_N = normal resistance of junction
e = electron charge
N_o = normal population density at Fermi energy
A = area of junction
d = thickness of junction.

Since the tunneling rate has to be as large as possible, one has to minimize the thickness of the oxide barrier and the thickness of the detector. The calculated value of the tunneling time τ_T (γ_T^{-1}) is approx. 2 µsec. The recombination rate is proportional to the density of the thermal excited quasiparticles[8], which in turn is strongly temperature dependent

$$N(T)_{thermal} \quad \alpha \quad \sqrt{T} \exp (-\Delta/kT)$$

where Δ = gap parameter
T = temperature
k = Boltzmann factor

This result implies, that the detector has to be operated at low temperature, ($\Delta/kT \gg 1$) such that the recombination rate becomes negligable compared to the tunneling rate (taking 1 µsec as the tunneling time, both rates become comparable in tin, roughly at a temperature of 0.8 K)[8].

Combining these equations we therefore get:

$$Q_{eff} = \frac{Q_o}{1 + \tau_T \cdot \gamma_R^o \sqrt{T} \exp(-\Delta/kT) + \tau_T \cdot \gamma_x}$$

In Fig. 5 we have plotted the temperature dependence of the alpha signal for 4 different junctions. The dashed curve represents the expected temperature dependence $Q_{eff}(T)$ for a tunneling time of 2 μsec and two values for γ_x, namely $\tau_T \cdot \gamma_x = 0$ and $\tau_T \cdot \gamma_x = 15$. For the recombination time γ_R^o was chosen such, that $\gamma_R^{-1} = 20$ nsec at 2.17 Kelvin (see ref. 8).

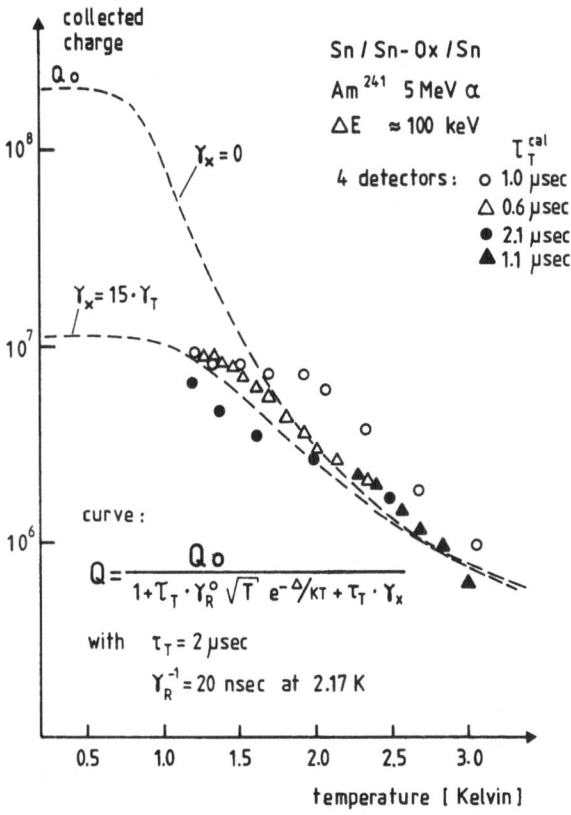

Fig. 4.   Temperature dependence of the Am-241 alpha signal in a Sn-junction. τ_T^{cal} is the calculated tunneling time of the junctions (equation (1)).

The agreement between the measured and the expected
temperature behaviour is satisfactory, although the various junctions differ rather strongly. Since these junctions
are not identical, one would expect different charge and
phonon loss processes.

Conclusion

Using two different types of radioactive sources,
we have shown that a superconducting tunneling junction
could be used as a high resolution detector for radiation
in the 20 keV energy domain, the effective gap was
measured to be 15 meV at T = 1.2 K. To achieve a resolution of 0.1 percent at 20 keV much has to be improved:
the noise of the electronic system has to be reduced by
an order of magnitude; reducing the temperature to 0.5 K
may still increase the signal. Also the geometric effects
of the detector have to be studied carefully in order to
control the energy deposition of the ionizing radiation
in the detector and to understand better the losses of
the produced quasiparticles.

REFERENCES

1. K.G. Mc Kay and K.B. Mc Alfee, Phys. Rev. 91:1079
    (1953).

2. G.H. Wood and B.L. White, Appl. Phys. Lett. 15:237
    (1969).

3. M. Kurakado and H. Mazaki, Phys. Rev. B 22:168 (1980).

4. G. Darbo et al., subm. to Nucl. Instr. and Methods
    (1984).

5. V. Zacek and A. Zehnder, June 1982, SIN internal
    report.

6. J.J. Simpson, Phys. Rev. D 23:649 (1981).

7. C.C. Chi et al., Phys. Rev. B 23:124 (1981).

8. P.W. Epperlein, thesis Universitaet Stuttgart (1977).

9. Preamplifier built by F. Pozar, SIN.

10. D.M. Ginsberg, Phys. Rev. Lett. 8:204 (1962).

NEW DETECTORS AND NEUTRINO MASS

S. Vitale*, G. Darbo*, G. Gallinaro*, S. Siri*, and
A. Barone[§]

*) Dipartimento di Fisica – Università di Genova
   I.N.F.N. – Sezione di Genova
§) Istituto di Cibernetica C.N.R. Arco Felice Napoli
   Istituto di Fisica, Università di Napoli

The different approaches to obtain new measures or experimental limits about the electron neutrino (or antineutrino) mass were already presented at this school by Lusignoli[1].

These experiments rely on the detection, in a nuclear decay, of small effects, due to neutrino mass on phase space dependence of transition rate when the available neutrino energy E_ν is coming to its minimum value. Atomic and molecular bonds on initial and final nucleus imply differences in energy, not at all negligible with respect to meaningful mass limits, that must be kept under control.

In experiments with external spectrometers, in which, for example, the β ray spectrum is measured, it is required first an excellent energy resolution and a "simple" source.

For calorimetric experiments, in which the total visible energy release in a transition, atomic effects included, has to be measured, the detector must also have an energy response uniform for fast electrons or atomic deexcitations.

Moreover the relaxation time, in which the final nucleus reaches his fundamental state in the detector material must be shorter than pulse formation time (this might be a problem at very low temperatures). The matching of source and detector has to be carefully investigated for the single experiment.

The performance of existing detectors, suitable for total absorption spectrometry, in this energy range, that is X ray detectors, is not satisfactory. Energy resolution is about an order of magnitude worse than electrostatic or magnetic spectrometers and uniformity in energy response cannot practically be investigated.

However Simpson[2] was able to set an upper limit to the neutrino mass of 55 ev with a detector energy resolution of about 500 ev F.W.H.M. but improvements of his limit are pratically impossible without a better resolving power.

In conclusion, feasibility investigations for new detectors with high resolution at low energies, under 20 kev, and suitable for total absorption spectrometry, seem to be worthwhile.

In principle a thermal calorimeter, in which the temperature increase due to a sudden energy release is measured, would be an ideal detector and will have an energy response, indipendent of the kind of energy loss. In practice such a detector is not out of question. At temperatures, below 1k, where thermal capacitances are very low a suitable semiconductor bolometer might worK with good performance. With such a detector at 4.2 k α particles produce signals very well above noise[3]. Anyway the question should be studied more in depth.

The energy resolution of a total absorption counter is limited by electronic noise due to the detector itself and to the amplifier and by statistical fluctuation in the number of primary processes involved in the pulse generation; that is, in a ionization chamber or a semiconductor detector by fluctuation in the number of electron charges collected for a given energy loss E.

The latter gives a lower limit for the energy resolution (Intrinsic energy resolution)

$$E_{FWHN} = 2.36 \left[F \, \mathcal{E}_0 \, E_0 \right]^{1/2}$$

where F is the Fano factor, and $\mathcal{E}_0$ is the average energy necessary to produce a signal corresponding to one electron charge. Spread due to electronic noise is equivalent to a number of electron charges and then proportional to $\mathcal{E}_0$.

Another relevant parameter is the minimum energy Δ necessary to produce an electron charge; in an intrinsic semiconductor Δ coincides with the energy gap between valence and conduction band. Low Δ values should favor indipendence of energy responce from the kind of excitation.

In present days and in the energy range of interest (under 20 KeV) the best suitable X ray detectors are:

a) Silicon or Germanium semiconductor detectors with a tipical energy resolution $E_{FWHM} \approx 500$ eV mostly due to electronic noise. Here $\varepsilon_o \approx 3$ e.v. and Δ is 1 ev for silicon and .7 ev for Germanium.

b) Gas scintillation proportional counters (Xenon). In this case $E_{FWHM} \approx 1$ KeV at $E_o = 20$ KeV and $E_{FWHM} \approx 350$ e.v. at $E_o = 2.5$ KeV. Here $\varepsilon_o = 22$ ev and $\Delta = 14$ ev ; the energy resolution is limited mainly by intrinsic statistical fluctuations. In fact in these detectors we "count" the light pulses due to the individual primary electrons and the noise level of such an amplifying system is very low.

To develope new better detectors two different approaches are possible.

First one can try to realize new detectors with a very small ε_o to have larger charge signals and smaller statistical fluctuations. This requires a small Δ value, but for a low level device thermal noise, Δ/KT must be high. So low temperatures are needed. Actually, quite often impurity and defects in materials introducing excitation energies $\Delta' \ll \Delta$ are the main source of noise .

Narrow gap semiconductors should be investigated and some cooled photodiode for infrared radiation, as InSb which has a $\Delta = .23$ ev. might be a good x ray spectrometer. However, because of tecnological problems, the quality of narrow gap semiconductors usually cannot be compared with that of silicon or germanium with negative effects on the expected detector performance.

As already discussed at this school by Barone[4] Superconducting tunnel junctions in which an energy gap of the order 1 m.e.v. is involved are quite actractive.

Experimental results have shown that values of $\mathcal{E}_o$ lower than 20 m.e.v. can be obtained with a detector thickness of $2-3\,\mu m$ of superconducting material (Lead or Niobium).

As tunneling probability depends on quasi particle density, $\mathcal{E}_o$ is increasing with detector thickness. Working at temperatures slightly over 1K the junctions are noisy and have a very low resistence. But at 1.1 K $\Delta/_{KT}$ is of the order of 10 only. Working at .3 K as it is possible with an He_3 cryostat, the junctions should drastically improve and a $\Delta E_{FWHM} \approx 50$ ev is foreseen.

A convenient choise might also be to work with superconducting materials with higher Δ such as V_3Si or Nb_3Ge at temperatures around $1-2K$.

The second approach that can be envisaged is to realize a semiconductor device, with normal gap materials and to provide it with some kind of internal gain. If the internal gain does not introduce much noise, as it happens in Gas Proportional Scintillation Counters, the device resolution should get near the limit due to statistical intrinsic fluctuation; for X ray energies of a few K.e.v this is a one order of magnitude gain. Phototransistor-like structures seem to be not reliable when high performance is essential. A suggestion comes from the recent developments of photoemissive surface tecnologies. A single crystal semiconductor with a very thin surface coating consisting of Cs and a small amount of oxigen has a very high photoemission efficiency. This is due to the fact that because of the surface treatment the conduction band is bent torward positive energies (Negative Electic Affinity, NEA)[5].

Providing in the semiconductor an electric field, electrons injected in the conduction band by the ionizing radiation will drift toward the NEA surface and then tunnel in the vacuum where they are accelerated to several tens K.e.v. and "counted", noise free by a silicon detector.

Extraction efficiency should be about 85%. The most suitable semiconductor seems to be GaAs that permit room temperature operation and has an $\mathcal{E}_o = 4.5$ ev. Silicon NEA surfaces are not impossible but require low temperature operation and present some technological problems.

In conclusion new X ray detectors with energy resolutions of the order of tens of e.v. should be possible. Due to the low $\triangle$ values their energy responce should be independent from excitation kind. If it is possible to solve in a convenient way the problems of source-detector matching in at least one case, a total absorption experiment on the neutrino mass, will be possible with resolution competitive to that of external spectrometer experiments[6].

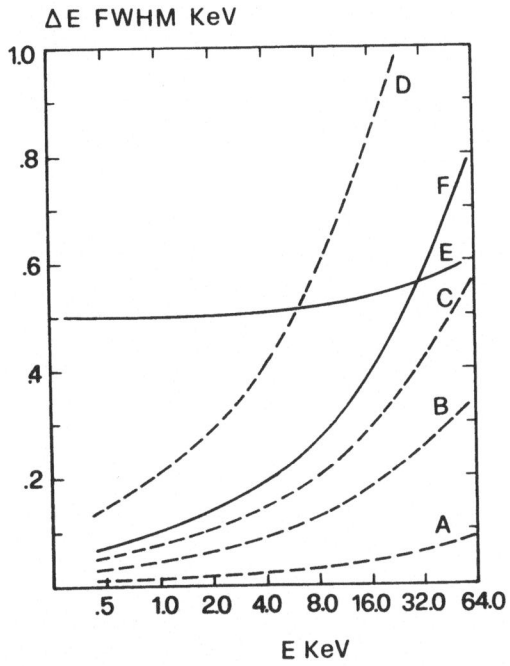

Fig. 1 - Limits on Energy resolution due to intrinsic statistical fluctuations (Dashed lines):
A-Superconducting Tunnel detector (ε_o = 20 meV),
B-Silicon semiconductor detector,
C-GeAs semiconductor detector,
D-Xenon gas ionization detector.
Total energy resolution (full lines):
E-semiconductor silicon detector
F-Expected resolution for a NEA GaAs detector and electron "counting".

REFERENCES

1. Lusignoli, "Neutrino mass," talk at this school.
2. J.J. Simpson Phys. Rev. D23, 649(1981).
3. G. Gallinaro, U. Valbusa, Private communication.
4. A. Barone, "Potential of superconducting tunnel junctions as detector in nuclear and particles physics" talk at this school.
5. For more detail in NEA tecnologies see for example: John S. Escher "NEA Semicondictor Photoemitters in: Semiconductors and Semimetals". Vol.15 R.K. Willerdson and A.C. Beer ed. Academic Press N.Y. 1981.
6. A. Lyubinov et al., Phys. Lett. 107B-19-(1981).

DIFFERENT FACETS OF MUONIC MOLECULES

Giovanni Fiorentini

Istituto Nazionale di Fisica Nucleare
Dipartimento di Fisica dell'Università
56100, Pisa, Italy

INTRODUCTION

Muonic molecules ($(X-\mu^{-}-X')$; X, X' = p,d or t) are in essence shrinked versions of the H_2^+ molecular ion, with a typical scale of length given by the muon Bohr radius, $a_\mu = a_o\, m_e/m_\mu \simeq 250$ fm, and with binding energies of up to few hundreds of eV. They are interesting by itself as a pure three body Coulomb problem, which can be studied, both theoretically and experimentally, to a high degree of precision. Beyond this, the study of muonic molecules is particularly important in connection with several fields of physics: the study of muon capture, $\mu^{-} + p \rightarrow \nu + n$, when the muon is bound in the (p-μ-p) molecule, is important in order to determine basic parameters of the weak interaction theory. Study of the two mirror fusion reactions,

1a) $d + d \rightarrow {}^{3}He + n$

1b) $d + d \rightarrow t + p$,

when the two nuclei are bound in the (d-μ-d) molecule can yield interesting information on charge symmetry violations in the p-wave d-d interaction near threshold. Also the spectroscopy of muonic molecules is a sensitive tool to investigate the tail of the nucleon nucleon potential. Finally, it has been pointed out that the fusion reactions occurring in muonic molecules - the so called muon catalyzed fusion - can be of interest for practical applications. In this respect it is worth mentioning the encouraging result of an experiment in progress at LAMPF.

There are several reviews covering the different aspects of the physics of muonic molecules [1-17] and I will mainly discuss the most recent developments after a brief introduction.

ENERGY LEVELS AND FORMATION PROCESSES OF MUONIC MOLECULES (THEORY)

The theoretical spectroscopy of muonic molecules is a field which has been mastered by Russian physicists since the beginning. Ya. Zeldovich in the early times, later S.S. Gershtein and more recently L. Ponomarev and his collaborators have developed methods of calculation of higher and higher accuracy. Particularly in the last few years the group of Ponomarev was able to describe the full spectrum of molecular levels with an accuracy of some meV on typical energies of order $e^2/a_\mu \sim 5000$ eV [18,19]. Some results are summarized in table I and II, where (J, v) denote the rotational and vibrational quantum numbers.

Table I. The mesomolecular spectrum according to: (a) the perturbative calculation; (b) the truncation method. The energies are in eV. (From ref. 18).

a

Jv	$(pp\mu)$	$(pd\mu)$	$(pt\mu)$	$(dd\mu)$	$(dt\mu)$	$(tt\mu)$
00	253.55	221.49	213.85	324.99	319.09	362.89
01	–	–	–	35.66	34.70	83.68
10	107.33	98.79	101.30 .	226.74	232.61	289.19
11	–	–	–	1.96	0.85	45.15
20	–	–	–	85.34	103.16	172.79
30	–	–	–	–	–	48.90

b

Jv	$(pp\mu)$	$(pd\mu)$	$(pt\mu)$	$(dd\mu)$	$(dt\mu)$	$(tt\mu)$
00	252.95	221.52	213.97	325.04	319.15	362.95
01	–	–	–	35.80	34.87	83.88
10	106.96	97.40	99.01	226.61	232.44	289.15
11	–	–	–	1.91	0.64	45.24
20	–	–	–	86.32	102.54	172.65
30	–	–	–	–	–	48.70

Table II. Hyperfine structure of the (d-μ-d) mesomolecule (a) and of the (d-μ-t) mesomolecule (b), from ref. 19. F is the total spin, the index N labels the different spin states with the same F, $\mathcal{E}_{Jv}$ is the hyperfine energy shift, in eV . w is the relative probability of forming the (J,v,F,N) state of the mesomolecule starting from a specific spin state of the mesoatom.

a

Jv	F	N	$\mathcal{E}^{FN}_{Jv}$,(eV)	$w^{FN}_{Jv}(\uparrow\downarrow)$	$w^{FN}_{Jv}(\uparrow\uparrow)$
00	$\frac{1}{2}$	1	0	0.1667	0.1667
	$\frac{3}{2}$	1	−0.0286	0.8333	0.0833
	$\frac{5}{2}$	1	0.0191	0	0.7500
01	$\frac{1}{2}$	1	0	0.1667	0.1667
	$\frac{3}{2}$	1	−0.0246	0.8333	0.0833
	$\frac{5}{2}$	1	0.0164	0	0.7500
10	$\frac{1}{2}$	1	−0.0169	0.2213	0.0560
	$\frac{1}{2}$	2	0.0070	0.0565	0.1384
	$\frac{3}{2}$	1	−0.0180	0.4436	0.1115
	$\frac{3}{2}$	2	0.0084	0.1119	0.2774
	$\frac{5}{2}$	1	0.0097	0.1667	0.4167
11	$\frac{1}{2}$	1	−0.0159	0.2222	0.0555
	$\frac{1}{2}$	2	0.0077	0.0555	0.1389
	$\frac{3}{2}$	1	−0.0161	0.4444	0.1111
	$\frac{3}{2}$	2	0.0079	0.1112	0.2778
	$\frac{5}{2}$	1	0.0082	0.1667	0.4167

b

Jv	F	N	$\mathcal{E}^{FN}_{Jv}$(eV)	$w^{FN}_{Jv}(\uparrow\downarrow)$	$w^{FN}_{Jv}(\uparrow\uparrow)$
00	0	1	0.0173	0	0.1111
	1	1	0.0282	0.0096	0.3301
	1	2	−0.1107	0.9904	0.0032
	2	1	0.0463	0	0.5556
01	0	1	0.0239	0	0.1111
	1	1	0.0312	0.0043	0.3319
	1	2	−0.1123	0.9957	0.0014
	2	1	0.0439	0	0.5556
10	0	1	0.0277	0.0007	0.0368
	0	2	−0.1039	0.1104	0.0002
	1	1	0.0162	0.0000	0.1111
	1	2	0.0249	0.0031	0.1101
	1	3	−0.1035	0.3303	0.0010
	1	4	0.0406	0.0000	0.1111
	2	1	0.0273	0.0056	0.1833
	2	2	−0.1041	0.5499	0.0019
	2	3	0.0447	0.0000	0.1852
	3	1	0.0433	0	0.2593
11	0	1	0.0445	0.0001	0.0370
	0	2	−0.1424	0.1110	0.0000
	1	1	0.0407	0.0000	0.1111
	1	2	0.0439	0.0002	0.1110
	1	3	−0.1422	0.3331	0.0001
	1	4	0.0501	0.0000	0.1111
	2	1	0.0443	0.0004	0.1851
	2	2	−0.1424	0.5552	0.0001
	2	3	0.0511	0.0000	0.1852
	3	1	0.0508	0	0.2593

A few points are to be remarked:

i) In contrast to the case of electronic molecules, there are very few rotational and vibrational levels. This is a consequence of the different ratio $m_{lepton}/m_{nucleus}$ in the two cases.

ii) The (p-μ-p) system has just two levels, with J=0 and J =1. These correspond, as a consequence of the Pauli principle, to "para" and "ortho" states respectively. The different nuclear spin content of the two states has important consequences when discussing the problem of muon capture.

iii) The (J,v) = (1,1) levels of the (d-μ-d) and (d-μ-t) have very small binding energies, in the range of eV, comparable to the typical energies of electronic systems. This suggests that resonant interactions between muonic and electronic molecules can occur, see below. It was very hard to establish theoretically the existence of these states, since it needed an accuracy of at least 0.1 eV in the calculations.

Formation of muonic molecules occurs through collisions:

(2)   (μ X) + X' → (X-μ-X') + energy.

The formation reactions are classified according to the way the binding energy of the muonic molecule is released. Besides the usual Auger process (Fig. 1a), it is also possible to have energy transfer through the excitation of the vibrational and rotational levels of ordinary molecules (Fig. 1b). This latter process is only possible

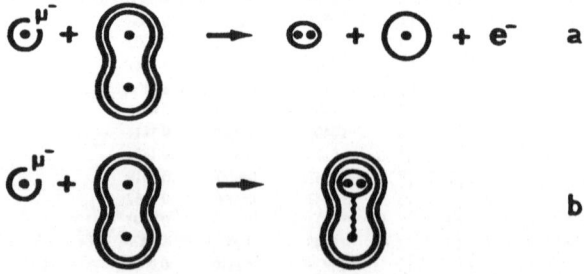

Fig.1 Formation of the mesomolecule   through (a)
      the Auger, non resonant, process and (b)
      the resonant process.

if the mesomolecule has weakly bound levels, with a binding energy E_b smaller than the dissociation energy of the Hydrogen molecule, a few eV. This is what occurs for the case of $(d-\mu-d)$ and $(d-\mu-t)$ systems, as we noted above. Due to the quantization of the vibrational and rotational energies this process can only occur if a resonance condition among E_b, the kinetic energy of the muonic atom, E_{kin}, and the quantum jumps of the electronic molecule, ΔE, is satisfied, see Fig. 2 :

(3a) $\quad E_{kin} = E_{res}$

(3b) $\quad E_{res} = \Delta E - E_b$.

It is intuitively clear that the resonant structure can be used as a tool in order to perform accurate determinations of the energy levels of muonic molecules. Also it is clear that under suitable conditions the resonant process can considerably increase the rate of formation of muonic molecules. Several recent developments in the field of muonic molecules are grounded on this idea.

The formation of muonic molecules via the two mechanisms has been studied in detail by Ponomarev and his collaborators[20,21]. The same group also calculated the rate of the ortho-para transition in the $(p-\mu-p)$ system, which is important for the study of muon capture[22].

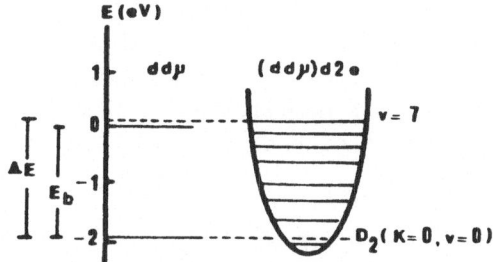

Fig. 2. Scheme of the resonant formation of the $(d-\mu-d)$ molecule.

THE (p-μ-p) MOLECULAR ION: A PROBE FOR THE STRUCTURE OF THE WEAK HAMILTONIAN.

The V-A theory predicts a strong dependence of the muon capture rate,

(4) $\bar{\mu} + p \rightarrow \nu + n$,

on the spin state of the muon relative to the proton. Measurements in gas deal essentially with muons bound around the proton in the singlet, 1s, atomic state, and thus give information on Λ_s, the capture rate in the singlet state. In order to get informations on the capture rate in the triplet, Λ_t , it is necessary to use high density targets, so that the (p-μ-p) molecule can be formed. The muon capture rates in the ortho and para state are, respectively [23]:

(5a) $\Lambda_{om} = 2 \gamma_0 (3/4 \Lambda_s + 1/4 \Lambda_t)$

(5b) $\Lambda_{pm} = 2 \gamma_p (1/4 \Lambda_s + 3/4 \Lambda_t)$,

where γ_0 and γ_p are factors connected with the overlap probability of the muon and one of the protons ($2\gamma_0 = 1.009 \pm 0.001$; $2\gamma_p = 1.143 \pm 0.001$). Clearly, in order to disentangle between the contributions of Λ_s and Λ_t it is necessary to know the molecular state of the muon when the measurement takes place. It is well known that the molecule is formed in the ortho state[24]. The transition probability to the para state, though expected to be small[22,25] , can however have significant effects on the interpretation of the results. Recently an experiment has beeen performed [26], with such an accuracy that it is possible to recognize in the data even the effect of the ortho-para transition.

The study of muon capture in atomic and molecular systems is now in such a stage that it is possible to get quite precise information on the structure of the weak Hamiltonian. I will simply refer to the lectures of Martino, Piccinini and Massa for a detailed discussion of this problem.

THE (d-μ-d) SYSTEM: A PROBE FOR THE STUDY OF NUCLEAR FORCES

As a consequence of the existence of the weakly bound states with $(J,v) = (1,1)$ it is possible to perform accurate experimental determinations of the binding energy [21]. One measures the formation rate of the muonic molecule as a function of the temperature of the target, i.e. as a function of the kinetic energy of the (d μ) atom involved in the reaction. By studying this dependence it is possible to deduce the resonant energy E_{res} of eqs. (3) and thus to derive the binding energy.

Actually, the experimental study of the molecular formation occurs through the detection of the nuclear fusion reaction which takes place in the molecule. The nuclei bound in the muonic molecule behave as if they were in a plasma with a density $\rho_{ef} \sim (a_\mu)^{-3}$, which is about 10^7 times the liquid hydrogen density, and with a temperature corresponding to their vibrational energy, $K T_{eff} \sim E_{vib} \sim 100$ eV. Under such extreme conditions, comparable to those inside a white dwarf, the nuclei fuse rapidly, the fusion rate being orders of magnitude larger than the muon decay rate and the rate of molecular formation. In summary, through the observation of the nuclear fusion reactions one gets informations about the energy levels of the muonic molecule.

In this way a few years ago it was possible to measure the binding energy of the $(J,v) = (1,1)$ level of the (d-μ-d) molecule[27]:

(6) $E_b = (2.196 \pm 0.003)$ eV .

More recently, effects associated with the hyperfine structure of the (d-μ-d) molecule have been observed by a SIN-Wien collaboration[28]. Through the analysis of their experiment it should be possible to derive the full spectrum of the hyperfine sublevels of the (1,1) level with an accuracy of about 1 meV.

Beyond their intrinsic interest, these measurements are of importance in order to establish the very long range part of the interaction potential between hadrons, i.e. at distances larger than the Compton wavelength of the pion. I would like to discuss this point in some detail. In the last few years several authors pointed out that long range interactions between hadrons could arise from the exchange of new (hypothetical) light particles or from the exchange of two gluons. Muonic molecules are of interest for the study of these anomalous interactions[29]. This occurs since the two hadrons are at distance **large** enough to neglect the contribution of the (not understood) short range part of the hadronic interaction and still **small** enough that the effect of the anomalous interaction can be detected. For example, measurements of the hyperfine structure of the (d-μ-d) molecule with an accuracy of the order of 1 meV can be sensitive to the exchange of a pseudoscalar particle with mass up to 1 MeV if its coupling constant is $g^2_{ps} < 10^{-1}$ [29].

Ponomarev and his collaborators [30] pointed out that the measurement of the branching ratio of reactions (1) when the deuterons are bound in the muonic molecule can be quite interesting in order to test the charge symmetry of nuclear reactions. Indications of violation of this symmetry in the p-wave interaction of two deuterons have been obtained in low energy d-d scattering. However, the extraction of the p-wave contribution to the cross section is quite difficult. On the contrary, this contribution can be <u>directly</u> measured when the fusion proceeds from the muonic molecule. The point is that: i) the muonic molecule is mainly formed in the (J,v)= (1,1) state as a consequence of the dominance of the resonant formation cross section, ii) electromagnetic transitions to lower states are slower than the rate for fusion, iii) since the total P-parity equals $(-1)^J$ the angular momentum L of the relative nuclear motion is odd for J = 1 and the dominant contribution arises from the p-wave. In conclusion, it looks that muonic molecules can be used as a "multipole meter" in the study of fusion reactions.

Recently an experiment performed in Gatchina[31] reported a large charge symmetry violation in the d-d muon catalyzed fusion. The same experiment also presented a nice measurement of the probability of the muon sticking to the fusion charged products. The relevance of this measurement will be made clear in the next section.

Also, the experimental study of the SIN-Wien group is of importance for the interpretation of experiments on muon capture in deuterium, particularly for the knowledge of the spin state of the (dμ) atom when the capture takes place. For the first time it was possible to accurately measure the rate of the hyperfine transition in collisions of the (dμ) atoms with deuterium and to deduce a value for collisions with hydrogen. Thus the long missing experimental information on the hyperfine populations of (dμ) atoms is provided (at least at low temperature), not only for pure D_2 but also for H/D mixtures.

THE (d-μ-t) MOLECULE. ENERGY FROM MUON CATALYZED FUSION?

The resonant formation (Fig. 1b) can occur also in this system and it was theoretically estimates, already in 1977 [32], that in a suitable mixture of deuterium and tritium a muon could catalyze about one hundred nuclear fusion reactions during its lifetime. In order to appreciate this number, let me remember that in the Alvarez experiment[33] where muon catalyzed fusion was observed for the first time the yield of detected fusions per muon was at the level of 10^{-2}. An experiment performed at Dubna[34] obtained a lower limit for the molecular formation rate $\Lambda_{mol} > 2 \; 10^8 s^{-1}$, which supports the idea that a muon can catalyze a lot of fusions in suitable conditions.

Recently an experiment has been performed at LAMPF[35], which provides very nice, quantitative data on the catalysis in deuterium tritium. By working at high deuterium-tritium density (up to 60 % of liquid hydrogen density) copious 14 MeV neutron production was observed, demonstrating up to 70 fusions per muon. On these grounds it was possible to derive that, under <u>optimal</u> conditions, a muon will catalyze about one hundred fusions, at least for temperature of the target < 543°K. The main results of the experiment are reported in table III and in Figs. 3 and 4.

Table III. Parameters critical to muon catalysis in deuterium tritium mixtures.

Parameter	Théory	Previous Experiment[34]	LAMPF Experiment[35]
Fusion rate	$10^{12}s^{-1}$	--	--
Rate of atomic capture in Hydrogen	$\sim 10^{11}s^{-1}$	--	--
Rate of transfer to tritium λ_{dt}	$2 \times 10^8 s^{-1}$	$(2.9 \pm 0.4) \times 10^8 s^{-1}$	$(2.8 \pm 0.1) \times 10^8 s^{-1}$
Mesomolecule formation rates:	$3 \times 10^4 s^{-1}$(non-resonant) $= 10^8 s^{-1}$(resonant, max) (temp. dependent)	$> 10^8 s^{-1}$ (no temp. dep. seen)	temp. dependent dtμ-d and dtμ-t parts:
$\lambda_{dt\mu-d}$	--	--	$(6.9 \pm 0.4) \times 10^8 s^{-1}$(534K)
$\lambda_{dt\mu-t}$	--	--	$(3.0 \pm 0.3) \times 10^8 s^{-1}$(534K)
Probability of sticking to He w_s	{ 0.0091 0.0086	--	$(7.6 \pm 0.5) \times 10^{-3}$
Relative He/H capture probability	~ 2	--	~ 6 (tentative)
Rate of transfer to Helium, λ_m	$5 \times 10^8 s^{-1}$	--	$\sim 4 \times 10^8 s^{-1}$ (tentative)
Optimal tritium concentration	~ 0.5	--	~ 0.5 (broad peak)
Optimal efficiency η_μ^{opt}	~ 50	--	90 ± 9 fusions/μ(534K)

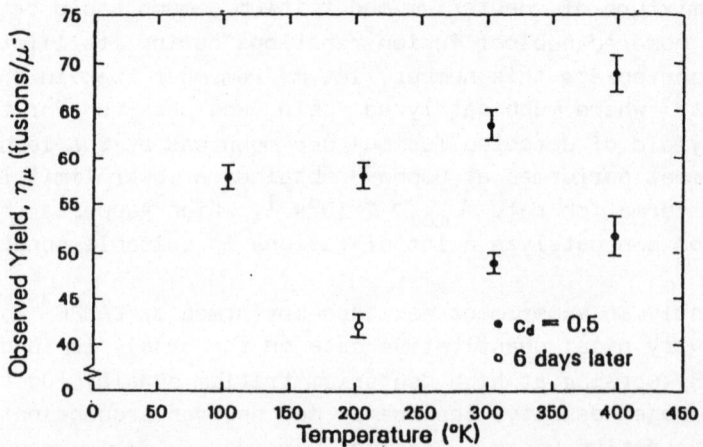

Fig.3 Muon catalyzed fusion yield (η_μ) for an equimolar
       d-t mixture at 0.6 liquid hydrogen density (ref.36)

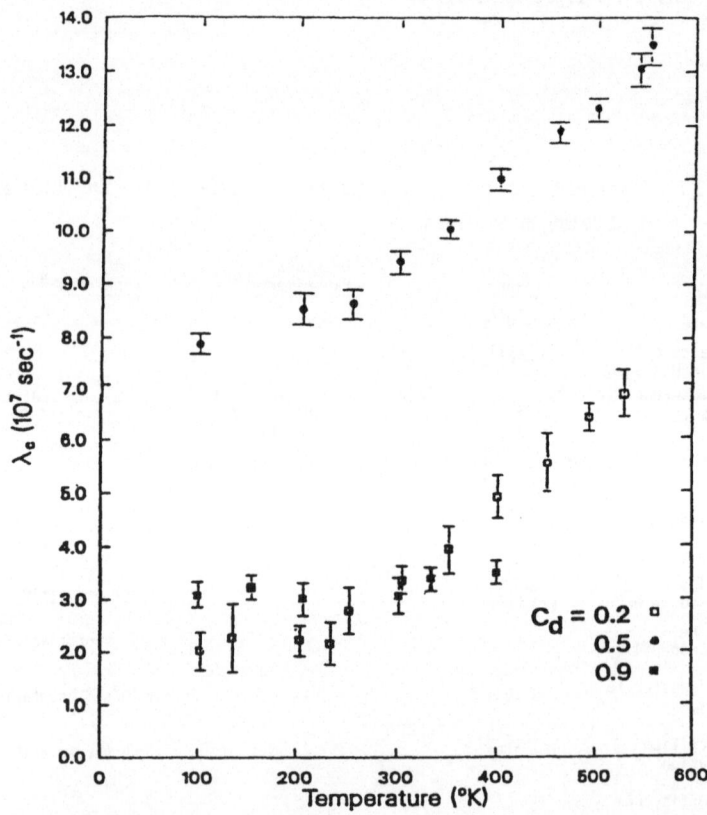

Fig.4 Muon catalysis cycle rate (λ_c) as a function of
       deuterium concentration and temperature (ref. 35)

It is worth observing that the bottle-neck of the chain is the sticking of the muon to the ^{4}He nucleus produced in the fusion. The measured value of the (effective) sticking probability, $w_s^{exp} =$ $(7.6 \pm 0.5)10^{-3}$ [35], which is in fair agreement with the theoretical prediction [36,37] , $w_s^{th} = (9 \pm 1) 10^{-3}$, implies that no matter how fast are all the other reactions one cannot hope for more than 150 fusions per muon.

It is worth reporting that it has been estimated that 100-200 fusions per muon would be enough in order to build a reactor with a positive energy balance[38].

Also independently of this, the LAMPF experiment is quite interesting since it opens the way to the study of the spectroscopy of the (d-μ -t) molecule, and more generally it gives informations on the kinetics of muons in deuterium-tritium mixtures. One can expect to have in the future a quite clear experimental picture thanks to the additional information which will arise from extensions of the LAMPF experiment and from other experiments planned and/or in preparation at SIN, Dubna and Leningrad. In this respect I would like to refer to the lecture by P. Kammel on the preliminary data obtained at SIN in a low density experiment.

CONCLUSIONS

Let me summarize the main points of this discussion:
i) essentially thanks to the efforts of Ponomarev's group one has now a detailed understanding of the spectroscopy of muonic molecules.
ii) The LAMPF experiment has produced interesting results on the kinetics of muon reactions in deuterium-tritium mixtures. Some of the results, particularly the temperature dependence of the formation rates of the (d-μ -t) molecule, are not understood theoretically. Additional experimental data, particularly at higher temperature and/or lower densities, will be quite useful. In this respect the SIN-Wien experiment is particularly interesting, since it explores a density interval which is complementary to that being investigated at LAMPF.

REFERENCES

1. Ya.B. Zeldovich and S.S. Gerstein, Usp. Fiz. Nauk. 71:581 (1960), (English transl. Sov. Phys. Uspekhi 3:593(1961).
2. S.S. Gerstein and L.I. Ponomarev, in "Muon Physics", V.W. H-ghes and C.S. Wu, eds. , Academic Press, New York (1975).
3. V.P. Dzhelepov, Atomnaja Energija 14:27 (1953).
4. E.H.S. Burhop, in "Electron and Ionic impact Phenomena", H.S.W. Massey, E.H.S. Burhop and H.G. Gilbody, eds., Oxford (1974).
5. A. Bertin, A. Vitale and A. Placci, Nuovo Cimento 5:423(1975).
6. A. Bertin et al. , Nuovo Cimento 72A:225 (1982).
7. "Mesons in Matter", Proceedings of the Int. Symposium on Meson Chemistry and Mesomolecular Processes in Matter, 7-10 June 1977.
8. L.I. Ponomarev, in "Proceedings of the 6th Int. Conference on Atomic Physics", Riga 17-22 August 1978, Plenum, New York (1979).
9. S.I. Vinitsky and L. I. Ponomarev, Physics of Elementary Particles and Atomic Nuclei 13:1336 (1982).
10. S. Tesh, Kernenergie 25:97(1981).
11. J. Meyer-ter-Vehn, Physik. Blätter 35:211 (1979).
12 L.I. Ponomarev, in Proceedings of the 10th European Conference on Controlled Fusion and Plasma Physics, Moscow, 14-19 September 1981.
13. L. Bracci and G. Fiorentini, Phys. Reports 86:170 (1982).
14. J. Rafelski, in "Exotic Atoms '79" , K. Crowe et al. , eds, Plenum, New York (1980).
15. W.H. Breunlich, Nucl. Phys. 353A:201 (1981).
16. G. Fiorentini, Nucl. Phys. 374A:607 (1982).
17. L. I. Ponomarev, to be published in the Proceedings of the 3rd International Conference on Emerging Nuclear energy systems, Helsinki, Finland, June 1983.
18. V. Melezhik et al., JETP 52:353(1981), and S. I. Vinitsky et al., JETP 55:400(1982). For a comprehensive review of the methods of calculation see S. Vinitsky and L. Ponomarev, Sov. Jour. of Part. and Nuclei 13:557(1982).
19. D.D. Bakalov et al. , JETP 52:1629(1981).
20. L.I. Ponomarev and M.P. Faifman, JETO 44:886(1976).
21. S. Vinitsky et al., JETP 47 : 444(1978).
22. D.D. Bakalov et al., Nucl. Phys. 384A:302(1982).
23. H. Primakoff, in "Muon Physics", C.S. Wu and W.H. Hughes, eds., Academic Press, New York (1975).

24. E. Zavattini, in "Muon Physics", C.S. Wu and W.H. Hughes, eds. Academic Press, New York (1975).

25. S. Weinberg, Phys. Rev. Letters 4:585(1976).

26. G. Bardin et al., Nucl. Phys. 352A:365(1981), G. Bardin et al., Phys. Letters 104B:320(1981). See also G. Bardin, Ph.D. Thesis, Univ. of Paris-Sud (1982) and J. Martino, Ph.D. Thesis, Univ. of Paris-Sud (1982).

27. V.B. Bystritsky et al., JETP 49:232(1979).

28. P. Kammel et al. , Phys. Letters 112B:319(1979) and P. Kammel et al. Phys. Rev. 28A:2611(1983).

29. L. Bracci, G. Fiorentini and R. Tripiccione, Nucl. Phys. 217B: 215(1983).

30. L.N. Bogdanova et al. Phys. Letters 155B:171(1982).

31. D.V.Balin et al. , Phys. Letters 141B:173(1984).

32. S.S. Gerstein and L.I. Ponomarev, Phys. Letters 72B:80(1977).

33. L.W. Alvarez et al. , Phys. Rev. 195:1127(1957). See also L.W. Alvarez, in Adventures in Experimental Physics" α:72 (1972).

34. V.M.Bystritsky et al. Phys. Letters 94B:476(1980).

35. S. Jones et al. ,Phys. Rev. Letters 51:1157(1983).

36. S.S. Gerstein et al., JETP 53:872(1981)

37. L. Bracci and G. Fiorentini,Nucl. Phys. 364A:383(1981).

38. Yu. V. Petrov, Nature 285:466(1980) and in Proceedings of the XIV LNPI Winter School, Leningrad (1978).

NEW EXPERIMENTAL RESULTS ON MUON CATALYZED

FUSION IN LOW DENSITY DEUTERIUM-TRITIUM GAS

presented by Peter Kammel

W. H. Breunlich[1], M. Cargnelli[1], P. Kammel[1], J. Marton[1]
P. Pawlek[1], J. Werner[1], J. Zmeskal[1], K. M. Crowe[2]
J. Kurch[2], R. H. Sherman[3], C. Petitjean[4], A. Janett[4]
H. Bossy[5], and W. Neumann[5]

Österreichische Akademie der Wissenschaften, Wien[1]
Lawrence Berkeley Lab., Univ. of California, Berkeley[2]
Los Alamos National Lab., Los Alamos[3]
Schweizerisches Inst. für Nuklearforschung, Villigen[4]
Physik Dept., Technische Univ., München, Garching[5]

INTRODUCTION

The general theoretical ideas for resonant formation of muonic hydrogen molecules have been discussed by G. Fiorentini at this school. Let us now take a closer look at this remarkable process, which is the starting point for the renewed interest in the field of muon catalyzed fusion. As an example Figure 1 presents the level scheme of the $d\mu t$ formation process

$$\mu t + D_2 \rightarrow [(d\mu t)dee]^* \qquad (1)$$

One should notice the different energy scales involved in this transition: <u>some hundred eV</u> on the muonic side (the extraordinary weakly bound $d\mu t$ state, responsible for resonant formation, has a binding energy of only -640 meV), <u>some hundred meV</u> for the energy spacing between vibrational levels on the electronic side. Resonant molecule formation is only possible if the value of the thermal kinetic energy of the initial μt atoms (dashed region in Figure 1) allows a transition between the indicated levels. As this thermal energy is of the order of <u>some meV</u> (at low temperatures) experimental physics has a precise method of determining the energies for resonant formation by observing the rate of mesomolecule formation as a function of temperature. Due to the high accuracy reached, tiny energy splittings, which usually are completely negligible on the muonic energy scale,

449

$$\mu t + D_2 \rightarrow [(d\mu t)dee]^*$$

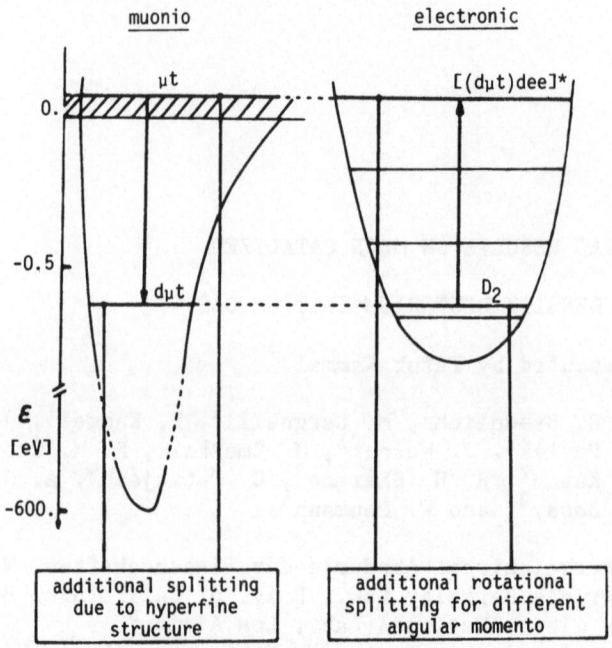

Fig. 1. Energy levels for resonant dμt formation. At resonance the energy gain for the transition from the scattering μt state to the dμt molecular level is spent on the excitation of the whole complex [(dμt)dee]. Additional energy splittings of some 10 meV influencing this energy balance are indicated for the muonic and the electronic part of this transition.

become important (see Figure 1). On the muonic side these small energy splittings are dominated by the hyperfine structure of muonic atoms and molecules. The sensitivity of resonant molecular formation to these splittings is particularly interesting, because hyperfine effects in muonic hydrogen have not been accessible to direct experimental observation before.

In the case of pure deuterium, where the resonant formation process was first discovered[1,2], the recent work of our group established muon catalysis as a powerful experimental method for the investigation of mesomolecular processes and bound muonic systems[3,4,5]. An accurate spectroscopy allows the determination of the binding energy and even the hyperfine structure of a dμd state within 1 meV or better (this corresponds to some ppm of the typical mesomolecular binding energy). These experiments also provide the long missing accurate experimental information about the transition rate between the hyperfine states of muonic deuterium. This detailed information about mesoatomic and mesomolecular processes[4,6] has considerable impact on the experiments on the elementary muon capture process in deuterium (discussed by M. Piccinini at this school).

Concerning the investigation of the $d\mu t$ fusion cycle, experimental research is just beginning. As mentioned by G. Fiorentini, very promising results have been obtained so far. Hyperfine effects, however, lead to considerable complication of the already complex sequence of muon induced reactions in deuterium-tritium mixtures and have been neglected in the first analysis of the two published experiments[7,8]. We have pointed out[9] that the consideration of hf effects is indispensable not only for a quantitative description of the kinetics, but also for an interpretation of these experiments in terms of basic processes. Therefore we have performed an experiment at SIN to understand the molecular formation mechanism as well as the processes of the $d\mu t$ fusion cycle in their full complexity.

THE EXPERIMENT AT SIN

Basic Considerations

The most important features of the kinetics of muon catalyzed reactions in D/T mixtures are sketched in Figure 2, including hyperfine effects both in molecular formation and in the various hyperfine transitions [for details see Reference 9]. A general description of the various steps of the reaction sequence is given in the recent review by L. I. Ponomarev[10].

As discussed in the introduction (Figure 1) the hyperfine components $\lambda_{dt\mu}^F$ of the $d\mu t$ molecular formation rate are expected to have quite different resonant behavior (F is the total spin of the μt atom). Due to the different binding energies the rates $\lambda_{dt\mu}^{F,D_2}$ and $\lambda_{dt\mu}^{F,DT}$ (for $d\mu t$ formation in collisions of μt atoms with D_2 and DT molecules, respectively) are also different. Thus, the rates observed experimentally are a superposition of these rates, weighted by the D_2 and DT molecular concentrations:

$$c_d \lambda_{dt\mu}^F = 2 \; c_{D_2} \lambda_{dt\mu}^{F,D_2} + c_{DT} \lambda_{dt\mu}^{F,DT} \tag{2}$$

The hyperfine transition can be induced by the processes:

$$\mu t \; (F=1) + t \rightarrow \mu t \; (F=0) + t \qquad (\text{rate } \lambda_t^{\mu t}) \tag{3}$$
$$\mu t \; (F=1) + d \rightarrow \mu t \; (F=0) + d \qquad (\text{rate } \lambda_d^{\mu t})$$

leading to an overall hyperfine transition rate

$$\lambda_{hf} = c_t \lambda_t^{\mu t} + c_d \lambda_d^{\mu t} \tag{4}$$

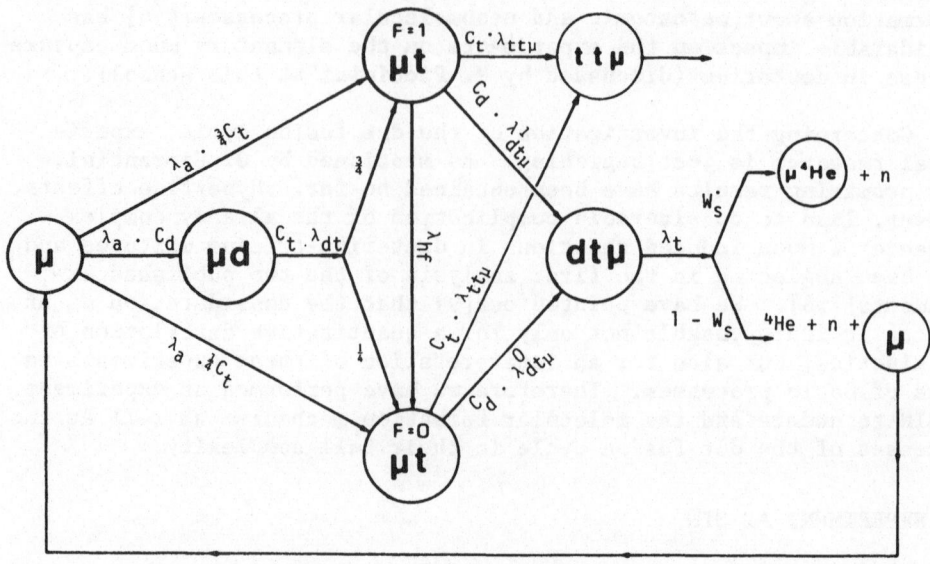

Fig. 2. Extended model of the kinetics of muon catalysis including
hyperfine effects. c_d and c_t denote atomic concentration of
deuterium and tritium, respectively. All rates normalized
to liquid hydrogen density, $\rho_o = 4.2.10^{22} cm^{-3}$. Details of
the dµd and tµt fusion branches have been omitted for
clarity.

A high, temperature independent value for $\lambda^{\mu t} = 9.10^8 s^{-1}$ is predicted theoretically[11]; no experimental results exist. The rate
$\lambda^{\mu t}_d$ is usually neglected[12].

Experimental Method

To disentangle all these rates a careful experimental strategy
is necessary. We chose the following target conditions: low temperature (The narrow thermal energy spread of the initial µt atoms
guarantees a high sensitivity to resonant structures in the dµt
formation); low target density (As most of the decisive rates are
proportional to the gas density, we are able to resolve reaction
rates which correspond to lifetimes of 1–10 ns in liquid hydrogen);
and a wide range of tritium concentrations (Varying c_t allows us to
unfold the two rates $\lambda^{F,D_2}_{dt\mu}$ and $\lambda^{F,DT}_{dt\mu}$).

The experimental setup was similar to that used in our previous
experiments [see Reference 4 for details]. The target cell was a
silver coated copper cylinder (vol. ∿1000 cm³) enclosed in an evacuated secondary containment which provided thermal insulation as well

as safety. The gas pumping and storage system was situated inside a glove box in the experimental area. Tritium and deuterium gas were passed through separate Pd-filters before being mixed in the target cell. 70 kCi of tritium was used in this experiment[13]. A mass spectrometer was connected directly to the target cell, and samples were taken before the after each run to determine the proportions of the different isotopic molecules. The target cell was surrounded by plastic scintillators, which had an overall efficiency for detecting electrons from muon decay of about 75%. Two neutron detectors with a total efficiency of ∿2% for 14 MeV neutrons, the signature of dt fusion, were used.

The measurements were performed in the μE4 beam of SIN in fall 1983. Most of the time we used a gas target of 1% liquid density. Typically 10% of the electronic stop signals corresponded to muons stopping in the target gas, and the true gas stop rate was ∿1 kHz. 18 data points in gaseous mixtures were measured and about 30000 fusion neutrons were collected per data point (see Table 1).

Experimental Results

We concentrate here on the subset of our data in which the tritium concentration was high (c_t = 88%). As can be seen from Figure 2, the complexity of mesomolecular processes is drastically reduced in this regime: complications due to muon transfer from deuterium to tritium become insignificant, since most of the muons are captured directly by tritium nuclei. Figure 3 shows data from one neutron detector for 88% tritium concentration and two target temperatures. These time spectra of dt fusion neutrons after muon stop were obtained by applying neutron/gamma discrimination and a high energy cut at 8.5 MeV neutron energy, which excludes most tt fusion neutrons (calculated neutron detector efficiency for 14 MeV

Table 1. Target Conditions for Data Runs

Density %	Tritium conc. %	Temperature			
		30 K	100 K	200 K	300 K
1	96	*			*
	88	*	*	*	*
	50	*	*	*	*
	40				*
	23				*
	10	*	*		*
	2	*			*
0.5	40				*

neutrons 0.38 ± .04%). Accidental background was less than 3% of the real events[14]. In spite of the low target density it was possible to obtain clean spectra of fusion neutrons starting immediately at the moment when the muon stops in the target. This is demonstrated by the characteristic shape of the energy spectra of 14 MeV neutrons. In Figure 3 capture neutrons due to wall stops are suppressed by a factor of ∿300 by requiring the detection of muon decay electrons in one of the electron detectors in a delayed time window (0.3–3.7 μs) after the fusion neutron. As a further test, we increased the suppression of capture neutrons compared to fusion neutrons by a factor of ∿8 by requiring a delayed coincidence signal of the two electron telescopes (each consisting of two plastic scintillators) instead of a single detector; the shape of the neutrons spectra did not change.

The spectra in Figure 3 show two distinct components with different decay rates. The amplitudes of the two components are not very sensitive to temperature variations, but the decay rate of the short-lived component increases strongly with higher temperatures. This behavior cannot be understood within the simple model used in previous works[7,8,15], which provides only one time constant at high c_t.

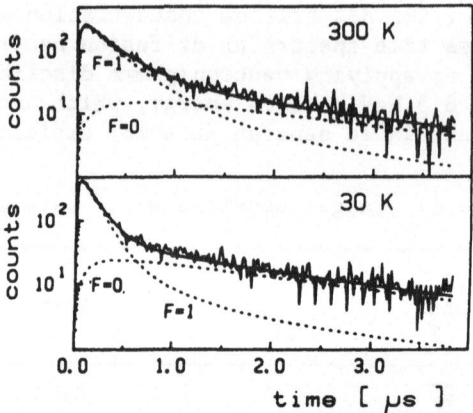

Fig. 3. Time spectra of dt fusion neutrons detected after muon stop (ρ=1% of liquid hydrogen, c_t=88%, bin width 16 ns). Neutron energy threshold of 8.5 MeV and detection of delayed decay electron were required. Data normalized to same number of muon stops. Solid curves correspond to fit with generalized reaction kinetics (Figure 2) using parameters given in Table 2.

To interpret our data we use the extended theoretical model of the muon induced fusion cycle discussed above (see Figure 2). In agreement with recent theoretical[10] estimates, thermalization times of muonic μt atoms were neglected. This model can adequately describe the observed data, as can be seen from the fit of our data with the exact solution of the reaction sequence of Figure 2. The build-up effect seen at the first 50 ns, an interesting phenomenon in itself, was included by a rate λ_a. These fits (Figure 3 and Table 2) indicate a surprising behavior of the observed rates, in particular an unexpected temperature dependence of the hyperfine transition rate λ_{hf}. At 300K we find,

$$\lambda_t^{\mu t} \leq \frac{\lambda_{hf}}{c_t} = (360 \pm 15) \cdot 10^6 s^{-1} \tag{5}$$

which is only one third of the theoretically predicted value[11] (This limit is deduced from Equation 3 making no assumption about $\lambda_d^{\mu t}$). The resonant $d\mu t$ formation rates, on the other hand, only show weak temperature dependence.

DISCUSSION AND OUTLOOK

We have presented results of our investigations of muon catalysis in deuterium/tritium mixtures. The first experimental observation of different components in the spectra of dt fusion neutrons at high tritium concentrations demonstrate the complexity of the muon induced fusion cycle. Our unexpected results prove that low density experiments are essential if one is to understand the basic processes of muon catalysis. An interpretation of our data in terms of a model including hyperfine effects is given.

For the hyperfine components $\lambda_{dt\mu}^F$ of the $d\mu t$ formation rate, first isolated in this experiment, high rates were found in the entire temperature range studied, with no dramatic temperature dependence. The formation rate from the upper hyperfine state of the

Table 2. Fit Results using Extended Kinetic Model including Hyperfine Effects. (Fitrange 0.048 μs – 4.0 μs, 239 Data Points). Important Fixed Parameters (all Rates Normalized to Liquid Density)

Temp.	Rates in $10^6 s^{-1}$			Events
	$\lambda_{dt\mu}^0$	$\lambda_{dt\mu}^1$	λ_{hf}	
30	30(3)	834(90)	642(27)	9700
300	45(4)	891(100)	317(13)	11000

μt atom, the highest mesomolecular formation rate yet observed, approaches $10^9 s^{-1}$. The importance of our experiment for the theoretical understanding of the resonance mechanism is demonstrated by a comparison with a recent calculation of dμt formation rates[16]. While the theoretical molecular formation rates[16] give a reasonable fit of the experimental data from reference 8, the same set of rates does not even qualitatively agree with the more detailed information, i.e. the hyperfine components of the dμt formation rate, observed directly in our experiment.

Evidence for qualitative new features in the hyperfine transition process was found. The present theoretical description[11] disagrees with our observed rates, both in magnitude and in temperature dependence. Refined theoretical models, including the possibility of a resonance mechanism for hyperfine transitions[17] have to be considered now.

Finally, we stress the preliminary nature of our results. The analysis of the region of high tritium concentration has already revealed surprising features and can be considered as a first step towards the full understanding of the d-μ-t fusion cycle. Obviously, the next step is a full description of the data over the entire range of tritium concentrations, disentangling the various components of molecular formation and hyperfine transition rates. Our data at low c_t (Figure 4) demonstrate that the characteristics of the observed neutron time spectra depend strongly on tritium concentration. The

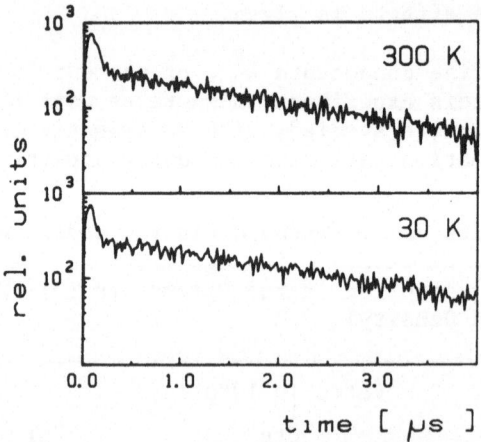

Fig. 4.   Time spectra of dt fusion neutrons after muon stop (ρ=1%) of liquid hydrogen, c_t=10%). Other parameters same as for Figure 3.

decay rate of the dominant time component is increasing with decreasing tritium concentration. If no additional effects turn up at low c_t this would suggest an unexpectedly high value of $\lambda_d^{\mu t}$ to induce additional hf transitions. Thus, a detailed verification of models has to await the analysis of our complete set of data.

Support by the following institutions enabled these investigations and is gratefully acknowledged: Austrian Academy of Sciences, Austrian Science Foundation, Swiss Institute for Nuclear Research, U.S. Department of Energy, and the German Bundesministerium für Forschung und Technik.

REFERENCES

1. S. I. Vinitskii, Sov.Phys.JETP, 47:444 (1978).
2. V. M. Bystritskii, Sov.Phys.JETP, 49:232 (1979).
3. P. Kammel, Phys.Lett., 112B:319 (1982).
4. P. Kammel, Phys.Rev., 28A:2611 (1983).
5. J. Zmeskal, Atomkernenergie/Kerntechnik, 43:193 (1983).
6. W. Bertl, Atomkernenergie/Kerntechnik, 43:185 (1983).
7. V. M. Bystritskii, Sov.Phys.JETP, 53:877 (1981).
8. S. E. Jones, Phys.Rev.Lett., 51:1757 (1983).
9. P. Kammel, Atomkernenergie/Kerntechnik, 43:195 (1983).
10. L. I. Ponomarev, Atomkernenergie/Kerntechnik, 43:175 (1983).
11. A. V. Matveenko, Sov.Phys.JETP, 32:871 (1971).
12. Note, however our recent experimental results for a similar process[6] and possible theoretical explanations[10].
13. The tritium was provided by the U.S. Department of Energy.
14. A time dependence of accidentals results from the delayed electron coincidence condition used and is discussed in [4].
15. S. S. Gershtein, Sov.Phys.JETP, 51:1053 (1980).
16. M. Leon, Phys.Rev.Lett., 52:605 (1984).
17. Compare the theoretical explanation for hyperfine transitions in H/D mixtures mentioned in [10].

PROGRESS IN MUON CATALYZED FUSION

AT JINR IN DUBNA (USSR)

Hubert Schneuwly

Institut de Physique
de l'Université
CH-1700 Fribourg, Switzerland

Up to now most of all experimental and theoretical work done
in the field of muon catalyzed fusion has been performed at the
JINR. For a review, see e.g. ref. 1. Coming back from a short visit
in Dubna, the organizers of this School asked me to report on what
is going on in this field at JINR.

EXPERIMENT

The Dubna machine producing the muons for these experiments
has been shut down in 1979 to be improved. Like to other places
the improvement program had to be revised during execution introducing some delay. The experimentalists hope that the accelerator
will produce at the end of 1984 at least the same muon current as
five years ago and increase it regularily during 1985.

The first experiment planned at the improved JINR accelerator
on muon catalyzed fusion is designed for the study of the $tt\mu$ system using a liquid tritium target. The experimental group is con-
ducted by Academician V.P. Dzhelepov. Among the members of his
group one finds well known people like V.M. Bystritsky, V.G. Zinov,
A. Gula, V.A. Stolupin, J. Woźniak and others.

The liquid tritium target with 35 cm^3 active volume and the
gas filling system is ready for work [2]. A detailed study of the
neutron registration efficiency in the experiments of the synthe-
sis reactions $tt\mu \rightarrow {}^4He + 2n + \mu$ and $dt\mu \rightarrow {}^4He + n + \mu$ has already been performed [3].

Like in earlier experiments the final neutron time distribution dN/dt will be measured and its analysis should allow the determination of the interesting parameters [4]. Whereas in pure deuterium the fusion rate is much larger than the $dd\mu$ molecule formation rate, the situation could be different in pure tritium where λ_f and $\lambda_{tt\mu}$ may have the same order of magnitude [5,6]. The formula describing the final neutron time distribution [4] is symmetric in λ_f and $\lambda_{tt\mu}$. Therefore any fit of the data will give two solutions corresponding to the transposition $(\lambda_f, \lambda_{tt\mu}) \leftrightarrow (\lambda_{tt\mu}, \lambda_f)$ which are indistinguishable on the basis of a χ^2 analysis alone.

The $tt\mu$ molecule formation rate $\lambda_{tt\mu}$ is however density dependent, whereas λ_f is not. Liquid tritium changes its density by nearly 20% by increasing its temperature from 20°K to 30°K. The measurement of the final neutron time distribution at two densities different by about 20% should permit to distinguish between the $tt\mu$ molecule formation rate $\lambda_{tt\mu}$ and the corresponding fusion rate λ_f if the two rates are different by more than about a factor of two [4].

THEORY

The muon capture in atomic hydrogen and helium isotopes is calculated by G.Ya. Korenman. Calculations in a semi-classical approximation using the stationary phase method with an effective potential in adiabatic representation have already been published [7,8]. The calculations are carried on using more refined methods.

Calculations by V.S. Meleshik, L.I. Ponomarev and M.P. Faifman on the elastic scattering processes of the type

$$p\mu(\uparrow\downarrow) + p \rightarrow p\mu(\uparrow\downarrow) + p$$

for all hydrogen isotopes have already been published [9]. Calculations of scattering cross-sections on molecules of the hydrogen isotopes have already be performed too [10].

The calculations of inelastic scattering processes like

$$
\begin{aligned}
d\mu(\uparrow\uparrow) + d &\rightarrow d\mu(\uparrow\downarrow) + d \\
t\mu(\uparrow\uparrow) + d &\rightarrow t\mu(\uparrow\downarrow) + d \\
d\mu(\uparrow\uparrow) + d &\begin{cases} \rightarrow d\mu(\uparrow\uparrow) + d \\ \rightarrow d\mu(\uparrow\downarrow) + d \end{cases} \\
d\mu \quad\quad + t &\rightarrow d + t\mu
\end{aligned}
$$

are in progress. V.S. Meleshik, A. Gula, J. Wozniak, M. Bubak, L.I. Ponomarev and others are the main authors of these calculations where one has to solve six to seven hundred more or less coupled equations. First results are expected this year. In the calculations of these inelastic cross-sections, the fact that one takes into account higher orders gives important corrections[2] to the approximate calculations.

The exact calculation of the weakly bound states $J = \nu = 1$ of the ddμ and dtμ systems is of great importance for the description of the whole muon catalyzed fusion process. The binding energies in this bound state are of the order of 1.91 eV for the ddμ system and 0.64 eV for dtμ. The precision the theoreticians want to reach [11] is $\Delta E = 10^{-3}$ eV which corresponds to $\Delta T \cong 10^{\circ}K$.

The shifts in energy of these weakly bound states caused by polarization of the electron-positron vacuum have been calculated by V.S. Meleshik [12] to 10 meV for the ddμ molecule and to 6.5 meV for dtμ. The matrix elements ignored in the calculation are estimated to make contributions no larger than about 0.1 meV.

The charge screening effects in the molecular complexes $\left[(dd\mu)d2e\right]$ and $\left[(dt\mu)d2e\right]$ have been evaluated by D.D. Bakalov and V.S. Meleshik [13] with the use of the mesic molecule wave functions calculated in the adiabatic representation of the three-body problem [14] with account of the hyperfine structure of the mesic molecule energy levels. In the weakly bound $J = \nu = 1$ states the screening effects shift the molecular-complex energy levels by 16 meV, resp. 25 meV. The quadrupole splitting has been evaluated to equal about 2 meV.

The relativistic corrections and the corrections for the electromagnetic structure of the nuclei to the nonrelativistic energy levels of the mesic molecules of all hydrogen isotopes have been calculated by D.D. Bakalov [15]. As an example, the correction due to the electromagnetic structure of the nuclei amounts to 2.2 meV for the $J = \nu = 1$ level in dtμ, whereas the recoil correction is 52.4 meV.

The hyperfine structure of the energy levels of the stationary states ($J \leq 1$, $\nu \leq 1$) of the μ-mesic molecules of the hydrogen isotopes have been calculated by D.D. Bakalov, S.I. Vinitsky and V.S. Meleshik [16] in first order perturbation theory in α^2 to an accuracy of about 0.1 meV. As an example, in dtμ the hyperfine splitting of the $J = \nu = 1$ state is of the order of $\Delta E = 193$ meV whereas the binding energy of the state is about E = 640 meV.

Because of the splittings of the levels of the μ–mesic molecules of the hydrogen isotopes one has about ten resonance conditions near to each other such that one does not expect a sharp
peak in the resonance formation rate of the molecular complexes
like $[(dt\mu)d2e]$. Up to now only the nuclear part of the level structure of these molecular complexes has been considered. Calculations
taking into account the electron part are presently in progress.

In mixtures of deuterium and tritium the calculations of the
muon recycling has been performed assuming that the muon transfer
from the dμ atom onto tritium proceeds from the dμ ground state.
In this case, the transfer rate λ_{dt} is of the order of $2 \cdot 10^8$ sec^{-1}.
The transfer can however also proceed from excited states where in
a collision with tritium the transfer rate can be comparable to the
deexcitation rate and exceed three or four orders of magnitude the
transfer from the dμ ground state. Taking into account the transfer
from excited states, the number of muon cycles can be increased by
reducing the tritium concentration in the mixture [17,18]. One can
imagine to measure this transfer rate from excited states and to
compare it to the deexcitation rate using the pion capture in mixtures of hydrogen and deuterium by varying the deuterium concentration.

I am grateful to P. Kammel and F. Kottman for clearifying comments and to D. Siradovic for her help in translations. I would
like to thank G. Fiorentini for a critical reading of this manuscript.

REFERENCES

1. L. Bracci and G. Fiorentini, Phys. Rep. 86:171 (1982) see also
   contributions to this volume by G. Fiorentini and P. Kammel
2. V. M. Bystritsky, J. Wozniak, A. Gula, V. P. Dzhelepov,
   V. K. Kapyshev, M. P. Malek, S. Sh. Mukhamet–Galeeva,
   L. A. Rivkis, V. A. Stolupin, V. A. Utkin, Sh. G. Shamsutdinov,
   JINR 13–83–636, Dubna 1983
3. V. M. Bystritsky, J. Wozniak, A. Gula, V. P. Dzhelepov,
   V. G. Zinov, JINR P1–83–515, Dubna 1983 (submitted to "Acta
   Physica Polonica")
4. V. M. Bystritsky, V. P. Dzhelepov, A. Gula, V.A. Stolupin,
   J. Wozniak, JINR E1–83–690, Dubna 1983 (submitted to "Acta
   Physica Polonica")

5. L. I. Ponomarev, M.P. Faifman, Zh. Eksp. Teor. Fiz. 71:1689 (1976) (Sov. Phys. JETP 44:886 (1976))
6. V. S. Melezhik, JINR 4-81-463, Dubna 1981
7. G. Ya. Korenman, Yad. Fiz. 35:390 (1982) (Sov. J. Nucl. Phys. 35: ... (1982))
8. G. Ya. Korenman, Kh. Tsookhuu, Yad. Fiz. 35:874 (1982) (Sov. J. Nucl. Phys. 35:... (1982))
9. V. S. Melezhik, L.I. Ponomarev, M.P. Faifman, Zh. Eksp. Teor. Fiz. 85:434 (1983) (Sov. Phys. JETP ...)
10. L. I. Menshikov, I.V. Kurtshatov Institute of Atomic Energy IAE-3811/12, Moscow 1983
11. S. I. Vinitsky, V. S. Melezhik, L. I. Ponomarev, Zh. Eksp. Teor. Fiz. 82:670 (1982) (Sov. Phys. JETP ...)
12. V. S. Melezhik, Pis'ma Zh. Eksp. Teor. Fiz. 36:101 (1982) (Sov. Phys. JETP Lett. 36:125 (1982))
13. D. D. Bakalov, V. S. Melezhik, JINR P4-81-835, Dubna 1981
14. L. I. Ponomarev, T.P. Puzynina, JINR E4-83-778, Dubna 1983
15. D. D. Bakalov, Zh. Eksp. Teor. Fiz. 79:1149 (1980) (Sov. Phys. JETP 52:581 (1980))
16. D. D. Bakalov, S. I. Vinitsky, V. S. Melezhik, Zh. Eksp. Teor. Fiz. 79:1629 (1980) (Sov. Phys. JETP 52:820 (1980))
17. L. I. Ponomarev, Atomenergie-Kerntechnik 43:175 (1983)
18. L. I. Menshikov, L. I. Ponomarev (in press)

TRIPLET STATE LIFETIMES AND

MUON CAPTURE IN GASEOUS HYDROGEN

I. Massa

Dipartimento di Fisica dell'Università di Bologna, and
Istituto Nazionale di Fisica Nucleare, Bologna, Italia

INTRODUCTION

It is now a long time since experiments exploiting the physical
phenomena which follow the slowing down of negative muon beams in
matter offer manifold opportunities of obtaining independent and sig-
gnificant informations on different fundamental physical fields[1];
for instance molecular and atomic physics, quantum electrodynamics
and nuclear physics. Nevertheless, despite its fundamental importance, it is today still absent a measurement of the muon capture
rate in hydrogen from the triplet spin state of mu-atomic systems.

When a negative muon is stopped in a hydrogen target, due to
the electrostatic attractive potential between μ^- and proton, the
replacement of an electron and the formation of a bound (μp) system
happen. This system, originally formed in an excited state, usually
promptly ($\lesssim 1$ ns) attains its 1s ground state through an electromagnetic cascade. It finally looks as an electrically neutral system, so
small that a strong muon-proton overlapping occurs. As a consequence,
the capture reaction

$$\mu^- + p \rightarrow n + \nu_\mu \qquad\qquad 1)$$

becomes $\sim 10^{-3}$ with respect to the other dominant disappeareance process, the muon decay

$$\mu^- \rightarrow e^- + \bar\nu_e + \nu_\mu \qquad\qquad 2)$$

whose rate ($\lambda_0 \simeq 4.5 \ 10^5 \ s^{-1}$) constitutes the reference scale.

It is by now currently accepted that process (1) is well theoretically described in the framework of V-A theory of weak interactions. Without entering into details[2], for our purposes it is interesting to underline few points :

i)   Theoretical description of process (1) and, in particular, calculation of the capture rate λ_c[3], requires (dropping second--class currents) to use four form factors : g_V , g_M , g_A , g_P (polar-vector, weak-magnetism, axial-vector and induced pseudo--scalar, respectively). The first two, moreover, are usually fixed exploiting other additional hypotheses as the Conserved Vector Current, so that it is possible to make analysis in terms of g_A and g_P only.

ii) If compared with other possible reactions, process (1) in hydrogen appears free from nuclear-physics complications and favourably sensitive to the two above mentioned quantities, in particular g_P . Actually, however, λ_c is expected to be strongly dependent on the total spin of the initially formed (μp) systems, both as regards the theoretical capture rate values in the triplet (λ_T) and in the singlet (λ_S) states (which are different of a factor ~ 70) and in relation to the different dependences of these quantities on the form factors. A measurement of λ_S , in fact, essentially allows us to determine g_A, whereas λ_T brings deeper informations on g_P.

iii) As will be explained later on, usual capture measurements (especially in gaseous hydrogen) are realized in such experimental conditions that essentially λ_S is measured. This is the reason why a measurement of λ_T has ever been considered a very important goal, through which a peculiar possibility of obtaining independent and deeper informations on the V-A structure of hadronic weak currents and, if possible, on g_P should be offered.

As will be seen in the following, the main experimental difficulty in measuring λ_T at usual pressure conditions is a consequence of the numerous scattering processes of (μp)'s against the surrounding gas molecules; due to these processes, in fact, at capture time the majority of mu-atoms are already landed onto the lower lying singlet state. Therefore, the "naif" method of preserving the initially formed triplet states is to stop the negative muons in very low-density hydrogen. Of course, more sophisticated techniques can be exploited, essentially along three directions :

i)   to use special trapping techniques (for instance a "muon bottle"
       apparatus[4]), to insure a high stopping efficiency of the incoming
       beam;

ii)   to set up particular techniques of sorting out the events due to
       muon nuclear capture in the triplet state, for instance by obser-
       ving the characteristic precession of the neutrons emitted in a
       magnetic field[5];

iii) to implement artificially the population of the triplet states
       (e.g. by laser pumping[6]).

    All these possibilities actually start from an initial situation,
which is the one provided by nature: the time of life of the imper-
turbed triplet states at a given density condition. At present, the
information gathered on this point derives from measurements of the
scattering cross sections of muonic hydrogen atoms against protons.
The object of this lecture is to discuss the essential lines of interpretation of such measurements along this direction, to present the
updated knowledge of the triplet state lifetimes derived thereby.
After a brief summary on some generalities (Sec. 2), the procedure to
obtain general information on the time of life of the triplet states
of (μp) atoms will be presented (Sec. 3). Sec. 4, finally, will be
devoted to the discussion of the theoretical predictions, by taking
into account the relevant experimental results, and to conclusions.

GENERALITIES

    We recall here that the story of the negative muons, after they
have been stopped in a H_2 target, is strongly influenced from the
hydrogen density value. In the case of <u>gaseous hydrogen</u>, (μp) systems
are formed in which the muon and proton spins can be parallel (total
spin F = 1, triplet state) or antiparallel (F = 0, singlet state); the
admixture of these two states is <u>initially</u> statistical, namely the
corresponding fractional populations are 3/4 and 1/4, respectively.
The kinetic energy at formation, E_0, can be considerably larger than
the thermal value at room temperature (~ 0.04 eV)[1].

    Due to their neutrality and reduced dimensions, the stable $(\mu p)_F$
systems easily diffuse throughout the medium in which they have been
produced, undergoing various types of elastic and inelastic processes,
like processes (1) and (2), or the formation of mu-molecular ions

$$(\mu p) + p \rightarrow (p\mu p) \qquad\qquad 3)$$

and the scattering processes

$$(\mu p)_{F_1} + p \rightarrow (\mu p)_{F_2} + p \qquad\qquad 4)$$

where F_1 and F_2 can both take the values 0 or 1. At a typical pressure of the gaseous hydrogen $\sim$10 atm, process (3) contributes only for some percent [7], so that reaction (4) represents the leading process. The corresponding collisions result in slowing the muonic atoms from their initial energy down to the thermal limit and in modifying the relative population of the hyperfine structure levels. The frequent scattering processes, in fact, in principle favour both the possible transitions between the two states; the dominant role of the triplet-singlet cross section [1], however, really determines a depopulation of the triplet state. Moreover, once the (μp) kinetic energy becomes lower than the energy gap separating the two states, $\Delta E_{\mu p} = 0.183$ eV, this depopulation becomes <u>irreversible</u>. The described development is more or less fast, according to the rates of processes (4), that is according to the scattering <u>cross sections</u> and hydrogen <u>pressure</u>.

Having in mind the possibility of measuring the capture rate λ_c with a maximum contribution from λ_T (to have a major sensitivity to g_P), the problem is, first of all, to know the cross section of processes (4) and, successively, to make out how low the hydrogen pressure must be to substantially preserve the statistical triplet proportion of (μp) atoms. As a reference we recall that at the experimental conditions of the measurement of Alberigi Quaranta et al. [8] (8 atm hydrogen pressure) it was verified that the capture process occurred completely in the singlet state; what means in a definite hyperfine structure state, without any perturbation due to nuclear structure and mu-molecular physics effects. As a consequence, as evidenced out by the subsequent analysis of Vitale et al. [9], the information which can be extracted on the pseudo-scalar form factor (by assuming muon-electron universality) has an accuracy of 30%.

The role of the mu-molecular effects becomes more important going to higher hydrogen pressures. In the extreme case of <u>liquid hydrogen</u>, in particular, the formation of molecular ($p\mu p$) systems is dominant. It is commonly accepted[10] that these molecular ions are initially formed predominantly in the (upper) orthostate and, until recently, it was assumed that they would remain in this state "forever" [11]. The recent experimental discovery of a long term transition from the orthostate to the parastate, by a Saclay-CERN--Bologna (SCB) collaboration [12], however, proves that in the long run the ($p\mu p$)'s can reach the lower-energy parastate. This implies the two following major consequences :

i)   The results of all the nuclear capture experiments performed in liquid hydrogen targets need to be reconsidered, to allow for the presence of a (reduced) admixture of ($p\mu p$) ions in the parastate, where the capture process occurs predominantly from muon-proton systems in the triplet spin states.

ii)   It is then conceivable the possibility of detecting muon capture events from the triplet state, if sufficient delays of observation are introduced with respect to the muon stopping time in the target.

It is interesting to note that, in the case of liquid hydrogen, the measurement of the ortho-para transition rate (λ_{OP}) has a significance similar to the physical information represented by the scattering cross section measurement for the gaseous hydrogen case.

THEORETICAL SURVIVAL TIMES OF THE TRIPLET STATES

Calculations

Predictions on the experimental conditions (especially as regards the hydrogen pressure) necessary to retain the statistical triplet proportion of (μp) atoms for a sufficiently long time, can be made through Monte Carlo simulation. The time distributions of the (μp) systems in the triplet state can be obtained as a function of gas pressure and few other parameters. In the computer program the formation of a high number of (μp)'s is considered, having initial energy E_0 and statistically distributed spins. The trajectorie of each mu--atom through the gas is then followed, by assuming a classical treatment of motion and scattering. Some assumptions may be underlined :

i)   in its moving through the gas, each (μp) has only one important interaction at a time;

ii)   the energy changes are evaluated on the basis of full two-body kinematics calculations, by taking into account the "Q-values" ($\pm \Delta E_{\mu p}$) attending in the cases of spin transitions, and the molecular H_2 motion (maxwellian velocity distribution);

iii) after each diffusion process, an isotropic angular distribution of the scattered (μp) atoms is assumed;

iv) besides muon decay and nuclear capture, other disappeareance channels, due to the presence of realistic quantities of deuterium and other impurities, are considered;

v) each (μp) atom is followed up to its disappeareance time, for a maximum time of 2.5 μs.

Scattering Cross Sections

It appears that the simulation of the diffusion process bases itself on an adequate knowledge of the scattering cross sections, which must be calculated, at any time, as a function of the experimental conditions (in particular the center-of-mass energy $E_{\mu p}$, at that time). By considering the information available at the present state of theoretical knowledge, this calculation is made through the following procedure.

a) For all the (μp) atoms in the triplet state and for those in the singlet state having $E_{\mu p} > \Delta E_{\mu p}$, the dependence of the various cross sections on $E_{\mu p}$ and on the scattering parameters is given by the expression formulated by Matveenko and Ponomarev[13]; that is, if the indexes i and j label the initial and final spin states of the scattering (μp) atom (which means

$$\sigma_{TS} = \sigma_{10} \ , \ \sigma_{TT} = \sigma_{11} \ , \ \sigma_{ST} = \sigma_{01} \ , \ \sigma_{SS} = \sigma_{00} \quad) :$$

$$\sigma_{ij}^{A} = \frac{4\pi}{k_i^2} \ \frac{\delta_{ij} \ D^2 + t_{ij}^2}{(D-1)^2 + (t_{11} + t_{00})^2} \qquad (i,j=1,0) \qquad 5)$$

where k_i is the momentum in the input channel of the reaction, t_{ij} are the elements of the reaction matrix, $D = t_{11} t_{00} - t_{10} t_{01}$ and δ_{ij} is the Kronecker symbol.

At the present collision energies ($\lesssim 2$ eV) the elements t_{ij} can be expressed as

$$t_{ij} = - a_{ij} \sqrt{k_i \ k_j} \qquad 6)$$

(here k_j is the momentum in the output channel and mu-atomic units are understood). In the low-energy limit the four scattering parameters a_{ij} are essentially constant, and can be expressed in terms of only two scattering lenghts, a_g and a_u :

$$a_{11} = (3 a_g + a_u)/4 \qquad\qquad a_{10} = \sqrt{3} (a_g - a_u)/4 \qquad 7)$$
$$a_{00} = (a_g + 3 a_u)/4 \qquad\qquad a_{01} = \sqrt{3} (a_u - a_g)/4$$

To extend the validity of eqs (7) to higher energies, the a_{ij}'s are replaced by their effective values

$$(a_{ij}*)^{-1} = (a_{ij})^{-1} + \frac{3\pi M k_i}{2a_{ij}^2} + \frac{3M k_i^2}{a_{ij}} \ln(\frac{9M k_i^2}{32}) \qquad 8)$$

where $M = (2 M_p + M_\mu)/4 M_\mu$, M_p and M_μ are the proton and muon mass, respectively, and the other symbols were already defined.

b) For (μp) atoms in the singlet state, with $E_{\mu p} < \Delta E_{\mu p}$, failing more refined corrections the resonant character of the process prevent from obtaining reliable values by the same procedure. Following the approach of Ponomarev et al.[14] , however, and exploiting the very weak energy dependence of σ_{00} on energy (excepted a narrow energy region), this cross section can be evaluated as in the early treatment by Gershtein[15]

$$\sigma_{00}^A = 4\pi (\frac{a_g + 3a_u}{4})^2 \qquad 9)$$

As a result of this treatment, the relevant cross sections are all expressed as functions of three common, independent parameters (a_u , a_g and $E_{\mu p}$). Their values, for a particular (reasonable) choice of a_u (3.7 a_μ) and $E_{\mu p}$ (0.2 eV, very little above the hyperfine structure splitting value), are shown in Fig. 1 (dashed curves) as a function of a_g .

A more realistic representation of the diffusion process is certainly obtained by describing the collisions of the mu-atoms against hydrogen molecules, rather than against free protons. The effect of molecular bond, that is of the internal degrees of freedom of the hydrogen molecules with which the (μp) atoms collide, may be taken into account by developing the method originally introduced by Matone, which considers the excitation of some rotational levels in the H_2 molecules (neglecting the possibility of excitation of vibrational levels)[16] . An interesting point is that, following this procedure, the molecular contributions to the cross sections can be factorized, so that each partial "molecular" cross section is expressed as :

$$\sigma_{J,I}^{J',I'} = [b_{ij} (I , I') B (J , J')]\sigma_{ij}^A \qquad 10)$$

where $b_{ij} (I , I')$ and $B (J , J')$ are suitable coefficients which depend, respectively, on the molecular total spin (I) and rotational momentum (J). The cross sections so calculated are then suitably

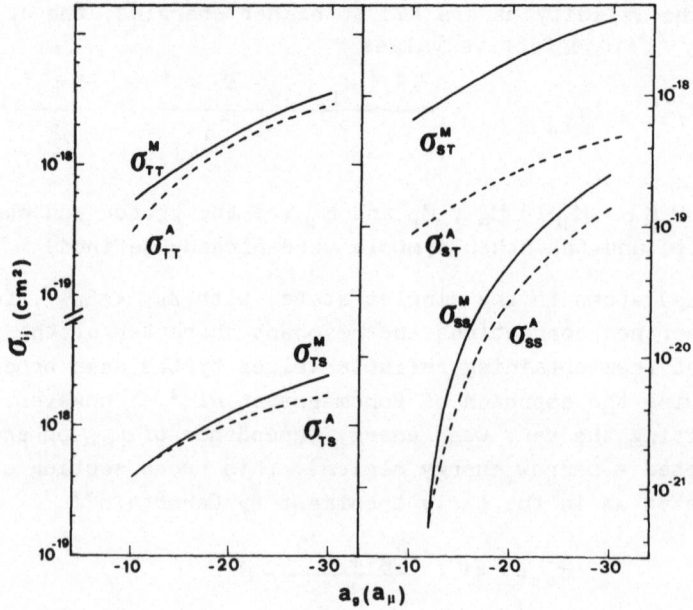

Fig. 1. Theoretical values of the cross sections relevant in the
scattering process of (μp) atoms against hydrogen atoms (A-
-model) or molecules (M-model). $a_u = 3.7\ a_\mu$ and $E_{\mu p} = 0.2$ eV
are assumed.

weighted and summed, to obtain the four effective σ_{ij}^M cross sections,
which are shown in Fig. 1 (full curves) besides the corresponding
"atomic" cross sections. It appears a common increase of the "molecular" cross sections, with respect to the "atomic" ones, which is
much more accentuated in the ST case. As a consequence, by going from
the A-model to the M-model, as concerns the lifetime of the triplet
states we may expect a slower depopulation, at least at early times
(i.e. until the (μp) system reaches an energy value $E_{\mu p} < \Delta E_{\mu p}$ and,
consequently, σ_{ST}^M vanishes).

Results

The fraction F_T of triplet states surviving at a time t following the (μp) atom formation can be assumed as an adequate parameter
to represent the behaviour of the triplet state population with time:

$$F_T(t) = N_T(t)/\{ N_T(t) + N_S(t)\} \qquad\qquad 11)$$

where N_T and N_S are the numbers of (μp) atoms in the triplet and singlet state, respectively, at time t. An example of F_T distribution, obtained with the described Monte Carlo procedure (for each model, A and M) is presented in Fig. 2. As expected, a substantial difference between the two curves mainly appears at early times of the distributions. Starting from the zero-time value 0.75, the A-curve shows in fact a very fast decreasing component before assuming, in first approximation, the shape of a simple negative exponential of lifetime τ_{TF}. The M-curve, on the contrary, while quite similar in the exponential decreasing, first exhibits a long equilibrium phase in which the number of (μp) systems in the triplet state do not substantially change, as a result of the high value of the ST with respect to the TS cross section. The curves in Fig. 2 are well representative of the general characteristics of the results obtained with the simulation program, in the whole range of each considered parameter (which range was chosen by taking into account the known experimental and theoretical values, summarized in Table I). We can then affirm that, even if we may assume (in first approximation) the lifetime τ_{TF} as characterizing the triplet depopulation rate, the initial flat behaviour in the molecular case makes the effective depopulation process somewhat slower than what is described by τ_{TF} itself.

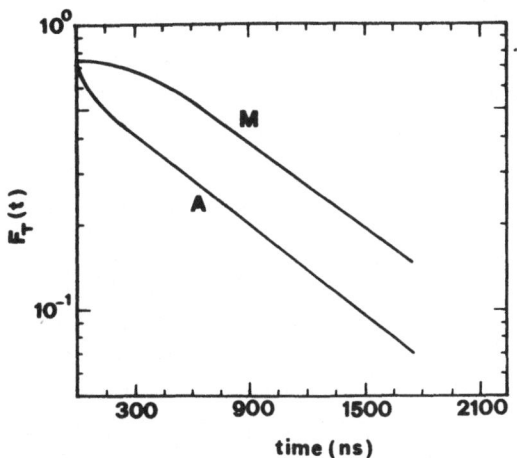

Fig. 2.   Fraction of triplet (μp) atoms, as a function of time, calculated through A- or M-model. P = 0.1 atm, E_0 = 0.8 eV, a_g = -18 a_μ and a_u = 3.7 a_μ are assumed.

Table I. Summary of the experimental and theoretical values of the (μp) elastic scattering parameters, relevant for the triplet lifetime calculations.

Authors	Year	σ_{00} (10^{-21}cm²)	E_0 (eV)	a_u (a_u)	a_g (a_u)	Ref
Dubna	1965	167±30	1 [a]	5 [a]	−33±2	Exp 17
Bologna-CERN	1967	7.6±0.7	0.55±0.20	5 [a]	−11.2±0.8	Exp 18
Bologna	1982	15±3	<0.8	3.7 [a]	−17.6±0.3	Exp 19
Dubna	1984	17.4±3.3	———	———	———	Exp 20
Zel'dovich et al.		1.2	———	5.25	−17.3	The 15
Cohen et al.		8.2	———	5	−11	The 21
Ponomarev		1.1	———	3.51	−13.5	The 22
Ponomarev et al.		99	———	3.51	−29.4	The 14
Monte Carlo range			$\{{}^{0.038}_{2}$	$\{{}^{3}_{7}$	$\{{}^{-30}_{-10}$	

[a] Assumed values

The results of simulation have simple behaviours which suggest the possibility of expressing in analytical form the τ_{TF} values, as a function of all the four parameters a_u, a_g, E_0 and P. This has been made, in the case of the M-model, with the aim of providing a useful tool to rapidly calculate at least the order of magnitude of τ_{TF}. The expression obtained has a simple analytical form, but it is not born from any attempt of physical interpretation :

$$\tau_{TF}^{M} = \exp\{6.82 - 0.97\ \ln P + 0.0924\ a_g - 0.122\ a_u + 0.167\ \ln E_0\} \qquad 12)$$

where lifetimes values in (ns) are obtained if P in (atm), scattering lenghts in (a_u units) and energies in (eV) are used. Expression (12) has been verified to provide τ_{TF}^{M} values which differ up to a maximum of 15% from the Monte Carlo values, in the whole considered ranges (see Table I).

DISCUSSION AND CONCLUSIONS

The results obtained through Monte Carlo simulation (we refer here to the case of the more complete and realistic M-model) must be analysed by taking into account the relevant experimental results. The scattering parameters and initial energy can be determined through measurements of the cross sections of process (4). A first set of measurements was carried out by a Dubna group[17] (diffusion chamber technique) and by a Bologna-CERN group[18] (counter technique). These results were affected by significant discrepancies between the measured cross sections and, consequently, the parameter values, as shown in Table I. A new experiment on process (4) has been performed more recently by a Bologna group[19] (counter technique), at the muon channel of the CERN Synchro-cyclotron. The results of this measurement, obtained by improved techniques as regards both the experimental and the analysis aspects, have solved the previous discrepancy on the cross sections and fixed a new set of parameter values (see Table I). Moreover these last results (which had also a very recent confirmation from a new measurement of a Dubna group[20]) clarified that measurements performed at 26 and 10 atm hydrogen pressure actually refer to diffusing (μp) atoms which were all in the singlet state.

As a matter of fact, the general characteristics of the Monte Carlo results show that it is necessary to go to an even lower gas pressure (< 1 atm) for a possible measurement of the capture rate in the triplet state. To describe the depopulation of the triplet state in the subatmospheric range, it becomes essential to evidence out the role of the flat top at early times of F_T , so that it is more realistic to evaluate the lifetime τ_T of the <u>absolute number</u> N_T of triplet states formed in a given pressure condition. This was done on the results of the simulation, allowing a variation of the scattering parameters as determined from the already cited Bologna experiment[19] and evaluating the lifetimes τ_T by the Peierls' formula[23] . One may mention here that in effect the inclusion of the muon decay contributes to reduce the behaviours of N_T vs. time to a decay curve which can approximately be described by a single exponential also in the low-pressure range. The results obtained in this way (down to P = 0.01 atm) are shown in Fig. 3, where also the lifetime region fixed from the known theoretical values is presented.

By examining Fig. 3 it appears even more evident that, to deal with triplet lifetimes suitable for observation (namely at least some hundredths of ns), the apparatus of measurement must be dimentioned to operate in a <u>substantially</u> subatmospheric range, even in the most optimistical hypothesis (see the dashed line of Fig. 3, which corresponds to a second solution suggested by the Bologna experiment, but which is hardly compatible with the existing theoretical models). This result obviously accentuates the already discussed experimental problems, beginning from the effects due to transfers to gas impurities and target walls, going to the difficulty of stopping a non-

-negligible part of the incoming beam in the hydrogen itself.

We must also recall that actually a measurement not in the tri-plet state but in the statistical mixture of the states would be generally realized. A measurement of this type should integrate the existing nuclear capture values at F_T = 0 (from high-pressure gas measurements) and F_T = 1/4 (from liquid measurements). As shown in

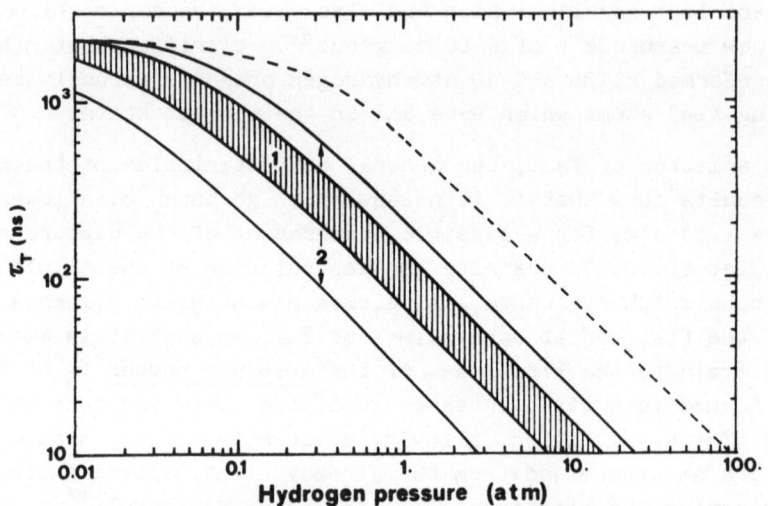

Fig. 3. Lifetimes of the absolute number of triplet states as a function of hydrogen pressure. a_u = 3.7 a_μ and (0.11 ≤ E_0 ≤ 0.8) eV are assumed. Region 1:(-18.6≤ a_g ≤ -16.6) a_μ, Bologna experiment. Region 2:(-29.4 ≤ a_g ≤ -11) a_μ, extreme theoretical values. Dashed line : a_g = -5 a_μ.

Fig. 4a, by hypothesizing a measurement of λ_{STA} with an absolute error $\varepsilon(\lambda_{STA})$ = ± 10 s^{-1}, a corresponding $\varepsilon(\lambda_T) \simeq$ 20 s^{-1} would be obtained on the triplet capture rate, by fitting procedure; a result which is mainly conditioned by the present large error on the gas measurements. It can be seen in Fig. 4b, in fact, that assuming to realize both gas measurements with the same absolute errors $\varepsilon(\lambda_S)$ = $\varepsilon(\lambda_{STA})$, much better information is obtained on λ_T, espe-

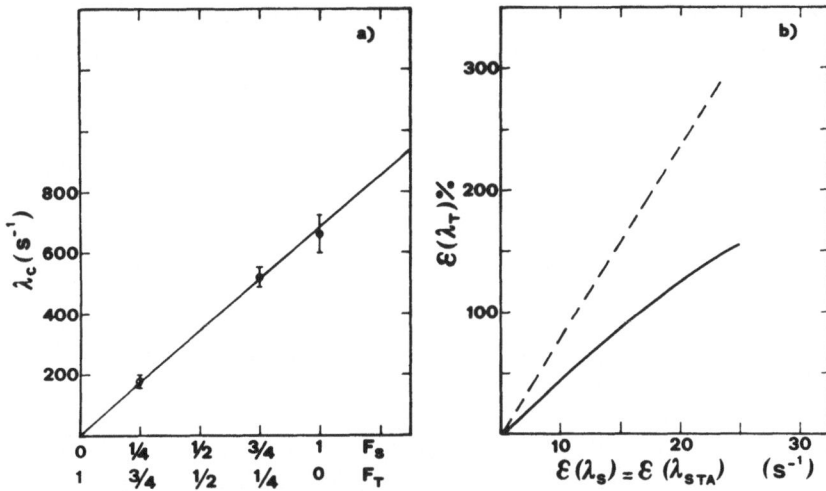

Fig. 4. a) Nuclear capture rate in hydrogen as a function of triplet (singlet) fraction. Point at $F_T = 0$: average of
results of gas measurements (λ_S). Point at $F_T = 1/4$:
result of SCB liquid measurement. Point at $F_T = 3/4$: pos-
sible measurement of λ_{STA} with absolute error ε (λ_{STA}) $= \pm$ 10
s^{-1} . The curve was obtained through fitting procedure.
b) Representation of the relative errors which are obtained
on λ_T (same procedure as in part a) hypothesizing equal
absolute errors for both the measurements in gaseous hydrogen. The dashed curve is obtained by exploiting the two gas
measurements only, whereas in the full curve also the liquid
measurement is taken into account.

cially if the liquid measurement is exploited also (continuous
curve). In conclusion, measurements of the muon capture rates in
low- and high-pressure gaseous hydrogen, with absolute errors $\sim$ 15
s^{-1} , comparable with the one of the recent SCB experiment in liquid
hydrogen (lifetime technique)[24] , should determine the triplet state
muon capture rate with a relative error lower than 100%. Such a
result should have a fundamental significance, due to the very different values expected for λ_S and λ_T, as a test of the V-A structure

of weak interactions. A similar result, moreover, should allow to improve of a factor ~ 2 the error on g_p , assuming muon-electron universality.

ACKNOWLEDGEMENTS

The author is indebted to A. Bertin and A. Vitale for fruitful discussions on the subjects of the present report.

REFERENCES

1. A. Bertin, A. Vitale and A. Placci, La Rivista del Nuovo Cimento, n. 5 (1975) 423.
2. See for example J. Martino, Muon capture in hydrogen, Lecture in this School.
3. H. Primakoff, in : Muon Physics II, Academic Press, Inc., New York, (1975) p. 3.
4. H. Anderhub et al., Phys. Lett. 101B (1981) 151.
5. J. H. Brewer, 8 - Icohepans, Vancouver, Canada (1979).
6. A. Bertin, G. Carboni, A. Placci, E.Zavattini, U. Gastaldi, G. Gorini, G. Neri, O. Pitzurra, E. Polacco, G. Torelli, G. Stefanini, A. Vitale, J. Duclos and J. Picard, Il Nuovo Cimento, 23B (1974) 489.
7. E. Zavattini, in : Muon Physics II, Academic Press, Inc., New York, (1975) p. 219.
8. A. Alberigi Quaranta, A. Bertin, G. Matone, F. Palmonari,
    · G. Torelli, P. Dalpiaz , A. Placci and E. Zavattini, Phys. Rev. 177 (1969) 2118.
9. A. Vitale, A. Bertin and G. Carboni, Phys. Rev. D11 (1975) 2441.
10. L. I. Ponomarev and M. P. Faifman, Sov. Phys. JETP 44 (1976) 886.
11. S. Weinberg, Phys. Rev. Lett. 4 (1960) 585.
12. G. Bardin, J. Duclos, A. Magnon, J. Martino, A. Richter, E. Zavattini, A. Bertin, M. Piccinini and A. Vitale, Phys. Lett. 104B (1981) 320.
13. A. V. Matveenko and L. I. Ponomarev, Sov. Phys. JETP 32 (1971) 871.
14. L. I. Ponomarev, L. N. Somov and M. P. Faifman, Sov. Jour. Nucl. Phys. 29 (1979) 67.
15. S. S. Gershtein, Sov. Phys. JETP 7 (1958) 318.
16. G. Matone, Lettere al Nuovo Cimento 2 (1971) 151; A. Bertin, I. Massa, M. Piccinini, G. Vannini, A. Vitale and G.Matone, Phys. Lett. 88B (1979) 185.

17. V. P. Dzhelepov, P. F. Ermolov and V. V. Fil'chenkov, Sov. Phys. JETP $\underline{22}$ (1966) 275.

18. A. Alberigi Quaranta, A. Bertin, G. Matone, F. Palmonari, A. Placci, P. Dalpiaz , G. Torelli and E. Zavattini, Il Nuovo Cimento $\underline{B47}$ (1967) 72.

19. A. Bertin, M. Capponi, I. Massa, M. Piccinini, G. Vannini, M. Poli and A. Vitale, Il Nuovo Cimento $\underline{72A}$ (1982) 225.

20. V. M. Bystritsky, Private Communication.

21. S. Cohen,, D. L. Dudd and R. J. Riddel, Phys. Rev. $\underline{119}$ (1960) 384.

22. L. I. Ponomarev, SIN preprint PR-77-011 (1977).

23. R. Peierls, Proc. R. Soc. London Ser. A $\underline{149}$ (1935) 467.

24. G. Bardin, J. Duclos, A. Magnon, J. Martino, A. Richter, E. Zavattini, A. Bertin, M. Piccinini, A. Vitale and D. Measday, Nucl. Phys. $\underline{A352}$ (1981) 365.

FORMATION OF THE LIGHTEST MUONIC ATOMS

AND THE 2S-LIFETIME OF THE $(\mu^{-4}\mathrm{He})^+$ - ION

Franz Kottmann

Institut für Hochenergiephysik
ETH-Hönggerberg
CH-8093 Zürich (Switzerland)

1. THE "MUON-BOTTLE" APPARATUS

There are many experiments which require the formation of muonic atoms in low density gases in order to
avoid collisions between the exotic atom and the host gas
atoms. Prominent examples are the nuclear muon capture in
the triplet state of μp_{1s} atoms (discussed by I. Massa in
this school), or the Lamb-shift in muonic hydrogen.

The conventional technique to stop a beam of negative
muons in a gas target is inadequate at pressures below
~ 1 atm, since the number of stopped muons drops approxi-
matively proportional with pressure, whereas the background level remains more or less constant. This problem
had been overcome with the "muon bottle", an apparatus
which was developed specifically for measurements on $\mu^- p$
and $(\mu^- \mathrm{He})^+$ atoms at gas pressures in the Torr- or even
sub-Torr region (1 Torr = 1.3 10^{-3} atm)[1,2,3]. The principle
of this technique is to inject negative pions of 40 MeV/c
momentum axially into a magnetic bottle filled with the
target gas. A quarter of the pions decays in flight within
the bottle. The muons from "backward decay" have very low
momenta in the lab system and are trapped on helical
orbits. These muons are then slowed down due to excitation
and ionization of target atoms, until they form a muonic
atom. The uv light from primary scintillation produced
during the slowing down process is detected by photomultipliers coated with a wavelength-shifter (Fig. 1,"PD").
At H_2-pressures below 2 Torr, the light production is
increased by producing secondary scintillation within

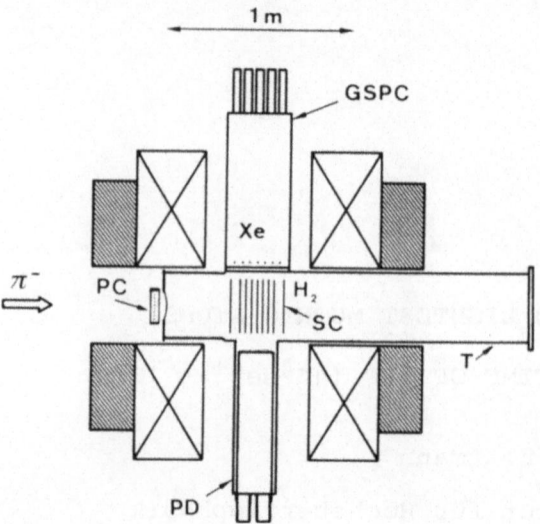

Fig. 1. Experimental set-up: PC = multiwire proportional
        chamber, SC = stop chamber, PD = photon detector
        (wavelength shifter, lucite light guide and RCA
        8854 photomultipliers), GSPC = xenon gas scin-
        tillation proportional chamber (wire planes,
        mirror light guide and 19 photomultipliers),
        T = stainless steel tank.

a set of HV wire planes ("stop chamber") in the target
gas[3]. For a trapped muon, typically 30 photons/μs are
detected, whereas pions as well as muons emitted at
forward angles do not produce a significant amount of
light and thus do not lead to a considerable background.
The detection of "stop-light" makes it possible to
determine precisely the time when the muonic atom is
formed and eventually to trigger a laser inducing
2S → 2P transitions (measurement of the Lamb-shift).

The stop rates of negative muons in H_2 or He gas
are 500/s at 1 Torr, 5000/s at 30 Torr, and 8000/s at
1 atm (120 μA primary proton beam, $3 \ 10^5$ π^-/s at 40 MeV/c)
At the lowest pressures, only muons of not too high initial
energies (~ 200 keV at 1 Torr) can be slowed down within
their lifetime. Above 30 Torr, the stop rate increases
very slowly with pressure because the geometry restricts
the maximum radius of the muon orbit to a value corresponding to an energy of about 2 MeV.

To detect the low-energy X-rays from the muonic cascade in the µp and µHe atoms, detectors with large efficient area were needed to cover a sizeable solid angle. This could only be realized with gaseous detectors; the energy resolution of usual $ArCO_2$ proportional chambers, however, is too poor to resolve the K_α-line (2P→1S, 1.90 keV) from the other K-lines of the µp atom (3P→1S: 2.25 keV, ..., 15P→1S: 2.52 keV). A better resolution was achieved by the use of xenon gas scintillation proportional chambers (GSPC). The known principles had to be adapted to the special requirements of an experiment in particle physics. The details are described in ref. 4 and 5. A relative energy resolution of 16 % (FWHM) and a time resolution of 80 ns (FWHM) was obtained at 2 keV. The Xe gas of the GSPC's is separated from the target volume by a 5 µm thick aluminized Mylar foil which is sustained by tungsten wires to hold a pressure difference up to 1 atm. The window diameter (efficient area) is 20 cm. The efficiency for the detection of soft X-rays, which is limited by the window thickness, is good down to an X-ray energy of 1.5 keV. Ionizing particles traversing the chamber are discriminated against X-rays with very high effectiveness by means of a pulse shape analysis.

Fig. 2 shows typical energy spectra of the X-rays from µp and µHe atoms demonstrating the good energy resolution and the low background level measured at very low gas densities with the muon bottle apparatus.

Another unconventional technique to stop mesons $(\mu, \bar{p})$ in low-density gases is the "cyclotron trap" presented by J. Missimer in this school. First tests showed that it also allows high stop densities at subatmospheric pressures[6].

2. FORMATION OF THE LIGHTEST MUONIC ATOMS AND MUONIC CASCADE PROCESSES

2.1 General Remarks

A decade ago, there existed a variety of model calculations about the formation of muonic atoms which lead to quite different conclusions. For instance, the estimates of the average muon energy for capture by a H-atom have varied from near thermal to several keV. In the meantime, more elaborate calculations have been performed, and new experimental information is available from X-ray intensity ratios measured on muonic atoms and molecular compounds.

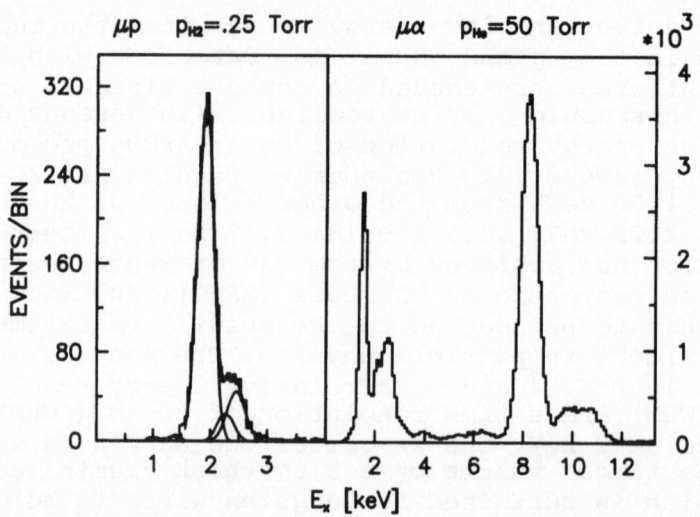

Fig. 2. Energy spectrum of μp K-line X-rays at 0.25 Torr
H$_2$, and of μHe K- and L-line X-rays at 50 Torr
He. In the μp-spectrum, the three fitted Gaussian
components K$_\alpha$(1.895 keV), K$_\beta$(2.246 keV), and
K$_{\gamma\delta}$...(≈2.45 keV) are shown.

The calculation of the interaction of muons and
muonic atoms with the surrounding gas atoms involves the
solution of many-body problems. It is therefore not surprising that even the most elementary case, the formation
of a μp-atom in atomic H$_1$-gas, is not understood in full
detail. Systematic investigations in this field are
complex and usually considered to be of minor relevance.
It turns out, however, that a reliable knowledge of the
interactions between the muonic atom and its environment
is often necessary for the correct analysis of important
experiments.

A particular difficulty arises in the case of
hydrogen gas when one wants to compare model calculations
with experimental results. Obviously, measurements have
to be performed with molecular hydrogen, whereas most
calculations only treat the simpler case of atomic H.
Naively, one would assume that a particular cross
section of an H$_2$-molecule is comparable to twice the

cross section of one H-atom, but such an approximation may be inappropriate in certain cases. For example, adiabatic ionisation of H_2-molecules by slow negative mesons does not occur as with H-atoms. With such arguments, some authors suggest that for the final slowing down and capture processes the cross section of an H_2-molecule has roughly the same size as that of one H-atom.

In what follows, various experimental investigations on the formation and cascade of μp and μHe atoms are briefly discussed.

2.2 Slowing Down of a Negative Particle in H_1 (H_2) - Gas

The energy loss dE/dx of a negative muon can be predicted accurately by the Bethe-Bloch formula at velocities $v_\mu > \alpha c$ ($\alpha = 1/137$; cf. Fig. 3). In the region of the Bohr velocity αc ($E_\mu = 2.8$ keV) and below, no precise calculations exist, and measurements were previously performed only for positive heavy particles (protons) for which the slowing down mechanisms are quite different.

With the "muon stop detector" of the muon bottle apparatus, it became possible to determine the mean value of dE_μ/dx in the velocity range $0.05\alpha c \leqslant v_\mu \leqslant \alpha c$ in H_2-gas. The scintillation light produced by the stopping muon is recorded as a function of time. The intensity of the light is almost proportional to the energy loss per unit time $dE/dt = v_\mu \cdot dE/dx$. In the experiment, a time t*

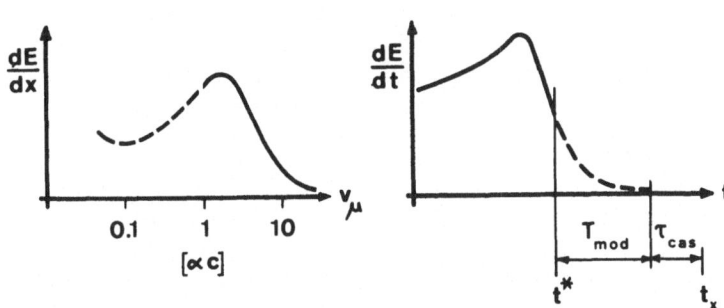

Fig. 3. Energy loss dE/dx of μ^- in H_2 gas as a function of the muon velocity, and the corresponding energy loss per unit time dE/dt as a function of time.

Table 1. Moderation times and upper
limits for the mean cascade
times, measured with μ^- in
H_2-gas

$p(H_2)$	T_{mod} [3]	τ_{cas} [7]
1.0 Torr	170 ± 40 ns	< 70 ns
0.25 Torr	670 ± 50 ns	< 200 ns

is defined when the intensity has fallen down to half of
its maximum value (Fig. 3). At this time, the muon velocity
is $v_\mu(t^*) \approx \alpha c$. After t^*, the muon is further slowed down,
and at the end of the "moderation time" T_{mod} it is
captured by an H_2-molecule and forms an excited μp atom.
At the end of the muonic cascade, a K X-ray is measured
(time t_x). The experimental results for T_{mod} and the
mean cascade time τ_{cas} are given in Table 1.

To compare these results with calculated values,
the cross section of an H_2-molecule is assumed to be
equal to that of one H_1-atom, as suggested by Leon[8]. The
moderation time calculated from the frequently used
stopping power given by Rosenberg[9] is $T_{mod} \approx 1400$ ns/p(Torr),
in clear disagreement with the data. Two recent calculations
by Cohen et al.[10] and Cohen[11], using completely different
methods, give $T_{mod} \approx 190$ ns/p(Torr), which agrees with
the measured values.

A more extensive discussion of different calculations
was given by Cohen[11]. Some of the older calculations which
disagree with the data must now be ruled out.

2.3 Atomic Capture (Coulomb Capture) in H_2 or He Gas or
    H_2 + He Mixtures

Various calculations have been performed on the
atomic capture of negative mesons. Surveys were given
by Gershtein and Ponomarev[12] and Schneuwly[13]. Leon[8]
pointed out that the capture process through electron
ejection is just the last ionization of the slowing down
and that the cross sections of both processes therefore
have to be calculated by the same method. Such consistent
calculations have been carried out for H atoms[10,11] and
He atoms[14]; they all show that capture occurs primarily

at energies about or below the ionization potential of the target.

Cohen et al. extended their "diabatic-state" method[10] in order to calculate the relative capture probabilities in a mixture of He and H atoms[15]. For the ratio of the muon capture probabilities, they get $W_{He}/W_H = 0.75$ in a 1:1 mixture. (Experimentally, this corresponds to a He-H_2 mixture with equal partial pressures, if one treats the molecule as one atom, as the authors suggest.) The same problem has been attacked in a different manner by Korenman and Rogovaya[16] who combined their calculation (based on the Born approximation) of the capture cross sections with the old dE/dx values of Rosenberg. It came out that the pions (or muons) are captured predominantly by He atoms at quite high energies ($\sim$ 80 eV); the capture ratio for a mixture with $p(He) = p(H_2)$ is $W_{He}/W_H \approx 3$ (these authors treated the molecule as two H atoms; otherwise the ratio would be ≈ 6).

Different measurements of the capture ratios in mixtures of hydrogen with other gases have been performed. Petrukhin and Suvorov[17] measured via the charge-exchange reaction $\pi^- + p \rightarrow \pi^0 + n$ the probability that a $\pi^- p$ atom is formed in a $H_2 + Z$ gas mixture. (Except for πp and $\pi^3 He$, there is no π^0 production in pionic atoms.) For $Z=2$, they found this probability to be $W(\pi^0) = 0.30$ at the partial pressures 40 atm H_2 + 40 atm He; this means that as much as 70 % of all pions form πHe atoms. This result seems to be in rough agreement with the calculation of Korenman and Rogovaya, but the authors interpreted their data (taken at various concentrations of several admixtures) in another way: The measured probability $W(\pi^0)$ has to be taken as the product of the probability P_H of initial atomic capture by hydrogen, and the probability that the pion is not transferred from H to He during the mesonic cascade, before the π^0 is produced. For the partial pressures mentioned above, their result is $P_H = 0.52$ and $W_{He}/W_{H2} = (1-P_H)/P_H = 0.92 \pm 0.05$.

A direct measurement of the Coulomb capture ratio was recently performed at SIN in the muon bottle at much lower densities. The K-line intensities of μp and μHe atoms, corresponding to the number of muonic atoms formed in the ground state, have been measured in the mixture 3 Torr H_2 + 3 Torr He (at room temperature). The result of a preliminary analysis is $N(\mu He_{1S})/N(\mu p_{1S}) = 0.8 \pm 0.2$. This number is an upper limit for the capture ratio W_{He}/W_{H2}, since transfer of the muon from H to He during the cascade cannot be ruled out a priori, although it seems unlikely at such a low density.

The difference between the raw data at low and high densities gives direct evidence that transfer during the cascade is very important (as supposed by Petrukhin and Suvorov), since the Coulomb capture ratio does not depend on pressure and the difference between π and μ is a minor effect. The final experimental results at low and high densities therefore agree with each other and also with the calculation of Cohen et al., whereas the prediction of Korenman and Rogovaya can now be ruled out.

2.4 Muonic Cascade Processes (μp, μHe)

Numerous measurements of the X-ray intensities of various muonic atoms have been performed, but not much data were available on the cascade of muonic hydrogen until quite recently. The formation of μp atoms has often been demonstrated, e.g. via the transfer process $\mu p_{1S} + Z \rightarrow \mu Z^* + p$, but the direct observation of the characteristic K-line X-rays (around 2 keV) was usually eluded since it requires more elaborate experimental techniques.

Nevertheless there is particular interest in the μp cascade which is quite different from all other muonic cascades. Muonic hydrogen is the only neutral muonic atom; for this reason, it is able to penetrate much deeper into the atomic shell of surrounding atoms, and the cross sections of collisional Stark mixing ($\mu p_{n,\ell}+H \rightarrow \mu p_{n,\ell'}+H$) and external Auger effect ($\mu p_n+H \rightarrow \mu p_{n'}+p+e$) are correspondingly larger. Internal Auger effect, which dominates the cascade of high-Z muonic atoms, is absent. - Several applications like the formation of μp_{2S} atoms, muon catalyzed fusion, or the study of πp and $\bar{p}p$ atoms, demand for a rather detailed understanding of the various processes occuring in the cascade of exotic hydrogen atoms.

New data are now available from X-ray measurements in the muon bottle with H_2 [18] and He [19] over a wide range of densities. Typical energy spectra are shown in Fig. 2. Additional (preliminary) results come from an experiment performed at SIN by a William and Mary - SIN collaboration who measured the K-line intensity ratios of muonic helium at pressures between 2.5 and 40 atm. The most interesting results are summarized in Fig. 4.

Several cascade calculations have been performed for muonic hydrogen (discussed in ref. 18). The intensity ratios are pressure-dependent since the rates of all

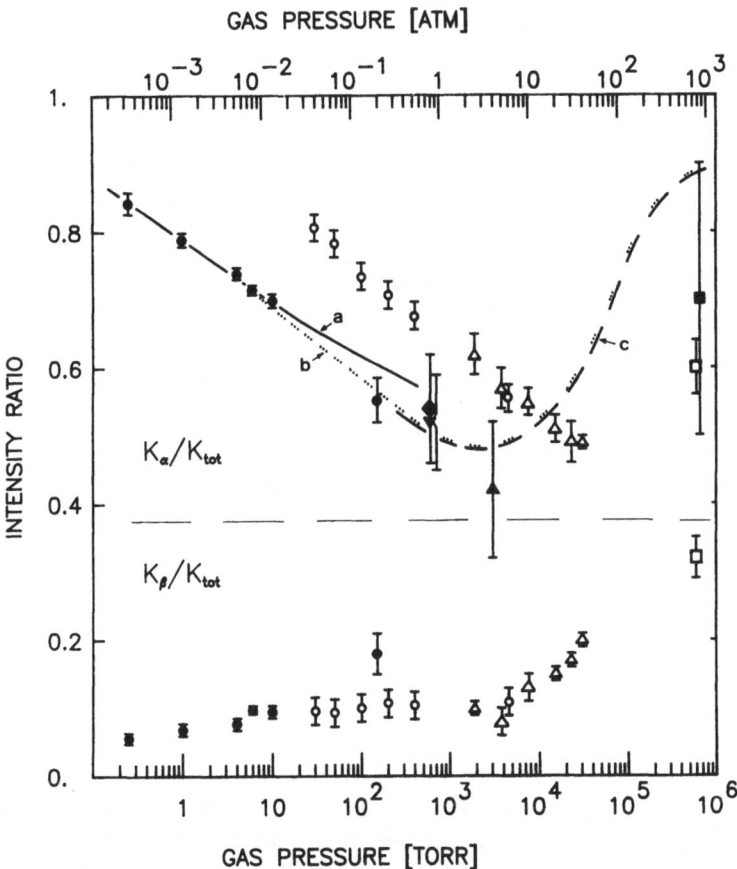

Fig. 4. The pressure dependence of the K-line intensity
ratios $I(K_\alpha)/I(K_{tot})$ and $I(K_\beta)/I(K_{tot})$ for μp
and μHe atoms.
μp: ●ref. 18; ◆ref. 24; ▼ref. 25; ▲ref. 26;
■ref. 27; calculated curves a: ref. 18;
b: ref. 28; c: ref. 29.
μHe: ○ref. 19; △ref. 20; □ref. 30.

collisional processes are proportional to the density, whereas the radiative transition rates remain constant. At H_2 pressures above 1 atm, the high rate of collisional Stark-mixing maintains a statistical population of the ℓ-sublevels for $n \gtrsim 10$. A cascade calculation treating the competition between the pressure-dependent processes and radiative transitions may therefore be started at intermediate n-levels with a known initial population. - The situation is completely different at pressures in the Torr-region where the cascade is almost purely radiative for $n < 10$; the intensity ratios are therefore sensitive to processes occuring at the highest n-levels.

At low densities, two particular reactions have to be taken into account in addition to the well-known external Stark- and Auger-effects: The "chemical reaction" $\mu p_n + H_2 \rightarrow \mu p_{n'} + H + H$ [21] and the "Coulomb deexcitation" $\mu p_n + H \rightarrow \mu p_{n'} + H$ (+ kin. energy) [22]. These effects speed up the muonic cascade at the highest n-levels. Quite high cross sections for the chemical reactions had to be adopted[18] to reproduce the short cascade times measured at 1 and 0.25 Torr (see Table 1).

For muonic helium, cascade calculations have been performed by Landua and Klempt[23] for the high-pressure region, but they have not yet been adapted to the lower densities.

3. THE 2S - LIFETIME AND THE CHEMISTRY OF THE $(\mu^- He)^+$ ION

3.1 The $(\mu^- He)^+_{2S}$ System

The muonic helium ion in the metastable 2S state $(\mu^- He)^+_{2S}$ is a preferred species in fundamental experiments proposed during the last years. Two prominent examples are the search for parity violation by weak neutral currents in light muonic systems[31,32], and measurements of the 2S - 2P energy differences in $\mu^4 He$ and $\mu^3 He$ yielding a precise test of vacuum polarization and determination of the nuclear rms charge radii[33]. Up to now, only the Lamb-shift of $\mu^4 He$ has been measured[34].

Preceding investigations of the metastable 2S muonic helium ion $(\mu^{-4} He)^+_{2S}$ at 7 atm [35] and at 10 to 50 atm [36] have shown only small deviations of the lifetime τ_{2S} from the "vacuum" value $\tau^0_{2S} \approx (\lambda_\mu + \lambda_{2X})^{-1} = 1.78$ µs. (λ_μ is the muon decay rate and $\lambda_{2X} = 1.06 \ 10^5 \ s^{-1}$ the

2S → 1S two-photon transition rate[37]. The muon capture and the M1(2S→1S) transition rates are negligibly small.) If $\tau_{2S} = \tau_{2S}^0$, i.e. in the absence of pressure-dependent 2S-decay channels, a fraction $\lambda_{2X}/(\lambda_\mu + \lambda_{2X}) = 0.19$ of the $(\mu^- He)_{2S}^+$ systems decays via emission of two X-rays. These X-rays have a continuous energy spectrum from 0 to 8.2 keV[37], the sum of their energies being equal to the energy difference $\Delta E(2S-1S) = 8.2$ keV.

The measured long lifetimes τ_{2S} in the pressure region 7 to 50 atm strongly disagree with theoretical predictions of the quenching rates induced by external interactions. Two possible cases are considered:

(i) The $(\mu^- He)_{2S}^+$-system is a bare ion undergoing collisions with the surrounding He atoms. The rate for radiative quenching of the muonic helium ion due to Stark mixing of the 2S and 2P states during collisions has been calculated by several authors[38,39,40] and is predicted to be at least two orders of magnitude larger than the experimental limit obtained at 7 atm[35].

(ii) Collisional quenching of the $(\mu^- He)_{2S}^+$ ion is prevented. Two mechanisms have been suggested:
The muonic helium ion could be neutralized by addition of an electron[40,41]. The interatomic potential would then be repulsive, like that of He-H, and Stark mixing consequently be very small. However, there are problems with this mechanism: The electron is not readily available since the ionization potential of He is greater than that of $(\alpha\mu^- e)$, and the internal Auger process would occur at a rate of $1.7 \cdot 10^8$ s^{-1}. Another mechanism was introduced and qualitatively discussed by Bertin et al.[36] and Carboni and Pitzurra[38]: Radiative quenching of the muonic 2S-state induced by collisions with the surrounding atoms may be prevented by a cluster of helium atoms around the muonic ion. From measurements of the mobility of ions in helium, it is known that big clusters are built up around an ion (up to 50 atoms in liquid helium). The electric field at the position of the ion in the centre of the cluster was supposed to be very weak for symmetry reasons and the Stark mixing rate correspondingly small. This qualitative argument, however, is in contradiction with a recent calculation[41], which indicates that molecular vibrations at the centre of a cluster cause an alternating electric field which is strong enough to rapidly quench the muonic 2S state.

3.2 Measurements

To clarify the puzzling situation discussed above, new experimental information was needed over a wider range of pressures. The importance of collisional quenching rate measurements at subatmospheric pressures has been long known[38].

In an experiment recently performed with the "muon bottle" apparatus at SIN, the μHe-2S-lifetime has been measured in a gaseous target at pressures between 50 Torr and 6 atm. A description of the experiment together with a discussion of the resulting X-ray - time- and energy distributions is given in ref. 19. Only the most important results will be summarized here.

At subatmospheric He pressures, delayed 2P → 1S transitions (8.2 keV) caused by radiative quenching of the metastable 2S state were detected. The time distribution of these delayed X-rays is well described by an exponential. The corresponding lifetimes $\tau_{2S}(p_{He})$ were fitted by means of a least squares procedure. The resulting values are confirmed by the observation of two-photon 2S → 1S transitions at pressures below 200 Torr. The two X-rays could be measured in coincidence since two detectors (Xe GSPC) were available.

At 6 atm, no significant effect was measured. The upper limit on the 2S lifetime, resulting from the absence of delayed two-photon transitions, is $\tau_{2S} < 250$ ns. An upper limit $\tau_{2S} < 25$ ns is obtained assuming radiative quenching as observed at lower pressures.

3.3 Results

From the measured total disappearance rate of the 2S state, $\lambda_{tot}(p_{He}) = [\tau_{2S}(p_{He})]^{-1}$, the pressure-independent decay rate $\lambda_o \approx \lambda_\mu + \lambda_{2X} = 5.6 \cdot 10^5$ s^{-1} was subtracted to get the 2S-quenching rate $\lambda_Q(p_{He}) = \lambda_{tot}(p_{He}) - \lambda_o$. The result is shown in Fig. 5. An excellent fit to $\lambda_Q(p_{He})$ was obtained with the sum of a linear and a quadratic term in the He density ρ_{He}:

$$\lambda_Q = k_1 \, \rho_{He} + k_2 \, \rho_{He}^2$$

with

$$k_1 = (1.5 \pm 0.6) \, 10^{-13} \; cm^3 \; s^{-1}$$
$$k_2 = (5.9 \pm 0.8) \, 10^{-32} \; cm^6 \; s^{-1}.$$

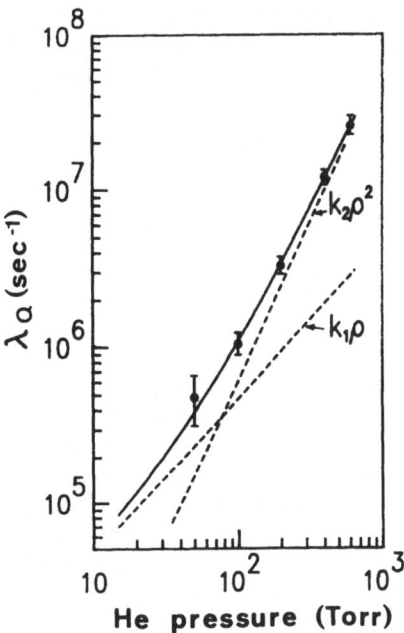

Fig. 5. 2S quenching rate λ_Q as function of He pressure. The solid line is the sum of a linear and quadratic term (broken lines) and corresponds to the fit mentioned in the text.

The dominant quadratic term is easily interpreted with the model of Cohen[41]: Molecular ions are formed in three-body collisions

$$(\mu He)^+_{2S} + He + He \rightarrow \{He-(\mu He)^+_{2S}\} + He,$$

and the 2S states are quickly quenched within the molecules. In the ground vibrational state of the diatomic molecular ion $He-(\mu He)^+_{2S}$, the calculated 2S quenching rate is $\lambda_{St}^{diat.} = 6 \cdot 10^7$ s^{-1}, independent of the gas pressure. Since the rate for the formation of molecular ions in three-body collisions depends quadratically on the pressure, the quadratic term of the measured quenching rate is to be interpreted as the rate of molecular ion formation. The value for k_2 is in close agreement with the measured three-body rate coefficient for the formation of He_2^+ in the helium afterglow[42] (as was presumed in ref. 41). Clearly, this quadratic dependence cannot be extrapolated to higher pressures, where the 2S quenching rate is limited by the rate λ_{St} of Stark transitions induced by the vibrating molecular ion. The result at 6 atm gives a lower limit for λ_{St} which is in accordance with the prediction[41].

The measured rate coefficient for the linear term k_1 is an order of magnitude larger than that calculated for "ordinary" collisional quenching[39,40]. It is also larger than the values obtained by Carboni and Pitzurra[38] for radiative quenching due to two-body formation of the unstable molecular ion $(\mu^{-4}He)^+_{2p}$-He, but may be compatible with this mechanism, since the theoretical values are rather uncertain.

The measurements show that the main 2S quenching process is accompanied by the emission of a K_α X-ray. However, it cannot be excluded that a minor part of all 2S states undergoes non-radiative quenching due to internal Auger effect in the molecular ion. This effect has only been calculated for higher muonic transitions[43]. A more elaborate analysis of the data will give an approximative value of the fraction λ_A/λ_{St} between non-radiative and radiative 2S-quenching processes:

$$\{\mu He^+_{2S}\text{-He}\} \underset{\lambda_{St}}{\overset{\lambda_A}{\rightleftarrows}} \begin{array}{l} \mu He^+_{1S} + He^+ + e^- \\ \{\mu He^+_{1S}\text{-He}\} + X(8.2 \text{ keV}) \end{array}$$

This will allow to determine λ_{St} since λ_A can be calculated.

3.4 Discussion

Previously, it was assumed that the formation of molecular ions lead to clusters preventing the 2S quenching. The data at subatmospheric pressures demonstrate, however, that molecular ion formation is by itself the main cause of 2S quenching, as was predicted by Cohen[41].

The formation of $(\mu He)^+$-He molecular ions has been supposed since a long time, but direct experimental evidence was only obtained with the observation of the quadratic pressure-dependence of the 2S quenching rate. Detailed theoretical studies by Russell[43] and Landua and Klempt[23] on the cascade processes in exotic helium atoms have shown that formation of molecular ions during the cascade is required to explain the anomalous K_β/K_α intensity ratio measured in liquid helium. During the cascade, the exotic ion is not thermalized, since it is accelerated by each external Auger effect. The rate of molecular ion formation for such ions has been calculated in ref. 23; the result is an order of magnitude lower than the value k_2 measured for the $(\mu He)_{2S}^+$ ions which are thermalized.

The long 2S lifetimes measured in the older experiments[35,36] are in contradiction with the new data. Fig. 6 shows the results of all experiments, including the preliminary upper limit $\tau_{2S} < 250$ ns of a recent measurement performed at 40 atm by the William and Mary SIN collaboration[20].

At subatmospheric pressures, the data are well described with the simple model of the formation of diatomic molecular ions, where the 2S-states are quenched rapidly. At pressures above a few atmospheres, the situation is more complicate since triatomic molecules $\{He-\mu He_{2S}^+-He\}$ may be formed before the 2S state is quenched in the diatomic molecule. The calculated Stark-mixing rate is an order of magnitude lower in the triatomic configuration [41]. It is therefore possible that τ_{2S} is in the order of 100 ns at high pressures, but, at present, no working model exists which is able

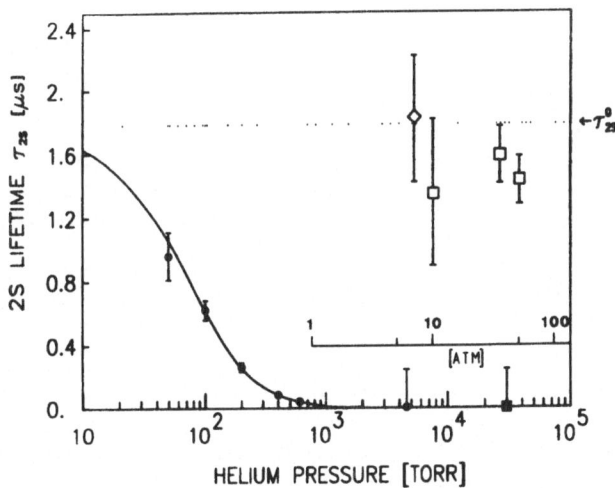

Fig. 6. 2S-lifetime as a function of He pressure.
● ref. 19; ■ ref. 20; ◇ ref. 35; □ ref. 36.
The line corresponds to the fit shown in Fig. 5.

to explain a 2S-lifetime as long as was necessary to perform the Lamb-shift experiment in μ^4He [34].

A new measurement of the 2S-2P energy difference in μ^4He has therefore been proposed at low pressure (p_{He}= 30 Torr), where the mechanisms of 2S quenching are comprehensible. This experiment will be performed in the muon bottle apparatus and is presently prepared by a ETH Zürich-Uni Basel collaboration.

For the formation of neutral muonic helium atoms ($e^-\mu^-$He), the high rate of molecular ion formation is a severe complication, as has been discussed by Souder et al.[44]. The cross section for the neutralization process must be very high to compete with molecular ion formation. Eventually, neutralization happens before the $(\mu He)^+_{1S}$ ion is slowed down to thermal energies. The cross section for neutralization has its maximum at a kinetic energy of a few eV[44], where the probability of molecular formation is probably much lower. Additional experimental information is necessary for a complete understanding of the neutralization process.

References

1. H. Anderhub, F. Kottmann, H. Hofer, P. Le Coultre, D. Makowiecki, O. Pitzurra, B. Sapp, P.G. Seiler, P. Shrager, M. Wälchli and P. Wolff, Phys. Lett. 60B:273 (1976).
2. F. Kottmann, in: Exotic Atoms, ed. by G. Fiorentini and G. Torelli (Servizio Documentazione dei Laboratori Nazionali di Frascati, 1977) p. 129.
3. H. Anderhub, J. Böcklin, M. Devereux, F. Dittus, R. Ferreira Marques, H. Hofer, H.K. Hofer, F. Kottmann, O. Pitzurra, P.-G. Seiler, D. Taqqu, J. Unternährer, M. Wälchli and Ch. Tschalär, Phys. Lett. 101B:151 (1981).
4. J. Böcklin, F. Dittus, R. Ferreira Marques, H. Hofer, F. Kottmann, R. Schären, D. Taqqu, M. Wälchli, W.-D. Herold and H. Kaspar, Nucl. Instr. and Meth. 176:105 (1980).
5. H.P. von Arb, J. Böcklin, F. Dittus, R. Ferreira Marques, H. Hofer, F. Kottmann, R. Schaeren, D. Taqqu and M. Wälchli, Nucl. Instr. and Meth. 207:429 (1983).
6. R. Abela, P. Blüm, D. Gotta, W. Kunold, K. Meissner, M. Schneider and L.M. Simons, SIN Newsletter 16:NL65 (1984).

7. H. Anderhub, H.P. von Arb, J. Böcklin, F. Dittus, R. Ferreira Marques, H. Hofer, F. Kottmann, D. Taqqu, J. Unternährer and Ch. Tschalär, "Search for the metastable 2S-state in muonic hydrogen at low gas pressures", to be published.

8. M. Leon, in: Exotic Atoms '79, eds. K. Crowe et al. (Plenum, New York, 1980) p. 141.

9. R.L. Rosenberg, Phil. Mag. $\underline{40}$:759 (1949).

10. J.S. Cohen, R.L. Martin and W.R. Wadt, Phys. Rev. $\underline{A24}$:33 (1981).

11. J.S. Cohen, Phys. Rev. $\underline{A27}$:167 (1983).

12. S.S. Gershtein and L.I. Ponomarev, in: Muon Physics III, ed. by V.W. Hughes and C.S. Wu (Academic Press, New York, 1975) p. 142.

13. H. Schneuwly, in: Exotic Atoms '79, eds. K. Crowe et al. (Plenum, New York, 1980) p. 147.

14. N.A. Cherepkov and L.V. Chernysheva, Yad. Fiz. $\underline{32}$:709 (1980) [Sov. J. Nucl. Phys. 32:366 (1980)].

15. J.S. Cohen, R.L. Martin and W.R. Wadt, Phys. Rev. $\underline{A27}$:1821 (1983).

16. G. Ya. Korenman and S.I. Rogovaya, Radiation Effects $\underline{46}$:189 (1980).

17. V.I. Petrukhin and V.M. Suvorov, Sov. Phys. JETP $\underline{43}$:595 (1976).

18. H. Anderhub, H.P. von Arb, J. Böcklin, F. Dittus, R. Ferreira Marques, H. Hofer, F. Kottmann, D. Taqqu and J. Unternährer, "Measurement of the K-line intensity ratios in muonic hydrogen between 0.25 and 150 Torr gas pressures", submitted to Phys. Lett. B.

19. H.P. von Arb, F. Dittus, H. Heeb, H. Hofer, F. Kottmann, S. Niggli, R. Schaeren, D. Taqqu, J. Unternährer and P. Egelhof, Phys. Lett. $\underline{136B}$:232 (1984).

20. A. Brodie, R. Dietlicher, M. Eckhause, D. Joyce, J.R. Kane, R.T. Siegel, D. Twerenbold, W.F. Vulcan, R.E. Welsh, R.J. Whyley, R.G. Winter and A. Zehnder, SIN-proposal R-82-11 (1982).

21. M. Leon and H.A. Bethe, Phys. Rev. $\underline{127}$:636 (1962).

22. L. Bracci and G. Fiorentini, Nuovo Cim. $\underline{43A}$:9 (1978).

23. R. Landua and E. Klempt, Phys. Rev. Lett. $\underline{48}$:1722 (1982).

24. H. Anderhub, H. Hofer, F. Kottmann, P. LeCoultre, D. Makowiecki, O. Pitzurra, B. Sapp, P.G. Seiler, M. Wälchli, D. Taqqu, P. Truttmann, A. Zehnder and Ch. Tschalär, Phys. Lett. $\underline{71B}$:443 (1977).

25. P.O. Egan, S. Dhawan, V.W. Hughes, D.C. Lu, F.G. Mariam, P.A. Souder, J. Vetter, G. zu Putlitz, P.A. Thompson, and A.B. Denison, Phys. Rev. $\underline{A23}$:1152 (1981).

26. A. Placci, E. Polacco, E. Zavattini, K. Ziock,
    G. Carboni, U. Gastaldi, G. Gorini and G. Torelli,
    Phys. Lett. 32B:413 (1970).
27. B. Budick, J.R. Toraskar and I. Yaghoobia, Phys. Lett.
    34B:539 (1971).
28. V.E. Markushin, Sov. Phys. JETP 53:16 (1981).
29. E. Borie and M. Leon, Phys. Rev. A21:1460 (1980).
30. G. Backenstoss, J. Egger, T. von Egidy, R. Hagelberg,
    C.J. Herrlander, H. Koch, H.P. Povel, A. Schwitter,
    and L. Tauscher, Nucl. Phys. A232:519 (1974).
31. J. Bernabeu, T.E.O. Ericson and C. Jarlskog, Phys.
    Lett. 50B:467 (1974).
32. G. Feinberg and M.Y. Chen, Phys. Rev. D10:190 (1974).
33. E. Zavattini, in: Lecture Notes in Physics 43 (Springer
    Verlag, Berlin, 1975) p. 370.
34. G. Carboni, U. Gastaldi, G. Neri, O. Pitzurra,
    E. Polacco, G. Torelli, A. Bertin, G. Gorini,
    A. Placci, E. Zavattini, A. Vitale, J. Duclos
    and J. Picard, Nuovo Cim. 34A:493 (1976);
    G. Carboni, G. Gorini, G. Torelli, L. Palffy,
    F. Palmonari and E. Zavattini, Nucl. Phys. A278:
    381 (1977);
    G. Carboni, G. Gorini, E. Iacopini, L. Palffy,
    F. Palmonari, G. Torelli and E. Zavattini, Phys.
    Lett. 73B:229 (1978).
35. A. Placci, E. Polacco, E. Zavattini, K. Ziock,
    G. Carboni, U. Gastaldi, G. Gorini, G. Neri and
    G. Torelli, Nuovo Cim. 1A:445 (1971).
36. A. Bertin, G. Carboni, A. Placci, E. Zavattini,
    U. Gastaldi, G. Gorini, G. Neri, O. Pitzurra,
    E. Polacco, G. Torelli, A. Vitale, J. Duclos and
    J. Picard, Nuovo Cim. 26B:433 (1975).
37. W.R. Johnson, Phys. Rev. Lett. 29:1123 (1972).
38. G. Carboni and O. Pitzurra, Nuovo Cim. 25B:367 (1975).
39. R.O. Mueller, V.W. Hughes, H. Rosenthal and C.S. Wu,
    Phys. Rev. A11:1175 (1975).
40. J.S. Cohen and J.N. Bardsley, Phys. Rev. A23:46 (1981).
41. J.S. Cohen, Phys. Rev. A25:1791 (1982).
42. C.P. de Vries and H.J. Oskam, Phys. Rev. A22:1429 (1980
43. J.E. Russell, Phys. Rev. A18:521 (1978).
44. P.A. Souder, T.W. Crane, V.W. Hughes, D.C. Lu, H. Orth,
    H.W. Reist and M.H. Yam, Phys. Rev. A22:33 (1980).

INDEX